# Fractals: Theory and Applications in Engineering

Springer
*London*
*Berlin*
*Heidelberg*
*New York*
*Barcelona*
*Hong Kong*
*Milan*
*Paris*
*Santa Clara*
*Singapore*
*Tokyo*

Michel Dekking, Jacques Lévy Véhel,
Evelyne Lutton and Claude Tricot (Eds.)

# Fractals:
# Theory and Applications in Engineering

Springer

Michel Dekking, Professor
Faculty of Technical Mathematics aus SC, Delft University of Technology,
Melalweg 4, 2628 CD Delft, The Netherlands

Jacques Lévy Véhel, Doctor
INRIA, Rocquencourt, B.P. 105, 78153 Le Chesnay Cedex, France

Evelyne Lutton, Doctor
INRIA, Rocquencourt, B.P. 105, 78153 Le Chesnay Cedex, France

Claude Tricot, Professor
Université Blaise Pascal, Département Mathématiques, 63177 Aubière Cedex, France

ISBN 1-85233-163-1 Springer-Verlag London Berlin Heidelberg

British Library Cataloguing in Publication Data
Fractals : theory and applications in engineering
 1.Fractals 2.Engineering mathematics
 I.Dekking, Michel
 620'.0015'14742
ISBN 1852331631

Library of Congress Cataloging-in-Publication Data
A catalog record for this book is available from the Library of Congress

Apart from any fair dealing for the purposes of research or private study, or criticism or review, as permitted under the Copyright, Designs and Patents Act 1988, this publication may only be reproduced, stored or transmitted, in any form or by any means, with the prior permission in writing of the publishers, or in the case of reprographic reproduction in accordance with the terms of licences issued by the Copyright Licensing Agency. Enquiries concerning reproduction outside those terms should be sent to the publishers.

© Springer-Verlag London Limited 1999
Printed in Great Britain

The use of registered names, trademarks, etc. in this publication does not imply, even in the absence of a specific statement, that such names are exempt from the relevant laws and regulations and therefore free for general use.

The publisher makes no representation, express or implied, with regard to the accuracy of the information contained in this book and cannot accept any legal responsibility or liability for any errors or omissions that may be made.

Typesetting: Camera ready by editors
Printed and bound at the Athenæum Press Ltd., Gateshead, Tyne & Wear
69/3830-543210 Printed on acid-free paper

# Foreword

The activity around the fractal models develops itself so rapidly that it is necessary, at regular times, to sketch a survey of the new applications and discoveries. Some old topics seem to come to a kind of maturity, like a local maximum; whilst other topics spring on a positive slope, bringing discussions and dynamics to the subject. The readers of the former book *Fractals in Engineering; From Theory to Industrial Applications* (Springer-Verlag 1997, ed. J. Lévy-Véhel, E. Lutton and C. Tricot) may find after two years some major differences with the present collection of works.

At first, a general impression is the following: Mathematics are more and more involved in the definition and use of fractal models. Let us mention two particularly active areas which both are strongly based upon theoretical arguments. Firstly, stochastic processes defined as extensions of fractional Brownian Motion functions become more and more sophisticated and adaptable to many experimental situations. Secondly, multifractal analysis has proved itself a powerful and versatile technique in Signal Processing. In both directions, new tools and new theorems are found regularly and the field of applications continues to grow.

The readers will find consequently several chapters on fractal stochastic processes in this book. Multifractal spectra are used in texture analysis and several new models are proposed. Some new achievements on IFS theory and image compression are given. Wavelets occur in different places (processes, fractal lattices, compression) and continue to play their fundamental role in signal analysis. Some intriguing vector calculus on fractal curves is sketched : This may well be a major topic in the near future. The chapters on fractal pores, fractal tunnels, chaotic flows and monolayer structures are good representatives of the wealth of activity occuring in physics and chemistry in connection with fractals. Note in particular a paper involving conformal invariance, quantum gravity and multifractal analysis that brings new insights into percolation phenomena and, among other results, proves the famous 4/3 Mandelbrot's conjecture on the boundary of a Brownian trajectory.

Finally, it seems that the time when the field was mainly concerned with a qualitative observation of fractal phenomena has definitely gone. In order to get operational and closer to the real world, the models are now deep-rooted in mathematics, with new theory behind.

We wish to thank all the authors who have generously contributed through recent and original work to the composition of this book. We also thank Mitzi Adams, Pierre Adler, Antoine Ayache, Christophe Canus, Marc Chassery, Nathan Cohen, Kenneth Falconer, Bertrand Guiheneuf, Stéphane Jaffard, Georges Oppenheim, Jacques Peyriere, Peter Pfeifer, Michel Rosso, for their help and careful reviews, and Nathalie Gaudechoux, for her Latex skill. Once more, the efficiency of our publisher Springer-Verlag is warmly thanked. We are also deeply grateful to INRIA and the University of Technology of Delft for their support.

Michel DEKKING,<br>
Jacques LÉVY VÉHEL,<br>
Evelyne LUTTON,<br>
Claude TRICOT.

# Table of Contents

## LOCALLY SELF SIMILAR PROCESSES

From Self-Similarity to Local Self-Similarity: the Estimation Problem .....  3
  *Serge Cohen*

Generalized Multifractional Brownian Motion: Definition and Preliminary Results .............................................................  17
  *Antoine Ayache, Jacques Lévy Véhel*

Elliptic Self Similar Stochastic Processes ............................  33
  *Albert Benassi, Daniel Roux*

Wavelets for Scaling Processes .......................................  47
  *Patrick Flandrin, Patrice Abry*

## MULTIFRACTAL ANALYSIS

Classification of Natural Texture Images from Shape Analysis of the Legendre Multifractal Spectrum .........................................  67
  *Piotr Stanczyk, Peter Sharpe*

A Generalization of Multifractal Analysis Based on Polynomial Expansions of the Generating Function .........................................  81
  *Antoine Saucier, Jiri Muller*

Local Effective Hölder Exponent Estimation on the Wavelet Transform Maxima Tree ........................................................  93
  *Zbigniew R. Struzik*

Easy and Natural Generation of Multifractals: Multiplying Harmonics of Periodic Functions ..................................................  113
  *Marc-Olivier Coppens, Benoit B. Mandelbrot*

## IFS

IFS-Type Operators on Integral Transforms ............................ 125
  *Bruno Forte, Franklin Mendivil, Edward R. Vrscay*

Comparison of Dimensions of a Self-Similar Attractor ................. 139
  *Serge Dubuc, Jun Li*

## FRACTIONAL CALCULUS

Vector Analysis on Fractal Curves ................................... 155
    *Massimiliano Giona*

Local Fractional Calculus: a Calculus for Fractal Space-Time ............ 171
    *Kiran M. Kolwankar, Anil D. Gangal*

## PHYSICAL SCIENCES

Conformal Multifractality of Random Walks, Polymers, and Percolation
in Two Dimensions ................................................. 185
    *Bertrand Duplantier*

Fractal Pores and Fractal Tunnels: Traps for "Particles" or "Sound Particles" 207
    *Jérôme Dorignac, Bernard Sapoval*

Fractal Pores and the Degradation of Shales ........................... 229
    *Luis E. Vallejo, Ann Stewart Murphy*

Continuous Wavelet Transform Analysis of Fractal Superlattices ......... 245
    *Hervé Aubert, Dwight L. Jaggard*

## CHEMICAL ENGINEERING

Mixing in Laminar Chaotic Flows: Differentiable Structures and Multifractal Features ....................................................... 263
    *Massimiliano Giona*

Adhesion AFM Applied to Lipid Monolayers. A Fractal Analysis. ........ 277
    *Gianina Dobrescu, Camelia Obreja, Mircea Rusu*

## IMAGE COMPRESSION

Faster Fractal Image Coding Using Similarity Search in a KL-transformed
Feature Space .................................................... 293
    *Jean Cardinal*

Can One Break the "Collage Barrier" in Fractal Image Coding? ......... 307
    *Edward R. Vrscay, Dietmar Saupe*

Two Algorithms for Non-Separable Wavelet Transforms and Applications
to Image Compression ............................................. 325
    *Franklin Mendivil, Daniel Piché*

# LOCALLY SELF SIMILAR PROCESSES

# From Self-Similarity to Local Self-Similarity: the Estimation Problem

Serge Cohen

Université de Versailles-St Quentin en Yvelines
45, avenue des Etats-Unis, 78035 Versailles FRANCE
or
CERMICS-ENPC
cohen@math.uvsq.fr

**Abstract.** In this article we review some methods used to identify the order $H$ of a fractional Brownian motion. This discussion is introduced to see how such techniques can be extended to locally self-similar processes. Moreover the model of the multifractional Brownian motion which is a locally self-similar process is further studied. In particular it is shown that its presentation given by J. Lévy Véhel and R. Peltier and the one given by A. Benassi, S. Jaffard and D. Roux are in some sense Fourier transformed of each other. Last some results for the estimation of the multifractionnal Brownian motion are recalled.

## Introduction

It is now a classical technique to estimate the scalar index that is relevant to describe the self-similarity property of a Gaussian process. Let us recall that in the Gaussian framework there is only one self-similar Gaussian process with stationary increments (see also [8] for a complete discussion of stationary self-similar Gaussian random fields defined in a generalized sense): the fractional Brownian motion (fBm) of index $H$ ($0 < H < 1$) which can be defined by $X_0 = 0$ a.s. and $\mathbb{E}(X_t - X_s)^2 = |t-s|^{2H}$. By computing the covariance of the fBm one can see easily that:

$$\forall \lambda > 0 \quad (X_{\lambda t})_{t \in \mathbb{R}} \stackrel{(d)}{=} \lambda^H (X_t)_{t \in \mathbb{R}} \tag{1}$$

where $\stackrel{(d)}{=}$ means that for every $(t_1, \ldots, t_n)$ the distribution of $(X_{\lambda t_1}, \ldots, X_{\lambda t_n})$ is the same as those of $\lambda^H (X_{t_1}, \ldots, X_{t_n})$. This means that the fBm is self-similar with index $H$. Estimating the parameter $H$ by experimental data is a classical but major issue for applications, and various methods has been proposed.

Actually the parameter $H$ governs at least two other properties of the fBm beside the self-similarity and I will classify these techniques by stressing which property is used for the estimation. Moreover the aim of this classification is to discuss whether these techniques can also be applied to locally self-similar processes. The need for localizing the self-similarity (1) is motivated by applications.

To understand this phenomenon let us recall that the fBm is used as a toy-model for simulation of the profile of a mountain. Actually one can show [21] that the sample paths of the fBm are almost surely Hölder continuous with parameter $H$ (at least if we consider the logarithmic factor negligible). Hence greater you take $H$, smoother is your profile. Even in this very basic model it seems highly desirable to have $H$ varying with the location because the geological nature of the mountain yields locally sharper profiles. From the mathematical point of view the problem is more delicate since the self-similarity is clearly global: it depends on the distribution of $X$ for every $t$. In this field the naïve approach which consists in replacing the index $H$ by a function in (1) :

$$\forall \lambda > 0 \quad (X_{\lambda t})_{t \in \mathbb{R}} \stackrel{(d)}{=} \lambda^{H(t)} (X_t)_{t \in \mathbb{R}} \tag{2}$$

is not really satisfactory since the value of the function $H$ at point $t$ is also governing the law of $X_s$ at time $s$ that are very far from $t$ since

$$X_s \stackrel{(d)}{=} \left(\frac{s}{t}\right)^{H(t)} X_t.$$

Even if one forgets the fact that (2) does not conveniently model all the phenomenons that are locally self-similar it seems very hard to characterize mathematically the processes satisfying (2) when $H$ is not constant. However if we write $X^H$ for the fBm of index $H$, it has been shown in Theorem 4 of [19] that the map $H \to X^H$ is continuous. This theorem can be seen as the first step toward local self-similarity.

Beside a popular model for image processing the fBm is also a very useful tool for the theory of developed turbulence. Since the articles of Kolmogorov [11–13] on the developed turbulence the fBm is used as a paradigm for a self-similar process that have an associated spectral density $\Pi(\xi) = \dfrac{1}{|\xi|^{1+2H}}$. In this particular situation the problem consists in estimating the fractional power of the spectral density. From a probabilistic point of view there is a paradox : *the fBm has stationary increments but it is not stationary*, hence it has no spectral density. In this article I recall that there are at least two ways to derive a stationary process from the fBm: first the fractional Gaussian noise (fGn) which is built with the increments of the fBm, second one can apply a wavelet transform to the fBm. In both cases the estimation of $H$ is well known and we wonder if such techniques can be easily extended to locally self-similar processes.

The article is organized as follows: in section 1 the fractional Gaussian noise is defined, its spectral density is computed and the problem of estimating $H$ as the fractional power of the spectral density is addressed. We remark that the estimation is hindered by long-range dependence properties of the fGn , and we present some hints to extend the estimation in the locally self-similar setting. In section 2 a wavelet transform of the fBm is used to estimate the parameter $H$ as the self-similarity index. It is pointed out how a convenient choice of the wavelet transform can circumvent the long-range dependence problem. The last section is devoted to the estimation of $H$ as the Hölder exponent of the sample paths of the fBm. In this direction two apparently different models have been proposed in [5] and [19] to model locally self-similar processes. In section 3 it is shown that these processes have the same law and that they are in some sense Fourier transformed of each other. Then we recall that these processes called multifractional Brownian motions (mBm) are actually locally asymptotically self-similar (lass) (see [5] and Proposition 5 in [19]) and we recall how to estimate the function $H$ of the mBm by generalized quadratic variations.

## 1 Fractional Brownian motion and spectral density.

In this section the harmonizable representation (cf [21]) of the fBm is first recalled to compute easily the spectral density of the fGn. Then the fGn is considered as a filtered fBm and it is shown that the order of the zero in frequency of the Fourier transform of the filter is responsible for the long-range dependence properties of the fGn. Furthermore we exhibit an estimator of $H$ based on the spectral density of the fGn in order to stress the problem of such methodology. Last we give a brief overview of the literature concerning locally stationary processes because it could be useful to define a "multifractional Gaussian noise" as a locally self-similar in a spectral sense generalization of the fBm.

A fBm can be represented as an integral of a white noise. In such framework the white noise is considered as a random Brownian measure $W(d\xi)$. (See [21] section 7.2.2 p 325-332 for the details.) An isometry between $L^2(\Omega)$ and $L^2(\mathbb{R})$ is done with $W(d\xi)$: for every $L^2(\mathbb{R})$-function $f$ one can define a Gaussian random variable denoted by

$$\int_{\mathbb{R}} f(\xi) W(d\xi) \qquad (3)$$

which is centered ($\mathbb{E}(\int_{\mathbb{R}} f(\xi) W(d\xi)) = 0$) and whose variance is $\|f\|^2_{L^2(\mathbb{R})}$. Hence

$$\|\int_{\mathbb{R}} f(\xi) W(d\xi)\|^2_{L^2(\Omega)} = \|f\|^2_{L^2(\mathbb{R})}$$

moreover the covariance of such integrals are easily computed with the formula

$$\mathbb{E}\left(\int_{\mathbb{R}} f(\xi) W(d\xi) \overline{\int_{\mathbb{R}} g(\xi) W(d\xi)}\right) = \langle f, g \rangle_{L^2(\mathbb{R})} \qquad (4)$$

for every function $f, g \in L^2(\mathbb{R})$. Then if we set

$$B_t^H = \frac{1}{C_H} \int_{\mathbb{R}} \frac{e^{it\xi} - 1}{|\xi|^{1/2+H}} W(d\xi), \qquad (5)$$

we check that

$$\mathbb{E}(B_t^H - B_s^H)^2 = \frac{2}{C_H^2} \int_{\mathbb{R}} \frac{1 - \cos((t-s)\xi)}{|\xi|^{1+2H}} d\xi \qquad (6)$$

because of (4). Since the right hand term of (6) is clearly an homogeneous function of order $2H$ in the variable $|t-s|$, $B^H$ is a fBm of order $H$ if

$$C_H^2 = 4 \int_0^{+\infty} \frac{1-\cos u}{|u|^{1+2H}} du.$$

The Fractional Gaussian noise (fGn) is then defined as the increments of the fBm with a step $\Delta > 0$

$$Y_t^{H,\Delta} \stackrel{def}{=} B_{t+\Delta}^H - B_t^H; \tag{7}$$

it is stationary Gaussian process with the covariance given by

$$\mathbb{E}\left(Y_t^{H,\Delta} Y_s^{H,\Delta}\right) = \int_\mathbb{R} \frac{e^{i(t-s)\xi}}{C_H^2 |\xi|^{1+2H}} \sin^2\left(\frac{\Delta\xi}{2}\right) d\xi \tag{8}$$

because of (4). Hence it yields the spectral density $\Pi_{Y^{\alpha,\Delta}}$ which is the Fourier transform of the covariance

$$\Pi_{Y^{\alpha,\Delta}}(\xi) = \frac{1}{C_H^2 |\xi|^{1+2H}} \sin^2\left(\frac{\Delta\xi}{2}\right) \underset{\xi \to 0}{\sim} \frac{\Delta^2}{4C_H^2} |\xi|^{1-2H}. \tag{9}$$

Let us remark that asymptotically for the large frequencies $|\xi| \to +\infty$ the spectral density $\Pi_{Y^{\alpha,\Delta}}$ fits the Kolmogorov's theory of turbulence. Nevertheless as stressed in formula (9) the spectral density has a pole at $\xi = 0$ when $H > 1/2$. This phenomenon which can also be presented as a divergence of the series $\sum_{j \in \mathbb{Z}} \mathbb{E}\left(Y_0^{H,\Delta} Y_j^{H,\Delta}\right)$ is known as long range dependence. It leads to a lack of decorrelation of the variables $Y_0^{H,\Delta} Y_j^{H,\Delta}$ as $j \to +\infty$ and it delays the convergence of the empirical covariance:

$$\lim_{N \to +\infty} \frac{1}{N} \sum_{p=1}^N Y_{\tau_p}^{H,\Delta} Y_{\tau_p + t}^{H,\Delta} = \mathbb{E}\left(Y_0^{H,\Delta} Y_t^{H,\Delta}\right). \tag{10}$$

Before illustrating the drawback of long range dependence for the estimation of $H$ let us explain why this happens for fGn. The fGn can be viewed as filtered fBm if it is written

$$Y_t^{H,\Delta} = B^H \star (\delta_\Delta - \delta_0)(t) \tag{11}$$

where $\delta_x$ is a Dirac measure at point $x$. In this framework the transfer function is

$$\widehat{(\delta_\Delta - \delta_0)}(\xi) = e^{i\Delta\xi} - 1 = O(\xi) \quad \text{when} \quad \xi \to 0$$

and the pole of $\Pi_{Y^{\alpha,\Delta}}$ is related to the zero of this transfer function. Hence one can guess that a filtered fBm with a transfer function that have a zero of order 2 at $\xi = 0$ will have no long range dependence. The consequence of this remark are developed in section 2.

Actually the identification of the spectral density of a stationary process is a classical problem. Since the spectral density is the inverse Fourier transform of the covariance, an estimator can be built by plugging the empirical covariance (10) into the inverse Fourier transform. As an example one can propose the Welch estimator [22]

$$\overline{\Pi}_{Y^{H,\Delta}}(\xi) = \frac{1}{N} \sum_{p=1}^{N} |\int_0^T \exp(-2\pi i \xi t) Y_t^{H,\Delta} w_\theta(t - \tau_p) dt|^2 \qquad (12)$$

where $w$ is a weight function which takes into account the process $Y^{H,\Delta}$ in the windows $(\tau_p - \frac{\theta}{2}, \tau_p + \frac{\theta}{2})$. Hence the estimator (12) is a continuous analog of the usual discrete periodogram which can be found for instance in [14]. The main issue with such time-frequency technique is the bias variance trade off that arises from the fact that the sample path is typically observed on a finite interval (i.e. $T$ is finite). Actually because of long dependence the time $\tau_{p+1} - \tau_p$ has to be greater than $\delta$ to have enough decorrelation between each term of the right hand side of (10). Since $T$ is finite $N$ is bounded above by $\frac{T}{\delta}$ and the mean square error in (10) can be important. The trade off consists in balancing these two different kinds of errors. A more complete discussion on this problem can be found in [2]. Moreover the extension of such estimators to locally self-similar processes makes the trade off even harder. Actually the width $\theta$ of the window has to be small enough to fit time variations of the spectral density. The most promising way to consider locally self-similar processes in this spirit seems to relate this problem to locally stationary processes as presented in [15] or in [6]. These two models are formally different, nevertheless they both aim to propose models for processes that are not stationary but for which the time varying spectrum can be estimated. In the framework of locally self-similar processes, the first step consists in generalizing the fGn to a locally stationary process which allows local variations of the spectrum. The second step in that direction is to define a time varying fractional Brownian motion whose increments will be the generalized fGn. Even if the relationship between locally self-similar processes and locally stationary processes is certainly an interesting starting point it is yet to be developed (see [17] for a related but distinct work).

## 2  Fractional Brownian motion and wavelets

In the previous section we have seen that the fBm can be transformed in a stationary process by taking its increments. Since the wavelets are particularly well suited to self-similar objects, it is not surprising that wavelet transform of the fBm is popular (see [9] or [2]). To compare this method with the estimation of $H$ as a spectral parameter of the fBm a brief summary of the multiresolution analysis is needed (See [16] for a comprehensive introduction to this theory). Since the sample paths are observed on a finite interval $[0,T]$ there is no loss in generality in considering a mulitresolution analysis of $L^2(0,1)$. An orthonormal basis of $L^2(0,1)$ can be built with a so-called mother-wavelet $\Psi$ and an important property of this $L^2$-function is the moment vanishing property:

$$\forall m \leq M \quad \int_0^1 t^m \Psi(t) dt = 0;$$

which is naturally equivalent to

$$\widehat{\Psi}(\xi) = O(|\xi|^M) \quad \xi \to 0,$$

where $\widehat{f}$ denotes in this article the Fourier transform of the function $f$. The multiresolution analysis is then given by

$$\Psi_{j,k}(t) = 2^{-j/2} \Psi(2^{-j} t - k) \quad \forall j \in \mathbb{Z} \quad k = 1 \text{ to } N_j \qquad (13)$$

which is an orthonormal basis of $L^2$. It is now classical to interpret the index $j$ as a scale parameter whereas $k$ is a location parameter. When this analysis is applied to the fBm coefficients

$$D_{j,k} = \int_0^1 \Psi_{j,k}(t) B^H(t) dt$$

are computed. Because of (13) $D_{j,k}$ can be viewed as a convolution of $B^H$ with $\Psi_{j,0}$, hence the same phenomenon as the one with the fGn happens if $M \geq 1$:

$$\forall j \in \mathbb{Z} \quad (D_{j,k})_{k \in \mathbb{Z}} \text{ is a stationary process.} \qquad (14)$$

Nevertheless if $M < 2$ these processes have long range dependence properties when $\frac{1}{2} < H < 1$. This drawback can be circumvented by choosing a mother-wavelet with $M = 2$, since in this case the spectral density of the process $(D_{j,k})_{k \in \mathbb{Z}}$ has no pole at $\xi = 0$ for every $0 < H < 1$. Once this choice has been done the estimation of $H$ is performed because of the self-similar property of $(D_{j,k})_{k \in \mathbb{Z}}$ which can be expressed as:

$$\forall j \in \mathbb{Z} \quad (D_{j,k})_{k \in \mathbb{Z}} \stackrel{(d)}{=} 2^{j(H+1/2)} (D_{0,k})_{k \in \mathbb{Z}}. \qquad (15)$$

Such identity can be found for instance in [7] where the estimation of $H$ is done with various functionals of the wavelet coefficients of the fBm. To understand the method let us take for example:

$$Y_j = \frac{1}{2} \log_2 \frac{1}{N_j} \sum_{k=1}^{N_j} D_{j,k}^2,$$

where $N_j$ is the number of coefficient $D_{j,k}$ which are available at the scale $j$: it is usually of order $2^{-j} N$ if $N$ is the cardinal of the discrete sample of the fBm. Since (14):

$$\lim_{N_j \to +\infty} \frac{1}{N_j} \sum_{k=1}^{N_j} D_{j,k}^2 = \mathbb{E} D_{j,0}^2,$$

and because of (15):

$$\mathbb{E}(D_{j,0})^2 = 2^{2j(H+1/2)} E(D_{0,0})^2.$$

Then it is not surprising that

$$\lim_{N_j \to +\infty} Y_j = j(H+1/2) + C,$$

as shown in [7]. Hence the estimation of $H$ is theoretically reduced to the estimation of the slope of a linear regression. Actually from the practical point of view the main problem is the computation of the coefficient $D_{j,k}$ when only a discrete sample of the fractional Brownian path is known. Following [2] this problem is reduced to the problem of knowing $(D_{0,k})_{0 \leq k \leq N_j}$ because the pyramidal algorithm (cf section 1.3 of [2]) yields quickly the $D_{j,k}$ once the algorithm is initialized with $D_{0,k}$. According to numerical experiments the estimation of $H$ is not perturbed when $B^H_{\frac{k}{N}}$ are used instead of $D_{0,k}$. At this point of the discussion an alternative to compute the $D_{j,k}$ is to performed a numerical integration, we refer to [3] for the details of such techniques.

As a conclusion of this section I stress once more the fact that wavelet coefficients are essentially useful to estimate the index governing the self-similarity property (1), which happens to be $H$ for the fBm. Hence if we want to extend such techniques to locally self-similar processes other generalizations than those of section 1 are possible. The interested reader can see [1] for an introduction to the problem. Even if it is not completely solved from a mathematical point of view, the main issue is clear. Whatever your model of locally self-similar process is, one expects that the process $(D_{j,k})_{k \in \mathbb{Z}}$ is not stationary anymore. Actually the aim is to carefully fix the range $[K^-(j), K^+(j)]$ in the following sum:

$$\frac{1}{2} \log_2 \frac{1}{K^+(j) - K^-(j)} \sum_{k=K^-(j)}^{K^+(j)} D_{j,k}^2$$

such that $K^+(j) - K^-(j)$ is big enough to allow convergence, and small enough to have small variations of the function $H$ in the intervals spanned by the supports of the functions $\Psi_{j,k}$ for $k \in [K^-(j), K^+(j)]$. The question seems open in the setting of wavelet techniques.

## 3 Multifractional Brownian motion(mBm)

In the last section I present a model of locally self-similar processes whose local Hölder regularity is identifiable. In this direction two apparently different processes have been independently proposed in [5] and [19]. Although the name "Multifractional Brownian motion" has been proposed for the first time by Jacques Lévy Véhel I will start from the harmonizable representation of the fBm (5). The authors of [5] define the multifractional Brownian motion (mBm) as :

$$B_t^h = \int_{\mathbb{R}} \frac{e^{it\xi} - 1}{|\xi|^{1/2+h(t)}} W(d\xi), \qquad (16)$$

where $h$ is a $(0,1)$-valued function with Hölder regularity $r$. In [5] one can find in Theorem 1.7 a complete study of the smoothness of the sample paths of the mBm under the additional assumption that $r > \sup h(t)$. These results are briefly summarized by the fact that $h(t)$ is almost surely the Hölder exponent of the sample paths of the mBm at point $t$. For a definition of the Hölder exponents I refer to Definition 9 of [19]. The same result of regularity is proved in Proposition 10 of [19] for a process which is also called multifractional Brownian motion although this process is defined as a non anticipative "moving average" of a white noise:

$$W_t^h = \frac{1}{\Gamma(h(t)+1/2)} \int_{-\infty}^{t} [(t-s)_+^{(h(t)-1/2)} - (-s)_+^{(h(t)-1/2)}] \widetilde{W}(ds), \qquad (17)$$

where $(x)_+$ equals $x$ if positive, 0 otherwise; where $\widetilde{W}$ is a Brownian measure and $\Gamma$ is the Gamma function. In [19] the roughness of the sample paths of the mBm is also expressed with the Hausdorff dimension of the local graph of the mBm. Let us recall the Proposition 8 of [19]

**Proposition 1.** *With probability one, for each interval $[a,b] \in \mathbb{R}^+$, the graph of the mBm $(W_t^h)_{t\in[a,b]}$ verifies the following property :*

$$dim_H\{W_t^h, t \in [a,b]\} = dim_B\{W_t^h, t \in [a,b]\} = 2 - \min\{h(t), t \in [a,b]\}$$

*where $dim_H$ means the Hausdorff dimension and $dim_B$ means the box dimension.*

In the last proposition we have seen for the first time the importance of the $\min_{[a,b]} h(t)$ which is actually also important for the estimation of $h$.

The following theorem shows that these processes with a convenient normalization have the same law and that $B^h$ is the Fourier transform of a well-balanced representation of the mBm as a moving average. Before proving this theorem the definition of the Brownian measure in (16) has to be more precisely defined. In [5] the white noise $\widehat{\widetilde{W}}(d\xi)$ is given by:

$$\widehat{\widetilde{W}}(d\xi) = \sum_{\lambda \in \Lambda} \eta_\lambda \widehat{\overline{\psi_\lambda}}(\xi) d\xi \qquad (18)$$

where $\Lambda$ is the set of dyadic points, $(\eta_\lambda)_{\lambda \in \Lambda}$ are independent identically standard Gaussian random variables, and where $(\psi_\lambda)_{\lambda \in \Lambda}$ is the Lemarié Meyer basis (cf [16]) of $L^2(\mathbb{R})$. It is worth emphasizing that the white noise can be presented with other orthonormal bases of $L^2(\mathbb{R})$ but the Lemarié Meyer basis is particularly convenient since the functions $\widehat{\overline{\psi_\lambda}}$ are vanishing in a neighborhood of $\xi = 0$.

Nevertheless the meaning of (18) is given as in (3) by taking a $\mathbb{C}$-valued function $f \in L^2(\mathbb{R})$ and letting

$$\int_{\mathbb{R}} f(\xi)\overline{\widetilde{W}}(d\xi) = \sum_{\lambda \in \Lambda} \eta_\lambda \int_{\mathbb{R}} f(\xi)\overline{\widehat{\psi}_\lambda} d\xi \qquad (19)$$

which is a centered Gaussian random variable with variance $\int_{\mathbb{R}} |f|^2(\xi)d\xi$. One can remark that the mBm can be written

$$B_t^h = \sum_{\lambda \in \Lambda} \eta_\lambda \langle f, \widehat{\psi}_\lambda \rangle_{L^2} \qquad (20)$$

but it is not clear that the mBm is a real valued process. Actually in this paper it will be a consequence of the proof of Theorem 1 (more generally one can read section 7.2.2 of [21] for a complete discussion). Before stating the theorem I would like to warn the reader against a possible confusion: even if $(\psi_\lambda)_{\lambda \in \Lambda}$ is a classical multiresolution analysis of $L^2(\mathbb{R})$ the construction of the mBm is not a part of the section 2 where wavelet coefficients of processes are used for estimation. Here the wavelets are a tool to define the mBm. Finally the link between moving average and harmonizable representation of the mBm is described by the following theorem.

**Theorem 1.** *Let a Brownian measure $\overline{\widetilde{W}}(d\xi)$ be given by (18) then the harmonizable representation of the multifractional Brownian motion:*

$$\int_{\mathbb{R}} \frac{e^{it\xi} - 1}{|\xi|^{1/2 + h(t)}} \overline{\widetilde{W}}(d\xi) \qquad (21)$$

*is almost surely equal up to a multiplicative deterministic function to the well balanced moving average:*

$$\int_{-\infty}^{+\infty} [|t - s|^{(h(t) - 1/2)} - |s|^{(h(t) - 1/2)}] W(ds), \qquad (22)$$

*where the Brownian measure $W(ds)$ is given by:*

$$W(ds) = \sum_{\lambda \in \Lambda} \eta_\lambda \psi_\lambda(s) ds. \qquad (23)$$

When $h(t) = 1/2$, $[|t - s|^{(h(t) - 1/2)} - |s|^{(h(t) - 1/2)}]$ is ambigous, hence the conventional meaning

$$[|t - s|^0 - |s|^0] \stackrel{def}{=} \left(\log(\frac{1}{|t - s|}) - \log(\frac{1}{|s|})\right)$$

is to be used. Conversely one can show that the non anticipative moving average representation of the mBm (17) is equal up to a multiplicative deterministic function to the harmonizable representation:

$$\int_{\mathbb{R}} \frac{e^{it\xi} - 1}{i\xi|\xi|^{h(t) - 1/2}} \overline{\widetilde{W}}(d\xi). \qquad (24)$$

*Hence the mBm given by the non anticipative moving average (17) has the same law as the mBm given by the harmonizable representation (16) up to a multiplicative deterministic function.*

The proof of the Theorem 1 is an application of the Parseval identity. Actually it is quite surprising that no one has already remarked that the two models [5] and [19] yield the same process since the proof is directly inspired by the corresponding proof for the fBm. Nevertheless I am convinced that this theorem is useful for further development of the theory. An easy consequence of Theorem 1 is that the process given by (21) is real valued since the process in (22) is real valued. At last one can remark that the law of the multifractional Brownian motions presented exactly as in [5] and in [19] are mutually singular because both processes have not been normalized in the same way. It also explains the multiplicative deterministic function that actually fits the normalizations.

*Proof:*
Since $\widehat{\psi_\lambda}$ is a $L^2$-function, Parseval formula yields

$$\langle \frac{e^{it\cdot}-1}{|\cdot|^{1/2+h(t)}}, \widehat{\psi_\lambda} \rangle_{L^2} = \langle \widehat{\frac{e^{it\cdot}-1}{|\cdot|^{1/2+h(t)}}}, \psi_\lambda \rangle_{L^2}. \tag{25}$$

Moreover $|\cdot|^{-(1/2+h(t))}$ is an homogeneous distribution, its Fourier transform is given by

$$\widehat{|\cdot|^{-(1/2+h(t))}}(s) = C(h(t)+1/2)|s|^{h(t)-1/2}$$

where $t$ is fixed when the Fourier transform is computed. If $h(t) \in (1/2,1)$ then $|\cdot|^{-(1/2+h(t))}$ shall be replaced by $Pf|\cdot|^{-(1/2+h(t))}$ in the sense of II,2 ; 26 in [20], it does not change the end of the proof since $\widehat{\psi_\lambda}$ is vanishing in some neighborhood of 0. Moreover the special case $h(t) = 1/2$ is easily explained by

$$\widehat{Pf\frac{1}{|\xi|}}(s) = C(1)\log(\frac{1}{|s|}) + D$$

(cf VII 7. 15 in [20]). Hence

$$\langle \widehat{\frac{-1}{|\cdot|^{1/2+h(t)}}}, \psi_\lambda \rangle_{\mathcal{E}',\mathcal{E}} = C(h(t)+1/2)\langle |\cdot|^{h(t)-1/2}, \widehat{\psi_\lambda} \rangle_{\mathcal{E}',\mathcal{E}}$$

where $\mathcal{E}$ is the space of fast decreasing functions and $\mathcal{E}'$ is the space of tempered distributions. Then

$$\widehat{\frac{e^{it\cdot}}{|\cdot|^{-(1/2+h(t))}}}(s) = C(h(t)+1/2)|s-t|^{h(t)-1/2}.$$

Moreover

$$\langle \widehat{\frac{e^{it\cdot}-1}{|\cdot|^{1/2+h(t)}}}, \psi_\lambda \rangle_{\mathcal{E}',\mathcal{E}} = \langle \widehat{\frac{e^{it\cdot}-1}{|\cdot|^{1/2+h(t)}}}, \psi_\lambda \rangle_{L^2}$$

since $s \to |t-s|^{(h(t)-1/2)} - |s|^{(h(t)-1/2)}$ is a $L^2$- function. Consequently the almost sure equality between (21) and (22) is proved.

Conversely one can show that (17) is equal to (24) up to a multiplicative deterministic function by showing that the Fourier transform of

$$s \to \mathbf{1}_{(-\infty,t]}(s)[(t-s)_+^{(h(t)-1/2)} - (-s)_+^{(h(t)-1/2)}]$$

is given by

$$\xi \to \frac{e^{it\xi} - 1}{i\xi|\xi|^{h(t)-1/2}},$$

an heuristic computation can be found in a note of [21] page 328.

Last we have to prove that the Gaussian processes

$$X_t = \int_\mathbb{R} \frac{e^{it\xi} - 1}{i\xi|\xi|^{h(t)-1/2}} \widehat{\widetilde{W}}(d\xi)$$

and

$$Y_t = \int_\mathbb{R} \frac{e^{it\xi} - 1}{|\xi|^{1/2+h(t)}} \widehat{\widetilde{W}}(d\xi)$$

have the same law. Actually the isometry (4) leads to

$$\mathbb{E}(X_s X_t) = \mathbb{E}(Y_s Y_t) = \int_\mathbb{R} \frac{(e^{it\xi} - 1)(e^{-is\xi} - 1)}{|\xi|^{1+h(t)+h(s)}}.$$

∎

Whatever definition you take for the mBm one can wonder if it is a convenient model for locally self-similar processes. Actually as explained in the introduction the mBm is only locally asymptotically self-similar (lass). As shown in the next theorem one can think of this property as the existence of a tangent fBm at each time $t$ where the mBm is defined. The parameter of the tangent fBm is then given by $h(t)$.

**Theorem 2.** *A mBm $X$ associated with a function $h$, is locally asymptotically self-similar (lass):*

$$\lim_{\varepsilon \to 0^+} \left( \frac{X(t+\varepsilon u) - X(t)}{\varepsilon^{h(t)}} \right)_{u \in \mathbb{R}} \stackrel{(d)}{=} \left( B^{h(t)}(u) \right)_{u \in \mathbb{R}} \qquad (26)$$

*where the equality is up to a multiplicative deterministic function of $t$, and where $B^{h(t)}$ is fBm with parameter $h(t)$.*

The proof can be found in [5] and to illustrate the meaning of the limit we check here that the fBm itself is lass. Since a $H$-fBm $X$ has stationary increments:

$$X(t+\varepsilon u) - X(t) \stackrel{(d)}{=} X(\varepsilon u)$$

and because it is self similar

$$X(\varepsilon u) \stackrel{(d)}{=} \varepsilon^H X(u),$$

hence in this particular case no limit is needed in the left hand side of (26) and the right hand side is the process $X(u)$ itself. However in the general case the fact that the function $h$ is governing only the self similar property of the process in neighborhood of $t$ is implemented by the limit when $\varepsilon$ goes to 0. In the previous theorem the limit is given for the law of the process in the $u$ variable. One can also find a version of this theorem in Proposition 5 of [19] when $u$ is fixed.

To estimate the function $h$ of a mBm with the process $X$ observed at sampling points $\frac{p}{N}, p = 0, \ldots, N$ the generalized quadratic variations are introduced:

$$\mathbf{V}_{\varepsilon,N}(t) = \sum_{p \in \mathcal{V}_{\varepsilon,N}(t)} \left( X(\frac{p+1}{N}) - 2X(\frac{p}{N}) + X(\frac{p-1}{N}) \right)^2. \qquad (27)$$

As pointed out in [10] the generalized variations have to be the sum of discrete second derivatives (i.e. $X(\frac{p+1}{N}) - 2X(\frac{p}{N}) + X(\frac{p-1}{N})$) to have $\sqrt{N}-$ rate of convergence for the estimator. Beside this technical remark it is clear that if the neighborhood

$$\mathcal{V}_{\varepsilon,N}(t) = \left\{ p \in \mathbb{Z}, \left| \frac{p}{N} - t \right| \leq \varepsilon \right\}$$

is such that $X$ is approximatively Hölder with exponent $h(t)$ on $\mathcal{V}_{\varepsilon,N}(t)$ then

$$\mathbf{V}_{\varepsilon,N}(t) \approx N^{1-2h(t)} \left\{ \frac{1}{N} \sum_{p \in \mathcal{V}_{\varepsilon,N}(t)} C^2(\frac{p}{N}, \omega) \right\} \qquad (28)$$

In the previous formula the random variables $C(\frac{p}{N})$ stand for the Hölder constants of the process $X$ at $\frac{p}{N}$. Unless the function $h$ is constant and equal to $1/2$ they are correlated. But because discrete second derivatives are used they behave as if they were uncorrelated. Hence an approximated Law of the Large Number and a Central Limit Theorem for weak dependent variables can be applied to the bracket in (28) if the neighborhood $\mathcal{V}_{\varepsilon,N}(t)$ is big enough. Then the ratio

$$\frac{\mathbf{V}_{\varepsilon,N/2}(t)}{\mathbf{V}_{\varepsilon,N}(t)}$$

is built to get rid of the

$$\lim_{N \to +\infty} \frac{1}{N} \sum_{p \in \mathcal{V}_{\varepsilon,N}(t)} C^2(\frac{p}{N}, \omega).$$

Following this heuristic analysis an estimator of the multifractional function $h(t)$ is proposed in [4]

$$\tilde{h}_{\varepsilon,N}(t) = \frac{1}{2} \left( \log_2 \frac{\mathbf{V}_{\varepsilon,N/2}(t)}{\mathbf{V}_{\varepsilon,N}(t)} + 1 \right). \qquad (29)$$

Please note that estimators of the same nature but built on variations of order $k$ :

$$S_n(k) = \frac{1}{n-1} \sum_{i=0}^{n-1} |X(\frac{p+1}{n}) - X(\frac{p}{n})|^k$$

have been proposed in [18] for the fBm (see Proposition 3.2 and Proposition 3.9).

At this point the question of the convergence of this estimator reminds us of the now familiar trade off between bias and variance. In this framework one can be more precise: $\varepsilon$ is the key factor. If $\varepsilon$ goes too quickly to 0 when $N$ goes to $+\infty$ then $\mathcal{V}_{\varepsilon,N}(t)$ is too small and the variance of the estimator is too big. On the contrary if $\varepsilon$ goes too slowly to 0 then $h(\frac{p}{N})$ is too far away from $h(t)$ and the bias is preponderant. As shown in [4] these two errors are balanced for $\varepsilon = N^{-1/3}$. These results are summarized in the following theorem.

**Theorem 3.** *Let us assume $X$ is a mBm with function $h$ at least $C^1$ and valued in $(0,1)$ and let $\varepsilon = N^{-\beta}$ with $0 < \beta < 1/2$. When $N \to \infty$,*

$$\widetilde{h}_{\varepsilon,N}(t) \to h(t) \quad (a.s.) .$$

*If $\varepsilon = N^{-1/3}$*

$$\mathbb{E}(\widetilde{h}_{\varepsilon,N}(t) - h(t))^2 \leq O(Log^2(N) N^{-2/3}) .$$

As a conclusion of this article the framework of multifractional processes seems more adapted to the estimation problem.

*Acknowledgement* The author is very grateful to Antoine Ayache for explaining that the finite part $Pf$ of distributions is to be used in the proof of Theorem 1 when $h(t) \in (1/2, 1)$.

# References

1. P. Abry and P. Gonçalvès (1998): Multiple-window wavelet transform and local scaling exponent estimation. Preprint.
2. P. Abry (1997): *Ondelettes et Turbulences.* Diderot Éditeur, Arts et Sciences.
3. J. M. Bardet (1997): *Tests d'autosimilarité des processus gaussiens. Dimension fractale et dimension de corrélation.* PhD thesis, Université d'Orsay.
4. A. Benassi, S. Cohen and J. Istas (1998): Identifying the multifractional function of a Gaussian process. *Statistic and Probability Letters*, 39: 337–345.
5. A. Benassi, S. Jaffard and D. Roux (1997): Gaussian processes and Pseudodifferential Elliptic operators. *Revista Mathematica Iberoamericana*, 13(1): 19–89.
6. R. Dahlhaus (1996): On the Kullback-Leibler information divergence of locally stationary processes. *Stochastic processes and their Applications*, 62:139–168.
7. L. Delbeke and P. Abry (1998): Wavelet-based estimators for the self-similarity parameter of fractional Brownian motion. Preprint.
8. R. L. Dobrushin (1979): Gaussian and their subordinated self-similar random generalized fields. *The Annals of Probability*, 7(1): 1–28.

9. P. Flandrin (1993): *Temps- fréquence*. Hermès.
10. J. Istas and G. Lang (1997): Quadratic variations and estimation of the local Holder index of a gaussian process. *Ann. Inst. Poincaré.*, 33(4): 407–437.
11. A. N. Kolmogorov (1941): Dissipation of energy in the locally isotropic turbulence. In S. K. Friedlander and L. Topper, editors, *Turbulence. Classic papers on statistical theory*, pages 156–158. Interscience publishers.
12. A. N. Kolmogorov (1941): The local structure of turbulence in incompressible viscous fluid for very large reynold's numbers. In S. K. Friedlander and L. Topper, editors, *Turbulence. Classic papers on statistical theory*, pages 151–155. Interscience publishers.
13. A. N. Kolmogorov (1941): On degeneration of isotropic tuburlence in an incompressible viscous liquid. In S. K. Friedlander and L. Topper, editors, *Turbulence. Classic papers on statistical theory*, pages 159–161. Interscience publishers.
14. J.-S. Leu and A. Papamarcou (1995): On estimating the spectral exponent of frational Brownian motion. *IEEE Transaction on information theory*, 41(1): 233–244.
15. S. Mallat, G. Papanicolaou, and Z. Zhang (1998): Adaptive covariance estimation of locally stationary process. *Annales of Statistics*, 26: 1–47.
16. Y. Meyer (1990): *Ondelettes et Opérateurs*, volume 1. Hermann, Paris.
17. G. Papanicolau and K. Sola (1998): Estimation of local power law processes. Available on http://georgep.Stanford.EDU:80/ papanico/.
18. R. F. Peltier and J. Lévy Véhel (1994): A new method for estimating the parameter of fractional Brownian motion. INRIA report 2396. Available on http://www-syntim.inria.fr/fractales/
19. R.F. Peltier and J. Lévy Véhel (1995): Multifractional Brownian motion: definition and preliminary results. Available on http://www-syntim.inria.fr/fractales/.
20. L. Schwartz (1978): *Théorie des distributions*. Hermann.
21. G. Samorodnitsky and M. S. Taqqu (1994): *Stable Non-Gaussian Random Processes*. Chapmann and Hall.
22. P. D. Welch (1967): The use of fast Fourier transform for the estimation of power spectra: a method based on time averaging over short modified periodograms. *IEEE Trans. on Audio*, AU-15: 70–73.

# Generalized Multifractional Brownian Motion: Definition and Preliminary Results

Antoine Ayache, Jacques Lévy Véhel

Projet Fractales, INRIA Rocquencourt,
Domaine de Voluceau, Rocquencourt,
BP 105, 78153 Le Chesnay Cedex - FRANCE
ayache@ceremade.dauphine.fr
Jacques.Levy_Vehel@.inria.fr

**Abstract.** The Multifractional Brownian Motion (MBM) is a generalization of the well known Fractional Brownian Motion. One of the main reasons that makes the MBM interesting for modelization, is that one can prescribe its regularity: given any Hölder function $H(t)$, with values in $]0,1[$, one can construct an MBM admitting at any $t_0$, a Hölder exponent equal to $H(t_0)$. However, the continuity of the function $H(t)$ is sometimes undesirable, since it restricts the field of application. In this work we define a gaussian process, called the Generalized Multifractional Brownian Motion (GMBM) that extends the MBM. This process will also depend on a functional parameter $H(t)$ that belongs to a set $\mathcal{H}$, but $\mathcal{H}$ will be much more larger than the space of Hölder functions.

## 1 Introduction

It is classical that there exists a unique gaussian process, self-similar of order $H$, vanishing at the origin and having stationary increments. This continuous, centered gaussian process is called the Fractional Brownian Motion (FBM). It has been introduced in 1968 by Mandelbrot and Van Ness [6]. This process can be defined as the stochastic integral, for $t \in \mathbb{R}$,

$$B_H(t) = \int_{-\infty}^{+\infty} \frac{e^{it\xi} - 1}{|\xi|^{H+1/2}} dW(\xi), \tag{1}$$

where $dW(\xi)$ is a Brownian measure (a good choice of $dW(\xi)$ leads to real valued FBM's). It is worthwhile noting that $\{B_{1/2}(t)\}$ is the well known Brownian Motion.

One of the main interests of the FBM in modeling is that one can prescribe the regularity of its paths. More precisely, the Hölder exponent of the FBM at any point $t_0$ is almost surely equal to $H$; for a random process $\{X(t)\}$ this exponent $\alpha(t_0)$ is defined as follows,

$$\alpha(t_0) = \sup\left\{\alpha, \lim_{h \longrightarrow 0} \frac{X(t_0 + h) - X(t_0)}{|h|^\alpha} = 0\right\}. \tag{2}$$

However, the fact that the parameter $H$ does not depend on $t_0$, is sometimes undesirable since it restricts the field of application. Some phenomena do not admit a constant Hölder exponent: for instance, the use of FBM for synthesizing artificial mountains does not allow to take into account erosion phenomena. The variation of the regularity may even contain an essential part of the signal information. For instance the variation of the Hölder exponent has been used for images segmentation [5].

An extension of the FBM, has been proposed independently by Lévy Véhel and Peltier [8] and by Benassi, Jaffard and Roux [3]. It is called the Multifractional Brownian Motion (MBM). The MBM can be defined by (1), except that the parameter $H$, is replaced by a Hölder function $H(t)$, with values in $]0, 1[$. This process shares many propreties with the FBM; for instance, at any point $t_0$, the Hölder exponent of the MBM is, almost surely, equal to $H(t_0)$ and the MBM is asymptotically locally self-similar of order $H(t_0)$. In addition the local Haussdorff and Box dimensions of the graph of the MBM at $t_0$, are both almost surely equal to $2 - H(t_0)$.

However, the continuity of the function $H(t)$ is sometimes undesirable, since it restricts the field of application. In some contexts (for example financial crash or image segmentation), it may seem more realistic that the Hölder exponent be a discontinuous function. Thus, it would be useful to construct a gaussian process extending the MBM that admits a discontinuous Hölder exponent, and this will be the aim of our paper. The method we use is inspired from the study of the Generalized Weierstrass function in [4].

The remainder of this article is organized as follows. In section 2, we define the Generalized Multifractional Brownian Motion (GMBM) of parameters the function $H(t) \in \mathcal{H}$ and the real $\lambda > 1$ where the space $\mathcal{H}$ is much more larger than the space of Hölder functions (see Definitions 1 and 2). In section 3 and in section 4, we will show that the GMBM shares some essential properties with the MBM: at any point $t_0$, the Hölder exponent of the GMBM is almost surely equal to $H(t_0)$ and the GMBM is asymptotically locally self-similar of order $H(t_0)$.

## 2 Definition of the GMBM

We have seen that the MBM depends on a funtional parameter that belongs to the space of Hölder functions. The GMBM will also depend on a functional parameter that belongs to a functional space $\mathcal{H}$. Let us first define $\mathcal{H}$.
We will say that a function $h : \mathbb{R} \longrightarrow \mathbb{R}$ is a $(\beta, c)$-*Hölder function* where $\beta > 0$ and $c > 0$ if and only if for all $t_1$, $t_2$ satisfying $|t_1 - t_2| < 1$, we have,

$$|h(t_1) - h(t_2)| \leq c|t_1 - t_2|^\beta.$$

**Definition 1.** $\mathcal{H}$ will be the set of the functions $H(t)$ defined on $\mathbb{R}$, such that $H(t) = \liminf\limits_{n \to \infty} H_n(t)$ where $(H_n(t))_{n \in \mathbb{N}}$ is a sequence of $(\beta, c_n)$-Hölder functions with values in $[a, b] \subset ]0, 1[$ that satisfy,

(a) for all $\epsilon > 0$ and $t_0$, there exist $n_0 = n_0(t_0, \epsilon)$ and $h_0 = h_0(t_0, \epsilon) > 0$ such that, for all $n \geq n_0$ and $|h| \leq h_0$ we have $H_n(t_0 + h) \geq H(t_0) - \epsilon$,
(b) for all $t$, $H(t) < \beta$ and $c_n = O(n)$.

**Definition 2.** Let $H(t) = \liminf\limits_{n \to \infty} H_n(t)$ be a function in $\mathcal{H}$ and let $\lambda > 1$. The Generalized Multifractional Brownian Motion (GMBM) of parameters the function $H(t)$ and the real $\lambda$ is the gaussian process $\{Y_{H,\lambda}(t)\}_{t \in \mathbb{R}}$ such that for all $t \in \mathbb{R}$,

$$Y_{H,\lambda}(t) = \sum_{n=0}^{\infty} \int_{D_n} \frac{e^{it\xi} - 1}{|\xi|^{H_n(t) + 1/2}} dW(\xi), \tag{3}$$

where $D_0 = \{|\xi| < 1\}$ and for all $n \geq 1$, $D_n = \{\lambda^{n-1} \leq |\xi| < \lambda^n\}$.

Note that the process (3) not only depends on the function $(H(t))$ but also on the sequence $(H_n(t))$.

It is clear that the GMBM is an extension of the MBM. Indeed when in (3) all the functions $H_n(t)$ are equal to a Hölder function $H(t)$, then the process $\{Y_{H,\lambda}(t)\}_{t \in \mathbb{R}}$ is the MBM of functional parameter $H(t)$.

It is worthwhile noting that when one merely replaces in (1) the parameter $H$ by a discontinuous function $H(t)$ one obtains a discontinuous process. Since our main aim is to control the Hölder exponent at each point, a definition such as (3) is needed.

One can also remark that the set $\mathcal{H}$ contains for instance all the functions that are both lower-semi-continuous (l.s.c.) and piecewise continuous (see Proposition 1); more interestingly, some very irregular functions, for instance functions of the type $b + (a-b)\chi_F(t)$ where $\chi_F(t)$ is the characteristic function of an arbitrary closed set $F$, as for example the Cantor set, belong also to $\mathcal{H}$. (see Proposition 2). At last, one can note that all the functions of $\mathcal{H}$ are l.s.c.

The following propositions give some insights on the set $\mathcal{H}$.

**Proposition 1.** Let $\{t_l\}_{l \in \mathbb{Z}}$ be an increasing sequence and let $H : \mathbb{R} \longrightarrow [a, b] \subset ]0, 1[$ be a function such that for all $t$,

$$H(t) = \sum_{l \in \mathbb{Z}} (a_l(t) - a)\chi_{]t_l, t_{l+1}[}(t) + (b_l - a)\sum_{l \in \mathbb{Z}} \chi_{\{t_l\}}(t) + a, \tag{4}$$

where, for all $l$, $a_l(t)$ is a continuous function on $[t_l, t_{l+1}]$ and $b_l \leq \min\{a_{l-1}(t_l), a_l(t_l)\}$. Then, the function $H(t)$ belongs to the set $\mathcal{H}$.

To prove Proposition 1 we need the following Lemma.

**Lemma 1.** *Let $a : [u, v] \longrightarrow [c, d]$ be a continuous function. For all $\epsilon > 0$, there exist a real $\eta_\epsilon > 0$ and a $C^1$ function $a_\epsilon : [u, v] \longrightarrow [c, d]$ satisfying the following properties:*

*(i) for all $t \in [u, u + \eta_\epsilon]$, $a_\epsilon(t) = a(u)$ and for all $t \in [v - \eta_\epsilon, v]$, $a_\epsilon(t) = a(v)$,*
*(ii) for all $t \in [u, v]$, $|a_\epsilon(t) - a(t)| \leq \epsilon$,*
*(iii) for all $t$, $|a'_\epsilon(t)| \leq e/\eta_\epsilon$ (we can take $e = 12$).*

**Proof** (of Lemma 1) Let $\epsilon > 0$, the function $a(t)$ being uniformly continuous on $[u, v]$, there exists a real $\alpha$, $0 < \alpha < (v - u)/16$, such that for all $t_1, t_2 \in [u, v]$, $|t_1 - t_2| \leq \alpha$, we have $|a(t_1) - a(t_2)| \leq \epsilon/2$.

Let $x_p = u + (p + 1)\alpha/4$, $p_m = \max\{p \in \mathbb{N}, x_p \leq \frac{u+v}{2}\}$ and let $y_q = v - (q + 1)\alpha/4$, $q_m = \max\{q \in \mathbb{N}, y_q \geq \frac{u+v}{2}\}$. For all $x$, $\beta \in [0, 1]$, we set

$$P_\beta(x) = -2\beta x^3 + 3\beta x^2. \tag{5}$$

We have, for all $x$ and for all $\beta$, $|P'_\beta(x)| \leq 3/2$.

The function $a_\epsilon(t)$ will be defined as follows.
For all $t \in [u, x_0 - \alpha/8]$

$$a_\epsilon(t) = a(u),$$

for all $t \in [y_0 + \alpha/8, v]$

$$a_\epsilon(t) = a(v),$$

for all $t \in [x_0 - \alpha/8, x_0]$

$$a_\epsilon(t) = \begin{cases} P_{(a(x_0)-a(u))}\left(\frac{8}{\alpha}[t - (x_0 - \frac{\alpha}{8})]\right) + a(u) & \text{if } a(x_0) \geq a(u) \\ P_{(a(u)-a(x_0))}\left(1 - \frac{8}{\alpha}[t - (x_0 - \frac{\alpha}{8})]\right) + a(x_0) & \text{else,} \end{cases}$$

$a_\epsilon(t)$ will be defined similarly on $[y_0, y_0 + \alpha/8]$,
for all $p \in \{0, \ldots, p_m - 2\}$ and for $t \in [x_p, x_{p+1}]$

$$a_\epsilon(t) = \begin{cases} P_{(a(x_{p+1})-a(x_p))}\left(\frac{4}{\alpha}[t - x_p]\right) + a(x_p) & \text{if } a(x_{p+1}) \geq a(x_p) \\ P_{(a(x_p)-a(x_{p+1}))}\left(1 - \frac{4}{\alpha}[t - x_p]\right) + a(x_{p+1}) & \text{else,} \end{cases}$$

$a_\epsilon(t)$ will be defined similarly on $[y_{k+1}, y_k]$ for all $q \in \{0, \ldots, q_m - 2\}$,
for all $t \in [x_{p_m-1}, y_{q_m-1}]$

$$a_\epsilon(t) = \begin{cases} P_{(a(y_{q_m-1})-a(x_{p_m-1}))}\left(\frac{1}{y_{q_m-1}-x_{p_m-1}}[t - x_{p_m-1}]\right) + a(x_{p_m-1}) \\ \quad \text{if } a(x_{p_m-1}) \leq a(y_{q_m-1}) \\ \\ P_{(a(x_{p_m-1})-a(y_{q_m-1}))}\left(1 - \frac{1}{y_{q_m-1}-x_{p_m-1}}[t - x_{p_m-1}]\right) + a(x_{q_m-1}) \\ \quad \text{else.} \end{cases}$$

∎

**Proof** (of Proposition 1). It follows from the previous Lemma, that for all $l \in \mathbb{Z}$ and $m \geq 1$, there exist a real $\eta_{l,m} > 0$ and a $C^1$ function $a_{l,m} : [t_l, t_{l+1}] \longrightarrow [a, b]$ such that,
- for all $t \in [t_l, t_l + \eta_{l,m}]$, $a_{l,m}(t) = a_l(t_l)$ and for all $t \in [t_{l+1} - \eta_{l,m}, t_{l+1}]$ $a_{l,m}(t) = a_l(t_{l+1})$,
- for all $t \in [t_l, t_{l+1}]$, $|a_{l,m}(t) - a_l(t)| \leq 1/m$,
- for all $t \in [t_l, t_{l+1}]$, $|a'_{l,m}(t)| \leq c/\eta_{l,m}$.

One can suppose that for each $l$, $(\eta_{l,m})_{m \in \mathbb{N}^*}$ is a decreasing sequence and that $\lim_{m \to \infty} \eta_{l,m} = 0$.

For all $k \geq 1$, the function $H_k(t)$ will be defined as follows.

$$H_k(t) = \sum_{l \in I_k} (a_{l,m(l,k)}(t) - a)\chi_{[t_l + 1/k, t_{l+1} - 1/k]}(t) + \sum_{l \in I_k} (\theta_{l,k}(t) - a)\chi_{[t_l - 1/k, t_l + 1/k]}(t) +$$
$$\sum_{l \in J_k} (\gamma_{l,k}(t) - a)\chi_{[t_l, t_l + 1/k]}(t) + \sum_{l \in L_k} (\delta_{l,k}(t) - a)\chi_{[t_l - 1/k, t_l]}(t) + a, \quad (6)$$

where

$$I_k = \{l \in \mathbb{Z}, \, (\exists m \, \eta_{l,m} > 8/k) \text{ and } (\exists m' \, \eta_{l-1,m'} > 8/k)\},$$

$$J_k = \{l \in \mathbb{Z}, \, l \notin I_k \text{ and } l + 1 \in I_k\},$$

and $L_k = \{l \in \mathbb{Z}, \, l \notin I_k \text{ and } l - 1 \in I_k\}$,
note that all these sets are finite;
for all $l \in I_k$, and $t \in [t_l - 1/k, t_l]$

$$\theta_{l,k}(t) = P_{(a_{l-1}(t_l) - b_l)}(1 - k[t - (t_l - 1/k)]) + b_l,$$

and for all $t \in [t_l, t_l + 1/k]$

$$\theta_{l,k}(t) = P_{(a_l(t_l) - b_l)}(k[t - t_l]) + b_l,$$

for all $l \in J_k$

$$\gamma_{l,k}(t) = P_{(a_l(t_l) - a)}(k[t - t_l]) + a,$$

for all $l \in L_k$

$$\delta_{l,k}(t) = P_{(a_l(t_l) - a)}(1 - k[t - (t_l - 1/k)]) + a,$$

and $m(l,k) = \max\{m, \eta_{l,m} > 8/k\}$. ∎

**Proposition 2.** *Let $H : \mathbb{R} \longrightarrow [a,b] \subset ]0,1[$ be the function defined by*

$$H(t) = \sum_{l \in \mathbb{N}} (a_l(t) - a)\chi_{]\alpha_l, \beta_l[}(t) + a, \quad (7)$$

*where*
- *for all $l, l' \in \mathbb{N}$, $l \neq l'$ we have $]\alpha_l, \beta_l[ \cap ]\alpha_{l'}, \beta_{l'}[ = \emptyset$,*
- *for all $l$, $a_l(t)$ is a continuous function on $[\alpha_l, \beta_l]$.*

Since any open set may be written as a countable disjoined union of open intervals, a consequence of the previous proposition is that all the functions of the type $(b-a)\chi_U(t) + a$, where $U \subset \mathbb{R}$ is open, belong to $\mathcal{H}$.

The proof of Proposition 2 is similar to the one of Proposition 1. ∎

Further properties of $\mathcal{H}$ will be investigated in a forthcoming work.

**Theorem 1.** *With probability one, $t \longmapsto Y_{H,\lambda}(t,\omega)$ is a continuous function*

To prove Theorem 1, we will need the following Lemma.

**Lemma 2.** *Let $\{X(t)\}_{t \in K}$ be a real valued gaussian process defined on a compact interval $K$ and suppose there exist 2 reals $c > 0$ and $\alpha > 0$, such that for all $t, t' \in K$*

$$E([X(t) - X(t')]^2) \leq c|t - t'|^\alpha. \tag{8}$$

*Then with probability one, $t \longmapsto X(t,\omega)$ is a continuous function.*

**Proof** (of Lemma 2). Let us show that the process $\{X(t)\}_{t \in K}$ satisfy the following property; there exists 3 reals $C > 0$, $\beta > 0$ and $\gamma > 0$ such that for all $t, t' \in K$

$$E(|X(t) - X(t')|^\beta) \leq C|t - t'|^{1+\gamma}. \tag{*}$$

When the inequality (*) is satisfied, it results from Kolmogorov criterium (see [10] page 57) that almost surely

$$\lim_{h \to 0} \sup_{\substack{t, t' \in K \\ |t - t'| < h}} |X(t) - X(t')| = 0.$$

Thus with probability one, $t \longmapsto X(t,\omega)$ is a continuous function.

To show (*), let us recall that (see [7] page 110), when $Z$ is a centered gaussian random variable, then for every real $r > 0$,

$$E(|Z|^r) = \frac{2^{r/2}\Gamma\left(\frac{r+1}{2}\right)}{\Gamma\left(\frac{1}{2}\right)}(E(|Z|^2))^{r/2}.$$

Thus taking in the previous equality $Z = X(t) - X(t')$ and $r = \frac{2(1+\gamma)}{\alpha}$ ($\gamma > 0$ being arbitrary), then using (8) we get (*). ∎

**Proof** (of Theorem 1) Let us first show that for all $l \in \mathbb{Z}$, with probability one $t \longmapsto Y(t, \omega)$ is a continuous function on $K_l = [\frac{l}{4}, \frac{l}{4} + \frac{1}{2}]$. It follows from Lemma 2, that it is sufficient to establish that: $\forall$, $t$, $t' \in K_l$

$$E([Y(t) - Y(t')]^2) \leq c|t - t'|^{2a}. \tag{9}$$

We have,

$$E([Y(t) - Y(t')]^2) \leq 2(A + B + G + H),$$

where

$$A = 2 \int_0^1 |e^{it\xi} - 1|^2 (|\xi|^{-H_0(t)-1/2} - |\xi|^{-H_0(t')-1/2})^2 \, d\xi,$$

$$B = 8 \sum_{n=1}^{\infty} \int_{\lambda^{n-1}}^{\lambda^n} (|\xi|^{-H_n(t)-1/2} - |\xi|^{-H_n(t')-1/2})^2 \, d\xi,$$

$$H = 2 \int_0^1 |\xi|^{-2H_0(t')-1} |e^{i(t-t')\xi} - 1|^2 \, d\xi,$$

$$G = 2 \sum_{n=1}^{\infty} \int_{\lambda^{n-1}}^{\lambda^n} |\xi|^{-2H_n(t')-1} |e^{i(t-t')\xi} - 1|^2 \, d\xi.$$

Let us give ad-hoc upper bounds of $A$, $B$, $H$ and $G$. The theorem on finite increments entails that for all $\xi \neq 0$, $t$, $t'$ and $n$, there exists $\tau \in [a+1/2, b+1/2]$, such that

$$||\xi|^{-H_n(t)-1/2} - |\xi|^{-H_n(t')-1/2}| = |\log|\xi|| |\xi|^{-\tau} |H_n(t) - H_n(t')|. \tag{10}$$

Then it follows from (10) and from Definition 1 (b) that,

$$A \leq 2 \left( \int_0^1 |e^{it\xi} - 1|^2 \frac{(\log \xi)^2}{\xi^{2b+1}} \, d\xi \right) |H_0(t) - H_0(t')|^2$$

$$\leq 2 \left( \left( \frac{|l|}{4} + \frac{1}{2} \right)^2 \int_0^1 \frac{(\log \xi)^2}{\xi^{2b-1}} \, d\xi \right) c_0^2 |t - t'|^{2\beta}, \tag{11}$$

and

$$B \leq 8 \sum_{n=1}^{\infty} \left( \int_{\lambda^{n-1}}^{\lambda^n} \frac{(\log \xi)^2}{\xi^{2a+1}} \, d\xi \right) c_n^2 |t - t'|^{2\beta}$$

$$\leq 8 \sum_{n=1}^{\infty} \lambda^{n-1}(\lambda - 1)(n \log \lambda)^2 \lambda^{-(n-1)(2a+1)} c_n^2 |t - t'|^{2\beta} \tag{12}$$

$$\leq 8(\lambda - 1)(\log \lambda)^2 \left( \sum_{n=1}^{\infty} n^2 \lambda^{-2a(n-1)} c_n^2 \right) |t - t'|^{2\beta}.$$

Moreover,

$$H \leq 2 \left( \int_0^1 \xi^{-2H_0(t')+1} \, d\xi \right) |t - t'|^2 \leq 2 \left( \int_0^1 \xi^{-2b+1} \, d\xi \right) |t - t'|^2. \tag{13}$$

At last we have,

$$G \leq 2\sum_{n=1}^{\infty} \int_{\lambda^{n-1}}^{\lambda^n} \xi^{-2a-1}|e^{i(t-t')\xi} - 1|^2\, d\xi \leq 2\int_1^{+\infty} \xi^{-2a-1}|e^{i(t-t')\xi} - 1|^2\, d\xi,$$

$$\leq 2\left(\int_1^{|t-t'|^{-1}} \xi^{-2a+1}\, d\xi\right)|t-t'|^2 + 8\left(\int_{|t-t'|^{-1}}^{+\infty} \xi^{-2a-1}\, d\xi\right), \tag{14}$$

$$\leq c|t-t'|^{2a}.$$

Thus (9) results from (11), (12), (13) et (14).

∎

## 3  Pointwise regularity of the GMBM

The following theorem shows that one can prescribe the regularity of the GMBM.

**Theorem 2.** *Let $\{Y_{H,\lambda}(t)\}_{t\in \mathbb{R}}$ be the GMBM of parameters the function $H(t)$ of $\mathcal{H}$ and the real $\lambda > 1$. Let $t_0$ be an arbitrary real and let $\alpha_Y(t_0)$ be the Hölder exponent of the process $\{Y_{H,\lambda}(t)\}_{t\in \mathbb{R}}$ at $t_0$ (see (2)).*
*Then*

$$\alpha_Y(t_0) = H(t_0),$$

*almost surely.*

To prove Theorem 2 we need the following Lemma.

**Lemma 3.** *Let $\{X(t)\}_{t\in I}$ be a real valued centered gaussian process defined on an interval $I$ and let $\alpha_X(t_0)$ be its Hölder exponent at $t_0$. $H(t)$ will be a deterministic funtion with values in $[a,b] \subset\, ]0,1[$.*

*(i) Suppose that for all $\epsilon > 0$, one can find two reals $c = c(t_0, \epsilon) > 0$ and $h_0 = h_0(t_0, \epsilon) > 0$ such that, for all $|h|, |h'| \leq h_0$,*

$$E([X(t_0+h) - X(t_0+h')]^2) \leq c|h-h'|^{2(H(t_0)-\epsilon)}. \tag{15}$$

*Then, $\alpha_X(t_0) \geq H(t_0)$ almost surely.*
*(ii) Suppose that for all $\epsilon > 0$, there exists a sequence $(h_n)$ satisfying $\lim_{n\to\infty} h_n = 0$ and a real $c = c(t_0, \epsilon) > 0$ such that, for all $n$,*

$$E([X(t_0+h_n) - X(t)]^2) \geq c|h_n|^{2(H(t_0)+\epsilon)}, \tag{16}$$

*then $\alpha_X(t_0) \leq H(t_0)$ almost surely.*

This lemma follows from general results on the regularity of gaussian processes. However, for sake of completness we prove it here by an elementary method.

**Proof** Let us fix $t_0 \in I$, $\epsilon > 0$ and suppose that $I = ]u, v[$. (15) entails that there exists 2 reals $c > 0$ and $h_0 > 0$ satisfying the following property: for all $h$, $h'$ such that $|h|, |h'| < h_0$ and $t_0 + h, t_0 + h' \in I$,

$$E([X(t_0 + h) - X(t_0 + h')]^2) \leq c|h - h'|^{2(H(t_0) - \epsilon/2)}. \tag{17}$$

Let $\eta = \frac{1}{2}\min(h_0, t - u, v - t)$ and $J = [-\eta, \eta]$. The gaussian process $\{Y_{t,\epsilon}(k)\}_{k \in J}$, will be defined on $J$ as follows

$$Y_{t_0,\epsilon}(k) = \begin{cases} \frac{X(t_0+k)-X(t_0)}{|k|^{H(t_0)-\epsilon}}, & \text{if } k \in J - \{0\} \\ 0 \text{ if } k = 0. \end{cases} \tag{18}$$

Suppose for a while that for all $h, h' \in J$

$$E(|Y_{t_0,\epsilon}(h') - Y_{t_0,\epsilon}(h)|^2) \leq C'|h - h'|^{\epsilon}. \tag{19}$$

It follows then from (18) and from Lemma 2 that almost surely $\lim_{k \to 0} Y_{t_0,\epsilon}(k) = 0$ which means that almost surely $\alpha_X(t_0) \geq H(t_0) - \epsilon$. Thus from now on our aim will be to show that the inequality (19) holds. It is clear that the inequality (19) is true when $h = 0$ or when $h' = 0$. For all $h, h' \in J - \{0\}$ we have

$$E(|Y_{t_0,\epsilon}(h') - Y_{t_0,\epsilon}(h)|^2)$$
$$\leq 2E\left(\frac{|X(t_0 + h') - X(t_0 + h)|^2}{|h'|^{2(H(t_0)-\epsilon)}}\right)$$
$$+ 2\left(\frac{1}{|h'|^{H(t_0)-\epsilon}} - \frac{1}{|h|^{H(t_0)-\epsilon}}\right)^2 E(|X(t_0 + h) - X(t_0)|^2).$$

It follows then from (17) that,

$$E(|Y_{t_0,\epsilon}(h') - Y_{t_0,\epsilon}(h)|^2) \leq 2c\frac{|h' - h|^{2(H(t_0)-\epsilon/2)}}{|h'|^{2(H(t_0)-\epsilon)}}$$
$$+ 2c\left(\frac{1}{|h'|^{H(t_0)-\epsilon}} - \frac{1}{|h|^{H(t_0)-\epsilon}}\right)^2 |h|^{2(H(t_0)-\epsilon/2)}.$$

From now on we will suppose that $0 < |h| \leq |h'|$, so we have $\frac{1}{|h'|^{2(H(t_0)-\epsilon)}} \leq \frac{2^{2(H(t_0)-\epsilon)}}{|h'-h|^{2(H(t_0)-\epsilon)}}$, and consequently,

$$E(|Y_{t_0,\epsilon}(h') - Y_{t_0,\epsilon}(h)|^2) \leq c2^{2H(t_0)+1}|h' - h|^{\epsilon} + 2c\left(1 - \left|\frac{h}{h'}\right|^{H(t_0)-\epsilon}\right)^2 |h|^{\epsilon}.$$

Let us study the case where $h' > 0$, the case where $h' < 0$ being similar.

When $h \in [-h', h'/2]$, since $|h' - h| \geq |h|$ the inequality (19) obviously holds. When $h \in ]h'/2, h']$, the Theorem on finite increments entails that

$$1 - \left|\frac{h}{h'}\right|^{H(t_0)-\epsilon} \leq 2(H(t_0) - \epsilon)\left(1 - \frac{h}{h'}\right),$$

consequently, we have

$$\left(1 - \left|\frac{h}{h'}\right|^{H(t_0)-\epsilon}\right)^2 |h|^\epsilon \leq 4\left(1 - \frac{h}{h'}\right)^2 h'^\epsilon,$$

$$\leq 4|h' - h|^\epsilon,$$

and then, we get (19) from this last inequality.

Now, we are going to show $(ii)$. For all $\epsilon > 0$, there exists $(h_n)$ a sequence of non vanishing reals converging to 0 and there exists $c > 0$ such that

$$E[(X(t_0 + h_n) - X(t_0))^2] \geq c|h_n|^{2(H(t_0)+\epsilon/2)}.$$

Since $\{X(t)\}$ is a gaussian process we have

$$\frac{X(t_0 + h_n) - X(t_0)}{|h_n|^{H(t_0)+\epsilon}} \rightsquigarrow \mathcal{N}(0, \sigma_n^2),$$

and the last inequality implies that $\lim_{n \to \infty} \sigma_n = +\infty$. Let us show that the sequence of random variables $\left(\frac{|h_n|^{H(t_0)+\epsilon}}{X(t_0+h_n)-X(t_0)}\right)$ converges in probability to 0. For all real $\eta > 0$, we have

$$P\left(\frac{|h_n|^{H(t_0)+\epsilon}}{|X(t_0 + h_n) - X(t_0)|} < \eta\right) = P\left(\frac{|X(t_0 + h_n) - X(t_0)|}{|h_n|^{H(t_0)+\epsilon}} > 1/\eta\right)$$

$$= \int_{(|x|>1/\eta)} \frac{1}{\sqrt{2\pi}\sigma_n} \exp(-x^2/2\sigma_n^2)\, dx$$

$$= \frac{1}{\sqrt{2\pi}} \int_{(|x|>1/\eta\sigma_n)} \exp(-x^2/2)\, dx.$$

Thus

$$\lim_{n \to \infty} P\left(\frac{|h_n|^{H(t_0)+\epsilon}}{|X(t_0 + h_n) - X(t_0)|} < \eta\right) = \frac{1}{\sqrt{2\pi}} \int_{-\infty}^{+\infty} \exp(-x^2/2)\, dx = 1.$$

Therefore, there exists a sub-sequence $k \mapsto n_k$ such that

$$\lim_{k \to +\infty} \frac{|X(t_0 + h_{n_k}) - X(t_0)|}{|h_{n_k}|^{H(t_0)+\epsilon}} = +\infty,$$

almost surely. This means that almost surely, $\alpha_X(t_0) \leq H(t_0) + \epsilon$. ∎

**Proof** (of Theorem 2).
**Step 1**: we will establish that for all $t_0$, $\alpha_Y(t_0) \geq H(t_0)$ almost surely. It follows from Lemma 3 $(i)$ that it is sufficient to show that for all $\epsilon > 0$, there exist 2 reals $c = c(t_0, \epsilon) > 0$ and $h_0 = h_0(t_0, \epsilon) > 0$, such that for all $|h|, |h'| \leq h_0$

$$E([Y(t_0 + h) - Y(t_0 + h')]^2) \leq c|h - h'|^{2(H(t_0)-\epsilon)}. \tag{20}$$

Let $h_0(t_0, \epsilon) \leq 1/2$ and $n_0 = n_0(t_0, \epsilon)$ be as in Definition 1 (a). For all $|h|$, $|h'| \leq h_0$, we have

$$E[(Y(t_0 + h) - Y(t_0 + h'))^2] \leq 2(R + S + T + U),$$

where

$$R = 2 \int_0^1 |e^{i(t_0+h)\xi} - 1|^2 (|\xi|^{-H_0(t_0+h)-1/2} - |\xi|^{-H_0(t_0+h')-1/2})^2 \, d\xi,$$

$$S = 8 \sum_{n=1}^{\infty} \int_{\lambda^{n-1}}^{\lambda^n} (|\xi|^{-H_n(t_0+h)-1/2} - |\xi|^{-H_n(t_0+h')-1/2})^2 \, d\xi,$$

$$T = 2 \int_0^1 |\xi|^{-2H_0(t_0+h')-1} |e^{i(h-h')\xi} - 1|^2 \, d\xi +$$
$$2 \sum_{n=1}^{n_0-1} \int_{\lambda^{n-1}}^{\lambda^n} |\xi|^{-2H_n(t_0+h')-1} |e^{i(h-h')\xi} - 1|^2 \, d\xi,$$

$$U = 2 \sum_{n=n_0}^{\infty} \int_{\lambda^{n-1}}^{\lambda^n} |\xi|^{-2H_n(t_0+h')-1} |e^{i(h-h')\xi} - 1|^2 \, d\xi.$$

Let us give ad-hoc upper bounds of $R$, $S$, $T$, and $U$. A similar method to that we have used to give upper bounds of $A$ and $B$, (see the proof of Theorem 1) leads to

$$R \leq 2 \left( (|t_0| + h_0)^2 \int_0^1 \frac{(\log \xi)^2}{\xi^{2b-1}} \, d\xi \right)^2 c_0^2 |h - h'|^{2\beta},$$
$$S \leq 8 \left( (\log \lambda)^2 (\lambda - 1) \sum_{n=1}^{\infty} n^2 \lambda^{-(n-1)2a} c_n^2 \right) |h - h'|^{2\beta}. \tag{21}$$

Moreover, as for all $t$ and $n$, $H_n(t) \in [a, b]$, we get

$$T \leq 2 \left( \int_0^1 \xi^{-2b+1} \, d\xi \right) |h - h'|^2 + 2 \left( \int_1^{\lambda^{n_0}} \xi^{-2a+1} \, d\xi \right) |h - h'|^2. \tag{22}$$

At last, Definition (a) implies that

$$U \leq \sum_{n=n_0}^{\infty} \int_{\lambda^{n-1}}^{\lambda^n} \xi^{-2H(t_0)+2\epsilon-1} |e^{i(h-h')\xi} - 1|^2 \, d\xi$$
$$\leq 2 \int_1^{+\infty} \xi^{-2H(t_0)+2\epsilon-1} |e^{i(h-h')\xi} - 1|^2 \, d\xi$$
$$\leq 2 \left( \int_1^{|h-h'|^{-1}} \xi^{-2H(t_0)+2\epsilon+1} \, d\xi \right) |h - h'|^2 + 8 \int_{|h-h'|^{-1}}^{+\infty} \xi^{-2H(t_0)+2\epsilon-1} \, d\xi \tag{23}$$
$$\leq c |h - h'|^{2(H(t_0)-\epsilon)}.$$

Thus, it follows from (21), (22) et (23) that the process $\{Y(t)\}$ satisfies (20).

**Step 2:** we will establish that for all $t_0$, $\alpha_Y(t_0) \geq H(t_0)$ almost surely. It follows from Lemma 3 (ii) that it is sufficient to show that for all $\epsilon > 0$, there exists a sequence $(h_n)$ of non vanishing real numbers converging to 0 and there exists $c = c(t_0, \epsilon) > 0$ such that for all $n$,

$$E([Y(t_0 + h_n) - Y(t_0)]^2) \geq c|h_n|^{2(H(t_0)+\epsilon)}. \tag{24}$$

As $H(t_0) = \liminf_{n \to \infty} H_n(t_0)$, there exists a subsequence $k \mapsto n_k$ such that for all $k$, $H_{n_k+1}(t_0) \leq H(t_0) + \epsilon$. From now on we will suppose that $n = n_k$. Let $h_n = \lambda^{-n}$, we have

$$\|Y(t_0 + h_n) - Y(t_0)\|_{L^2(\Omega)} = \left(E[(Y(t_0 + h_n) - Y(t_0))^2]\right)^{1/2}$$
$$\geq \left(\int_{\lambda^n}^{\lambda^{n+1}} |\xi|^{-2H_{n+1}(t_0)-1} |e^{i(t_0+h_n)\xi} - e^{it_0\xi}|^2 \, d\xi\right)^{1/2}$$
$$- \left(\int_{\lambda^n}^{\lambda^{n+1}} \left(|\xi|^{-H_{n+1}(t_0)-1/2} - |\xi|^{-H_{n+1}(t_0+h_n)-1/2}\right)^2 |e^{i(t_0+h_n)\xi} - 1|^2 \, d\xi\right)^{1/2}$$
$$\geq \left(\int_{\lambda^n}^{\lambda^{n+1}} |\xi|^{-2H(t_0)-2\epsilon-1} |e^{ih_n\xi} - 1|^2 \, d\xi\right)^{1/2}$$
$$- 2 \left(\int_{\lambda^n}^{\lambda^{n+1}} \left(|\xi|^{-H_{n+1}(t_0)-1/2} - |\xi|^{-H_{n+1}(t_0+h_n)-1/2}\right)^2 \, d\xi\right)^{1/2}.$$

Thus setting $u = h_n \xi$, we get

$$\begin{aligned}\int_{\lambda^n}^{\lambda^{n+1}} |\xi|^{-2H(t_0)-2\epsilon-1} |e^{ih_n\xi} - 1|^2 \, d\xi \\= h_n^{2(H(t_0)+\epsilon)} \int_1^\lambda u^{-2H(t_0)-2\epsilon-1} |e^{iu} - 1|^2 \, du.\end{aligned} \tag{25}$$

Moreover, it follows from (10) that

$$\int_{\lambda^n}^{\lambda^{n+1}} \left(|\xi|^{-H_{n+1}(t_0)-1/2} - |\xi|^{-H_{n+1}(t_0+h_n)-1/2}\right)^2 d\xi$$
$$\leq \left(\int_{\lambda^n}^{\lambda^{n+1}} (\log \xi)^2 \xi^{-2a-1} \, d\xi\right) c_{n+1}^2 h_n^{2\beta}$$
$$\leq \left((n+1)^2 \lambda^n (\lambda - 1)(\log \lambda)^2 \lambda^{-(2a+1)n} \lambda^{-2(\beta - H(t_0) - \epsilon)n}\right) c_{n+1}^2 h_n^{2(H(t_0)+\epsilon)}$$
$$= \left((n+1)^2 (\log \lambda)^2 (\lambda - 1) \lambda^{-2(a+\beta - H(t_0) - \epsilon)n}\right) c_{n+1}^2 h_n^{2(H(t_0)+\epsilon)}. \tag{26}$$

At last, (25) and (26) entail that

$$\|Y(t_0 + h_{n_k}) - Y(t_0)\|_{L^2(\Omega)} \geq \left\{\left(\int_1^\lambda u^{-2H(t_0)-2\epsilon-1} |e^{iu} - 1|^2 \, du\right)^{1/2} \right.$$
$$\left. - C_\lambda \lambda^{-(a+\beta - H(t_0) - \epsilon)n_k} (n_k + 1) c_{n_k+1}\right\} h_{n_k}^{H(t_0)+\epsilon},$$

and then Definition 1 (b), implies that for $k$, big enough,

$$\|Y(t_0 + h_{n_k}) - Y(t_0)\|_{L^2(\Omega)} \geq \tfrac{\sqrt{2}}{2} \left(\int_1^\lambda u^{-2H(t_0)-2\epsilon-1} |e^{iu} - 1|^2 \, du\right)^{1/2} h_{n_k}^{H(t_0)+\epsilon}.$$

∎

## 4  Local self-similarity of the GMBM

**Definition 3.** *We will say that a process $\{X(t)\}$ is asymptotically locally self-similar of order $\alpha > 0$ at $t_0$, if and only if the process $\left\{\frac{X(t_0+\rho u)-X(t_0)}{\rho^\alpha}\right\}_u$ converge in distribution to a non trivial limit as $\rho \to 0^+$.*

**Proposition 3.** *Let $\{Y_{H,\lambda}(t)\}_{t\in\mathbb{R}}$ be the GMBM of parameters the function $H(t) = \liminf_{n\to\infty} H_n(t)$ of $\mathcal{H}$ and the real $\lambda > 1$. Let $t_0$ be a real satisfying the following properties.*

1. *There exist 2 reals $c = c(t_0) > 0$ and $\delta = \delta(t_0) > 1 - a$ such that for all $n$,*
$$|H_n(t_0) - H(t_0)| \leq c\lambda^{-2\delta n};$$

2. *These exist $n_0 = n_0(t_0)$ and $h_0 = h_0(t_0) > 0$ such that, for all $n \geq n_0$ and $|h| \leq h_0$ we have $H_n(t_0 + h) \geq H(t_0)$.*

*Then the process $\{Y_{H,\lambda}(t)\}_{t\in\mathbb{R}}$ is asymptotically locally self-similar of order $H(t_0)$ at $t_0$. More precisely,*

$$\lim_{\rho\to 0^+} \text{Law}\left\{\frac{Y(t_0+\rho u) - Y(t_0)}{\rho^{H(t_0)}}\right\}_{u\in\mathbb{R}} = \text{Law}\left\{\sigma(t_0)B_{H(t_0)}(u)\right\}_{u\in\mathbb{R}},$$

*where $\{B_{H(t_0)}(u)\}_{u\in\mathbb{R}}$ is the standard fractional brownian motion of parameter $H(t_0)$ and where*

$$\sigma^2(t_0) = 4\int_{-\infty}^{+\infty} \frac{\sin^2 u/2}{|u|^{2H(t_0)+1}} du.$$

It is worthwhile noting here that there exist many "interesting" functions in $\mathcal{H}$ that satisfy 1. and 2. for all real $t_0$. Indeed, following the same lines as in the proof of Proposition 1 one can show that:

- functions that are both lower-semi-continuous and piecewise constant;
- functions of the type $b + (a - b)X_F(t)$, where $0 < a < b < 1$ and $X_F(t)$ is the characteristic function of any arbitrary closed set $F$,

satisfy 1. and 2..

Proposition 3 is a consequence of the following Lemmas.

**Lemma 4.** *For all $t_0, u, v$ and $\rho$ small enough, we have*

$$E[(Y(t_0 + \rho u) - Y(t_0))(Y(t_0 + \rho v) - Y(t_0))]$$
$$= \sum_{n=0}^\infty \int_{D_n} \frac{(e^{i\rho u\xi}-1)(e^{-i\rho v\xi}-1)}{|\xi|^{2H_n(t_0)+1}} d\xi + 0(\rho^{2H(t_0)})$$

**Proof** Let

$$N_n(\xi, x, y) = \frac{e^{i(t_0+x)\xi} - 1}{|\xi|^{H_n(t_0+y)+1/2}}$$

$$E[(Y(t_0 + \rho u) - Y(t_0))(Y(t_0 + \rho v) - Y(t_0))]$$
$$= \sum_{n=0}^{\infty} \int_{D_n} [N_n(\xi, \rho u, \rho u) - N_n(\xi, 0, 0)][\overline{N_n(\xi, \rho v, \rho v)} - \overline{N_n(\xi, 0, 0)}] d\xi$$
$$= \sum_{n=0}^{\infty} \Bigl\{ \int_{D_n} [N_n(\xi, \rho u, \rho u) - N_n(\xi, \rho u, 0)][\overline{N_n(\xi, \rho v, \rho v)} - \overline{N_n(\xi, \rho v, 0)}] d\xi$$
$$+ \int_{D_n} [N_n(\xi, \rho u, \rho u) - N_n(\xi, \rho u, 0)][\overline{N_n(\xi, \rho v, 0)} - \overline{N_n(\xi, 0, 0)}] d\xi$$
$$+ \int_{D_n} [N_n(\xi, \rho u, 0) - N_n(\xi, 0, 0)][\overline{N_n(\xi, \rho v, \rho v)} - \overline{N_n(\xi, \rho v, 0)}] d\xi$$
$$+ \int_{D_n} [N_n(\xi, \rho u, 0) - N_n(\xi, 0, 0)][\overline{N_n(\xi, \rho v, 0)} - \overline{N_n(\xi, 0, 0)}] d\xi \Bigr\}.$$

At last, by a similar method to that we have used to give upper bounds of $A, B, H$ and $G$ in the proof of Theorem 1, we can show that

$$E[(Y(t_0 + \rho u) - Y(t_0))(Y(t_0 + \rho v) - Y(t_0))]$$
$$\leq \alpha_1 \rho^{2\beta} |u|^\beta |v|^\beta + \alpha_2 \rho^{\beta+H(t_0)} |u|^\beta |v|^{H(t_0)} + \alpha_3 \rho^{\beta+H(t_0)} |u|^{H(t_0)} |v|^\beta$$
$$+ \sum_{n=0}^{\infty} \int_{D_n} \frac{(e^{i\rho u\xi} - 1)(e^{-i\rho v\xi} - 1)}{|\xi|^{2H_n(t_0)+1}} d\xi,$$

where $\alpha_1, \alpha_2$ and $\alpha_3$ are 3 constants. ∎

**Lemma 5.** *If the sequence $(H_n(t_0))$ satisfies the condition 1. of Proposition 3, then we have*

$$\lim_{\rho \to 0+} \frac{1}{\rho^{2H(t_0)}} \sum_{n=0}^{\infty} \int_{D_n} \frac{(e^{i\rho u\xi} - 1)(e^{-i\rho v\xi} - 1)}{|\xi|^{2H_n(t_0)+1}} d\xi$$
$$= \int_{-\infty}^{+\infty} \frac{(e^{iu\eta} - 1)(e^{-iv\eta} - 1)}{|\eta|^{2H(t_0)+1}} d\eta$$

**Proof**

$$\int_{-\infty}^{+\infty} \frac{(e^{iu\eta} - 1)(e^{-iv\eta} - 1)}{|\eta|^{2H(t_0)+1}} d\eta = \frac{1}{\rho^{2H(t_0)}} \int_{-\infty}^{+\infty} \frac{(e^{i\rho u\xi} - 1)(e^{-i\rho v\xi} - 1)}{|\xi|^{2H(t_0)+1}} d\xi.$$

Thus, we have

$$\left| \frac{1}{\rho^{2H(t_0)}} \sum_{n=0}^{\infty} \int_{D_n} \frac{(e^{i\rho u \xi} - 1)(e^{-i\rho v \xi} - 1)}{|\xi|^{2H_n(t_0)+1}} - \int_{-\infty}^{+\infty} \frac{(e^{iu\eta} - 1)(e^{-iv\eta} - 1)}{|\eta|^{2H(t_0)+1}} d\eta \right|$$

$$\leq \frac{1}{\rho^{2H(t_0)}} \sum_{n=0}^{\infty} \int_{D_n} \left| (e^{i\rho u \xi} - 1)(e^{-i\rho v \xi} - 1) \right| \left( \frac{1}{|\xi|^{2H_n(t_0)+1}} - \frac{1}{|\xi|^{2H(t_0)+1}} \right) \right| d\xi$$

$$\leq 2\rho^{2(1-H(t_0))} |u||v| \left[ \int_0^1 \left( \frac{1}{\xi^{2H_0(t_0)-1}} + \frac{1}{\xi^{2H(t_0)-1}} \right) d\xi \right.$$
$$+ \sum_{n=1}^{\infty} \lambda^{2n} \int_{\lambda^{n-1}}^{\lambda^n} \left| \frac{1}{\xi^{2H_n(t_0)+1}} - \frac{1}{\xi^{2H(t_0)+1}} \right| d\xi \right].$$

This last series is convergent, since the Theorem on finite increments and 1. (see Proposition 3) imply that

$$\sum_{n=1}^{\infty} \lambda^{2n} \int_{\lambda^{n-1}}^{\lambda^n} \left| \frac{1}{\xi^{2H_n(t_0)+1}} - \frac{1}{\xi^{2H(t_0)+1}} \right| d\xi,$$

$$\leq 2\lambda^{2a}(\lambda - 1) \log \lambda \sum_{n=1}^{\infty} n \lambda^{2(1-a)n} |H_n(t_0) - H(t_0)|,$$

$$\leq c \sum_{n=1}^{\infty} n \lambda^{2(1-a-\delta)n} < \infty.$$

∎

**Lemma 6.** *We have*

$$E[(Y(t_0 + \rho u) - Y(t_0))(Y(t_0 + \rho v) - Y(t_0))]$$
$$= \left( \int_{-\infty}^{+\infty} \frac{(e^{iu\eta} - 1)(e^{-iv\eta} - 1)}{|\eta|^{2H(t_0)+1}} d\eta \right) \rho^{2H(t_0)} + O(\rho^{2H(t_0)}).$$

**Proof** This Lemma is a straightforward consequence of Lemmas 4 and 5. ∎

**Proof** (of Proposition 3) Lemma 6 yields the convergence of the finite dimensional distribution of the process $\left\{ \frac{Y_{H,\lambda}(t_0 + \rho u) - Y_{H,\lambda}(t_0)}{\rho^{H(t_0)}} \right\}_{u \in \mathbb{R}}$ to those of $\{\sigma(t_0) B_{H(t_0)}(u)\}_{u \in \mathbb{R}}$. To have the convergence in distribution for the topology of the uniform convergence on compact sets a tightness result is required.

Following the same lines as in the proof of Theorem 2 and using 2. one can show that, for integer $l = 2$ and for every compact $K$ there exists a finite constant $c(t_0, l)$ such that for $\rho > 0$ small enough,

$$\forall u, v \in K \quad E\left( \frac{Y(t_0 + \rho u) - Y(t_0 - \rho v)}{\rho^{H(t_0)}} \right)^2 \leq c(t_0, l)|u - v|^{lH(t_0)}.$$

Since the processes are Gaussian this inequality can be extended to $l$ large enough to get $H(t_0).l > 1$. Hence one can classically deduce that the sequence of the laws of $\left(\frac{Y(t_0+\rho u)-Y(t_0)}{\rho^{H(t_0)}}\right)_{\rho>0}$ is relatively compact.

∎

## Acknowledgement

The authors would like to thank Albert Benassi and Serge Cohen for their valuable comments on earlier versions of this papers.

## References

1. A. Benassi, P. Bertrand, S. Cohen and J. Istas (1999): Identification d'un processus gaussien multifractionnaire avec des ruptures sur la fonction d'échelle. To appear in C. R. Sci. Paris.
2. A. Benassi, S. Cohen and J. Istas (1998): Identifying the multifractional function of a Gaussian process. Statistics & Probability Letters 39, 337–345.
3. A. Benassi, S. Jaffard and D. Roux (1997): Gaussian processes and pseudodifferential Elliptic operators. Rev. Mat. Iberoamericana 13 (1), 19–89.
4. K. Daoudi, J. Lévy Véhel and Y. Meyer (1998): Construction of continuous functions with prescribed local regularity. Const. Approx. 014 (03), 349–385.
5. J. Lévy Véhel (1996): Introduction to the multifractal analysis of images. Y. Fisher Editor, Springer Verlag.
6. B. B. Mandelbrot and J. Van Ness (1968): Fractional brownian motion, fractional noises and applications. Siam Review Vol.10, 422–437.
7. A. Papoulis (1991): Probability, Random variables and stochastic processes. Mc-Drawhill.
8. R. F. Peltier and J. Lévy Véhel (1995): Multifractional Brownian Motion: definition and preliminary results. Rapport de recherche INRIA, No. 2645.
9. G. Samorodnistsky and M. S. Taqqu (1994): Stable non-gaussian random processes. Chapman & Hall.
10. E. Wong and B. Hajek (1985): Stochastic processes in engineering systems. Springer-Verlag.

# Elliptic Self Similar Stochastic Processes

Albert Benassi and Daniel Roux

Laboratoire de Mathématiques Appliquées, CNRS URA 1501
Clermont-Ferrand 2, 63177 Aubière cedex - FRANCE
{benassi,roux}@ucfma.univ-bpclermont.fr

**Abstract.** In this paper we define the elliptic stochastic processes for some constant coefficients pseudo differential operator and with respect to some stochastic measure. We characterize the elliptic processes which are with stationary increments and self-similar (SISS). We are doing this by a characterization of stochastic measures entering in the problem. We also give examples of SISS elliptic processes with non stable laws.

## 1 Introduction

The scale invariance property plays a fundamental role in many areas of theoritical and applied sciences. A single word can unify different occurrences of this concept: the word fractal, introduced by Mandelbrot in ([9]). Roughly speaking a fractal $F$ presents the same aspect whatever is the scale of observation: every little part of $F$ looks like the entire object after zooming. A mathematical definition of this property is "self-similarity".

A stochastic process $X = \{X(x), \ x \in \mathbb{R}^d\}$ is self-similar with scale parameter or index $H$ ($H > 0$) if it satisfies the following identity

$$\forall \eta > 0, \quad \text{law } \{\eta^{-H} X(\eta \, x), \ x \in \mathbb{R}^d\} =_{Law} \{X(x), \ x \in \mathbb{R}^d\}$$

where $=_{Law}$ means equality in law for processes.

As an example let us mention the well known fractional brownian motion ([10]). It is the centered gaussian process with covariance function

$$\mathbb{E}(B_H(x) B_H(y)) = \frac{1}{2}(|x|^{2H} + |y|^{2H} - |x-y|^{2H}).$$

The fractional brownian motion admits an integral represention

$$B_H(x) = \int_{\mathbb{R}} (|x-y|^{H-1/2} - |y|^{H-1/2}) \, W(dy)$$

where $W$ is the real gaussian white noise, i.e. the derivative in the distribution sense of the standard brownian motion. Another equivalent representation is given by

$$B_H(x) = c_H \int_{\mathbb{R}} \frac{\exp(ix\xi) - 1}{|\xi|^{H+1/2}} \widehat{W}(d\xi)$$

where $\widehat{W}$ denotes the Fourier transform of $W$. The link between the two representations is a Plancherel identity using the invariance in law of $W$ (up to a constant $c_H$) under Fourier transform.

This means that the fractional brownian motion is a solution of the elliptic equation
$$LX = W, \quad X(0) = 0$$
where $L$ is the pseudo-differential operator defined by
$$\widehat{L\varphi}(\xi) = \frac{c_H}{|\xi|^{H+1/2}}\widehat{\varphi}.$$

This is the point of view we adopt in the present article to study self similar processes. In a previous work ([2]) this was the key of several results on elliptic gaussian processes: wavelet decomposition, local and uniform regularity, and local self-similarity. Here we will consider non gaussian processes.

Let us recall that in the gaussian case, the class of processes with stationary increments which are self-similar with index $H$ - we shall write $SI - SS(H)$ for short - is entirely described in the paper of Dobrushin ([4]). In the stable non gaussian case the situation is more intricate and many examples are known, see Kono-Maejima ([7]). Let us suppose that $d = 1$ and the process is $\alpha-$ stable symmetric. Then in the gaussian case, $\alpha = 2$, the knowledge of two parameters, the index parameter $H$ and the variance parameter $\sigma^2$, is sufficient to determine the law of the process. In the non gaussian case, $\alpha < 2$, there exists numerous different processes which admit the same parameters $H, \sigma$, where $\sigma^2$ is a substitute for the variance, see Taqqu et al ([13]) or ([6]). At the present time there exists no exhaustive classification.

In order to better understand the self-similarity property we choose to study the class of "elliptic processes". A stochastic process is said to be elliptic when there exists a stochastic measure $M$ on $\mathbb{R}^d$ and a pseudo-differential elliptic operator $L$ such that
$$LX = M, \quad \mathbb{P} \text{ a.s.} . \tag{1}$$
The self similarity property of $X$ will impose
$$X(0) = 0, \quad \mathbb{P} \text{ a.s.} . \tag{2}$$
Precise meaning for this equation will be given later.

To define completely the class of self-similar processes under scope we require symmetry of the laws and stationarity of the increments.

The paper is organized as follows. In part 2 we solve a non-random version of equation (1)-(2). In part 3 and 4 we shall describe the class of stochastic measures we need and then we solve (1)-(2) in the random case. Then we characterize the self-similar processes among the elliptic ones. In the part 5 we give a wavelet decomposition of these processes, and we use this decomposition in order to get regularity results. In a last part we present an explicit example of elliptic

$SI - SS(H)$ and non stable process: it is build as the perturbation of a stable one by using percolation ideas.

With the exception of the last example, we give only sketches of proofs for most results of this paper. The reader can find all details of proofs and complements in ([1]).

## 2 Elliptic equations

Let $(\mathbb{R}^d, \mathcal{R}^d)$ the $d$-dimensionnal real space and its Borel $\sigma$- field. The canonical euclidean norm will be used throughout the paper and denoted by $|\ |$.

The self-similarity index will always be denoted by $H$ and in this paper we will consider only the case $0 < H < 1$.

If $\alpha \geq 1$ we denote by $\alpha^*$ its conjugate exponent,
$$\frac{1}{\alpha} + \frac{1}{\alpha^*} = 1.$$

We can now introduce the elliptic operators $L$ which we need. Let $\beta > 0$ and $m_\beta := \frac{d}{\beta} + H$. Let $s$ be some real valued measurable function on the unit sphere $\Sigma_{d-1}$ and $S$ be the associated function defined by $S(\xi) = s(\frac{\xi}{|\xi|})$, $\forall \xi \neq 0$. We define a $m_\beta$-homogeneous symbol $\rho_{\beta,S}$ on $\mathbb{R}^d$,

$$\rho_{\beta,S}(\xi) := |\xi|^{m_\beta} S(\xi), \tag{3}$$

and the associated operator $L_{\beta,S}$

$$L_{\beta,S} f(x) = \int_{\mathbb{R}^d} \exp(ix\xi) \rho_{\beta,S}(\xi) \widehat{f}(\xi) \frac{d\xi}{(2\pi)^d} \tag{4}$$

where $\widehat{f}(\xi) = \int_{\mathbb{R}^d} \exp(-ix\xi) f(x) \, dx$. Both notations $\mathcal{F}(f)$, $\widehat{f}$ will be used to denote the Fourier transform of $f$.

Let us recall some properties of the Fourier transform on $\mathbb{R}^d$ we shall need.

If $T$ is a integrable function (or a distribution) on $\mathbb{R}^d$ we have

$$\forall \lambda > 0,\ \forall y \in \mathbb{R}^d, \quad \mathcal{F}(T(\lambda\ .))(\xi) = \lambda^{-d} \widehat{T}(\frac{\xi}{\lambda}),$$
$$\mathcal{F}(T(y + \ .))(\xi) = \exp(-iy\xi) \widehat{T}(\xi). \tag{5}$$

If $\mu \notin 2\mathbb{N}$,

$$\mathcal{F}(|.|^\mu)(\xi) = \frac{1}{|\xi|^{d+\mu}} \tag{6}$$

and for every function $Q(x) = q(\frac{x}{|x|})$ there exits some function $r$ such that

$$\mathcal{F}(|.|^\mu Q(.))(\xi) = \frac{1}{|\xi|^{d+\mu} R(\xi)}, \quad \text{with } R(\xi) = r(\frac{\xi}{|\xi|}). \tag{7}$$

This defines a one to one transfom $\tau_\mu$, $\tau_\mu(Q) = R$.

Let us introduce the kernels $R_{\beta,S}$ and $K_{\beta,S}$ by

$$R_{\beta,S}(x,\xi) := \frac{\exp(ix\xi) - 1}{|\xi|^{\frac{d}{\beta}+H} S(\xi)} \quad (8)$$

$$K_{\beta,S}(x,y) := |x-y|^{H-\frac{d}{\beta}} Q(x-y) - |-y|^{H-\frac{d}{\beta}} Q(-y) \quad (9)$$

where $S = \tau_{H-\frac{d}{\beta}}(Q)$. It is easy to deduce the following result from (5), (7).

**Lemma 1.** *If $1 < \beta \leq 2$ we have*

$$\forall x, \xi \in \mathbb{R}^d, \quad \mathcal{F}(K_{\beta,S}(x,.))(\xi) = R_{\beta^*,S}(x,\xi).$$

We denote by $\mathcal{R}_{\beta,S}$, $\mathcal{K}_{\beta,S}$ the operators with respective kernels $R_{\beta,S}(x,\xi)$ and $K_{\beta,S}(x,y)$,

$$\mathcal{R}_{\beta,S} f(x) = \int_{\mathbb{R}^d} R_{\beta,S}(x,\xi) f(\xi)\, d\xi, \quad \mathcal{K}_{\beta,S} f(x) = \int_{\mathbb{R}^d} K_{\beta,S}(x,y) f(y)\, dy. \quad (10)$$

In the general case the kernels $R_{\beta,S}$ and $K_{\beta,S}$ are distributions on $\mathbb{R}^d$ and the integrals in (10) are taken in the distribution sense.

From now on, the function $s$ is fixed and we forget to write the dependance on $s$ (or $S$); we shall always suppose that $s$ is a positive valued measurable function on $\Sigma_{d-1}$ which satisfies

$$s(-\xi) = s(\xi) \text{ and } \exists c \in (0,1), \ c \leq s(\xi) \leq \frac{1}{c}, \ \forall \xi \in \Sigma_{d-1}. \quad (11)$$

Let us now introduce the Banach spaces we need to solve (1)-(2). We denote by $\Delta$ be the $d$-dimensional laplacian and by $W_0^{s,\beta}$ the Sobolev homogeneous space

$$W_0^{s,\beta} = \{u, \ (-\Delta)^{s/2} u \in L^\beta(\mathbb{R}^d) \ , \ u(0) = 0\}$$

equipped with the semi norm $\|u\|_{s,\beta} = \left(\int_{\mathbb{R}^d} |(-\Delta)^{s/2} u(x)|^\beta\, dx\right)^{1/\beta}$.

**Lemma 2.** *Let $\alpha \in (1,2]$, $H \in (0,1)$ and $m = \frac{d}{\alpha} + H$, $m_* = \frac{d}{\alpha^*} + H$. Then $W_0^{m,\alpha}$ and $W_0^{m_*,\alpha^*}$ are Banach spaces.*

This result is valid for $W_0^{s,\beta}$ under the general condition $0 < s - \frac{d}{\beta} < 1$, see Meyer ([11]).

We consider the following two problems:
find functions $u$ such that

$$(P_*) \quad L_{\alpha^*} u = \widehat{f}, \quad u(0) = 0$$

find functions $v$ such that

$$(P) \quad L_\alpha v = g, \quad v(0) = 0.$$

**Theorem 1.** *We suppose that $1 < \alpha \le 2$, $s$ satisfies (11) and $f, g \in L^\alpha(\mathbb{R}^d)$.
Then
(i) $\mathcal{R}_{\alpha^*}(f)$ is well defined and is the unique element in $W_0^{m_*, \alpha^*}$ solution of problem $(P_*)$.
(ii) $\mathcal{R}_\alpha(\widehat{g})$ is well defined and is the unique element in $W_0^{m, \alpha}$ solution of problem $(P)$.
(iii) Furthermore we have Plancherel like formulas*

$$\mathcal{R}_{\alpha^*}(f) = \mathcal{K}_\alpha(\widehat{f}), \quad \mathcal{R}_\alpha(\widehat{g}) = \mathcal{K}_{\alpha^*}(g). \tag{12}$$

Idea of the proof
(i) It is easy to check that $\forall x \in \mathbb{R}^d$, $R_{\alpha^*}(x, .)$ belongs to $L^{\alpha^*}(\mathbb{R}^d)$. Thus $u := \mathcal{R}_{\alpha^*}(f)$ is well defined for every $f \in L^\alpha(\mathbb{R}^d)$. Using properties of Fourier transform we see that $u$ belongs to $W_0^{m_*, \alpha^*}$ and is a solution of problem $(P)$.
(ii) Here we need the result ([14])

$$\forall \alpha \in [1, 2], \quad g \in L^\alpha(\mathbb{R}^d) \Rightarrow \widehat{g} \in L^{\alpha^*}(\mathbb{R}^d).$$

Then we can proceed as for the proof of (i).
(iii) These formulas are consequence of (7).

## 3  Elliptic stochastic processes

The notion of linear process has been defined by Gelfand ([5]). When it is based on $\mathcal{D}(\mathbb{R}^d)$, the space of smooth functions with compact support in $\mathbb{R}^d$, a linear process is also called a generalized process or a random distribution. This will be the case here.

Let $M : \mathcal{D}(\mathbb{R}^d) \to L^0(\Omega, \mathbb{P})$ be such a linear process. In order to define several invariance properties we introduce operators $t_x, u_x, r_\mu$ by setting $\forall x, y \in \mathbb{R}^d$, $\forall \mu \in \mathbb{R} \setminus \{0\}$,

$$t_x(f)(y) = f(y - x), \quad u_x(f)(y) = e^{ix \cdot y} f(y), \quad r_\mu f(x) = f(\mu x).$$

A linear process is said to enjoy

1. (SM1), stationarity if
$$Law(\{M(t_x\varphi), \varphi \in \mathcal{D}(\mathbb{R}^d)\}) = Law(\{M(\varphi), \varphi \in \mathcal{D}(\mathbb{R}^d)\}),$$
$\forall x \in \mathbb{R}^d$
2. (SM2), unitarity if
$$Law(\{M(u_x\varphi), \varphi \in \mathcal{D}(\mathbb{R}^d)\}) = Law(\{M(\varphi), \varphi \in \mathcal{D}(\mathbb{R}^d)\}),$$
$\forall x \in \mathbb{R}^d$
3. (SM3), scaling property of order $K$ if
$$Law(\{M(r_\mu\varphi), \varphi \in \mathcal{D}(\mathbb{R}^d)\}) = Law(\{|\mu|^{-K-d}M(\varphi), \varphi \in \mathcal{D}(\mathbb{R}^d)\}),$$
$\forall \mu > 0$
4. (SM4), symmetry if
$$Law(\{-M(\varphi), \varphi \in \mathcal{D}(\mathbb{R}^d)\}) = Law(\{M(\varphi), \varphi \in \mathcal{D}(\mathbb{R}^d)\}).$$

**Definition 1.** *A linear process $M$ based on $\mathcal{D}(\mathbb{R}^d)$ is a stochastic measure with control measure $\mu$ if $M(\varphi)$ extends to functions $\varphi = 1_A$ with $A \in \mathcal{R}^d$, $\mu(A) < \infty$ and satisfies the following $\sigma-$ additivity property: for every sequence $(A_i)_{i \in \mathbb{N}}$ of Borel sets which are pairwise disjoints and such that $\mu(\cup_{i \in \mathbb{N}} A_i) < \infty$ we have*
$$M(\sum_i 1_{A_i}) = \sum_i M(A_i), \quad \mathbb{P} \text{ a.s.} .$$

*A linear process is said of type $(\alpha, p)$ with control measure $\mu$ if*
$$\exists c > 0, \ \forall \varphi \in \mathcal{D}(\mathbb{R}^d), \ \|M(\varphi)\|_{L^p(\Omega, \mathbb{P})} \leq c \|\varphi\|_{L^\alpha(\mathbb{R}^d, d\mu)}. \tag{13}$$

We shall denote $SL^{\alpha,p}(\mu)$ the space of linear processes $M$ for which (13) is true.

Notice that if $M$ satisfies scaling property of order $\frac{d}{\alpha}$ then $\widehat{M}$ satisfies scaling property of order $\frac{d}{\alpha_*}$. We shall restrict our study to the $\alpha'$s such that $1 < \alpha \leq 2$.

Now we have to introduce the spaces of stochastic processes which are useful for the setting of elliptic stochastic problems.

**Definition 2.** *We shall say that a linear stochastic process $X$ belongs to $SW_0^{m,\alpha,p}(\mu)$ if*
$$(-\Delta)^{m/2} X \in SL^{\alpha^*,p}(\mu) \quad \text{and} \quad X(0) = 0$$
*where we define the vanishing in 0 condition by using gaussian densities $g_\sigma(x) = (\sigma\sqrt{2\pi})^{-d} \exp(-\frac{x^2}{2\sigma^2})$,*
$$X(0) = 0 \ \Leftrightarrow \ \lim_{\sigma \to 0} <X, g_\sigma> = 0, \quad \mathbb{P} \text{ a.s.} .$$

The semi norm $\|X\|_{m,\alpha,p,\mu}$ is the best constant $c$ in (13) for the random distribution $(-\Delta)^{m/2} X$.

We recall that $\alpha \in (1,2]$ and $s$ satisfies (11). Let $M$ be a stochastic measure of type $(\alpha, p)$ with control measure $\mu$ and $L_\alpha, L_{\alpha^*}$ be the elliptic operators given in (4). We consider the two equations

$$(SP_*) \quad L_{\alpha^*} X = M, \quad X(0) = 0$$

$$(SP) \quad L_\alpha Z = \widehat{M}, \quad Z(0) = 0.$$

**Theorem 2.** *We suppose that the measure $\mu$ is absolutely continuous with respect to the Lebesgue measure on $\mathbb{R}^d$. Then there exists a unique pair $(X, Z)$ in $SW_0^{m_*, \alpha_*, p}(\mu) \times SW_0^{m, \alpha, p}(\mu)$ such that $X$ (resp. $Z$) solves $(SP_*)$ (resp. $(SP)$). These solutions are given by*

$$X = \mathcal{R}_{\alpha_*}(\widehat{M}), \quad Z = \mathcal{R}_\alpha(M).$$

*Furthermore we have Plancherel like formulas*

$$\mathcal{R}_{\alpha_*}(\widehat{M}) = \mathcal{K}_\alpha(M), \quad \mathcal{R}_\alpha(M) = \mathcal{K}_{\alpha_*}(\widehat{M}). \tag{14}$$

Idea of the proof
   We use Theorem 1 and a regularization argument.

*Remark 1.* Recall that $s$ is supposed to be symmetric. If the linear process $M$ is also symmetric (condition SM4) then the process $Z$ is real valued.

## 4  Self-similar Processes

It seems more relevant to study scaling properties for the pair $(X, Z)$ than separately for each process. Therefore, we introduce now the notion of inverse scaling property.

**Definition 3.** *$(X, Z)$ satisfies the inverse scaling invariance property with index $H$ - or shorter the ISI(H) - if*

$$\forall \lambda > 0, \quad Law(\frac{1}{\lambda^H} X(\lambda.), \lambda^H Z(\frac{.}{\lambda})) = Law(X(.), Z(.)). \tag{15}$$

*Remark 2.* If the pair $(X, Z)$ is ISI(H) then each process is separately H-self-similar and any linear combination $aX + bZ$ satisfies the property

$$\forall \lambda > 0, \quad Law(a\frac{1}{\lambda^H} X(\lambda.) + b\lambda^H Z(\frac{.}{\lambda})) = Law(aX(.) + bZ(.)).$$

We can now study ISI(H) property for a pair $X, Z$ which is solution of $SP_*, SP$.

**Theorem 3.** *Let us suppose that hypotheses of Theorem 2 are valid. Then the solution $X, Z$ of $SP_*, SP$ is inverse scale invariant with index $H$ if and only if the linear process $M$ satisfies conditions SM1 to SM4 with $K = \frac{-d}{\alpha^*}$. In this case $\mu$ is proportional to Lebesgue measure.*

Idea of the proof

The inverse scaling invariance follows from (5),(6) and a change of variables in the expressions of $L_\alpha$ and $L_{\alpha^*}$. Formal computations are justified by using regularizations of $M$.

In the gaussian case the SI-SS(H) processes $X$ and $Z$ are with proportional laws; it means that their laws are equal up to a mulptiplicative factor.

In the more general $\alpha$−stable case this problem is studied in the book of Samorodnitski-Taqqu ([13]), ch. 7; the processes $X, Z$ are the so-called harmonizable and moving average fractional stable motions. The laws of $X$ and $Z$ are not proportional if $\alpha < 2$, and the proof of this result given in ([13]) rests upon codifference arguments. This is generalized in the following result, with a proof which makes use of operators $L_\alpha$ and $L_{\alpha^*}$.

**Corollary 1.** *Under the conditions of Theorem 2 and if $\alpha < 2$ the processes $X, Z$ are with no proportional laws.*

Idea of the proof

From $(SP), (SP_*)$ we get

$$L_\alpha L_{\alpha^*} X = L_\alpha M, \quad L_{\alpha^*} L_\alpha Z = L_{\alpha^*} \widehat{M}.$$

If the laws of $X$ and $Z$ would be equal up to the multiplicative factor $c$ we should get

$$L_\alpha M =_{Law} cL_\alpha * \widehat{M}$$

and then $\mathcal{R}_{\alpha_*} L_\alpha M =_{Law} c\widehat{M}$.

This would give $\mathcal{R}_{\alpha_*} L_\alpha = c\mathcal{F}$ and then $\alpha = \alpha_*$. It is just the gaussian case, $\alpha = 2$.

## 5  Wavelet decomposition, regularity

Let us recall some basic facts we need about the wavelet bases of Lemarié-Meyer.
Such a wavelet basis $(\Psi_\lambda)_\lambda$ is indexed by

$$\Lambda = \{(j, k, l), \ j \in \mathbb{Z}, \ k \in \mathbb{Z}^d, \ l \in E_d = \{0, 1\}^d \setminus \{(0, \cdots, 0)\}\}$$

and we have
$$\Psi^\lambda(x) = \Psi^{(l)}(2^j x - k), \ \Psi_\lambda(x) = \frac{\Psi^\lambda(x)}{\|\Psi^\lambda\|_{L^2}}.$$

Let us define
$$\Psi_{\lambda,\alpha}(x) = 2^{jd/\alpha}\Psi^{(l)}(2^j x - k).$$

**Lemma 3.** *If $1 < \alpha \leq 2$ then $(\Psi_{\lambda,\alpha})_{\lambda \in \Lambda}$ is an unconditional basis of $L^\alpha(\mathbb{R}^d)$.*

Let us define a family of functions $(\Phi_{\lambda,\alpha})_{\lambda \in \Lambda}$ and a family of r.v. $(\eta_\lambda)_{\lambda \in \Lambda}$ by

$$\Phi_{\lambda,\alpha}(x) = \mathcal{R}_{\alpha_*}(\widehat{\Psi^{(l)}})(2^j x - k) > -\mathcal{R}_{\alpha_*}(\widehat{\Psi^{(l)}})(-k), \ \eta_\lambda = M(\Psi_{\lambda,\alpha})$$

With the help of the family $(\Phi_{\lambda,\alpha})_{\lambda \in \Lambda}$ we can now express the process $X$, solution of $(SP_*)$.

**Theorem 4.** *Under the conditions of Theorem 2 we have*

$$X(x) = \sum_{\lambda \in \Lambda} 2^{-jH}\Phi_{\lambda,\alpha}(x)\eta_\lambda. \tag{16}$$

*The family $(\eta_\lambda)_\lambda$ satifies some "partial" stationarity properties:*
*(i) for every fixed $(j,l)$ the sequence $(\eta_\lambda)_{k \in \mathbb{Z}^d}$ is stationary and*
*(ii) the sequence $(\eta_\lambda)_{j \in \mathbb{Z}}$ is stationary.*
*Therefore we can express $X$ as the following sum*

$$X(x) = \sum_{j \in \mathbb{Z}} 2^{-jH} X_j(2^j x) \tag{17}$$

*where $(X_j(x))_j$ is a stationary sequence of processes with stationary increments.*

The same result is true for $Z$, the solution of problem $(SP)$.

Let us mention that in ([12]) Pesquet-Popescu gives wavelet decompositions of self similar non gaussian processes.

Idea of the proof
The result follows from the wavelet decomposition of $M$,

$$M(dx) = \sum_{\lambda \in \Lambda} \Psi_{\lambda,\alpha^*}(x) M(\Psi_{\lambda,\alpha}) \ dx$$

and the equality

$$2^{-jH}\Phi_{\lambda,\alpha}(x) = \int_{\mathbb{R}^d} \frac{\exp(ix\xi) - 1}{|\xi|^{\frac{d}{\alpha_*}+H} S(\xi)} \widehat{\Psi_{\lambda,\alpha^*}}(\xi) \ d\xi.$$

**Theorem 5.** *Under the conditions of Theorem 2 and if we suppose that there exists positive constants c,c' such that*

$$P\{|\eta_\lambda| > t\} \leq ct^{1/\alpha}$$

*and*

$$P\{\max_{|k|\leq 2^j,\,l} |\eta_{k,j,l}| < t\} \geq c' P\{|\eta_\lambda| < t\}^{2^j d}$$

*then the condition $H > \frac{d}{\alpha}$ implies that with probability 1 the paths of X are continuous.*

The conditions on the $\eta$'s are satisfied in the case of symmetric $\alpha$-stable r.v.; the result in this case is well known ([13]).

## 6 Examples

In this paragraph we shall give examples of elliptic self-similar processes with stationary increments.

- 1) $\alpha = 2$, $(= \alpha_*)$ and $M$ is the gaussian white noise. In this situation $X$ and $Z$ are essentially the same process: their laws are equal up to a multiplicative factor.
- 2) $1 < \alpha < 2$ and $M$ is the symmetric $\alpha$-stable measure. In this case, $X$ and $Z$ are fractional stable processes which are thoroughly studied in ([13]).
- 3) In the third example the law of the process will be non stable. We follow ideas of [3] where a model of intermittency is proposed. It rests upon the following remark: the set $\Lambda$ which indices the wavelet basis can be written as

$$\Lambda = T_{2^d} \times E_d$$

if we denote by $T_q$ the regular $q$-tree (from each vertex $(q+1)$ edges are drawn). We restrict ourselves to the case $d = 1$ in which case $E_1 = \{1\}$ and $\Lambda$ reduces to $T_2$. Using the 2-adic tree $T_2$ we can encode the dyadic cells of $\mathbb{R}$, in such a way that the significative values of $\psi_\lambda$ are taken on the cell $C_\lambda$. Now let $X$ an elliptic symmetric self-similar process with stationary increments. The decomposition given in (16) can be now written as

$$X(x) = \sum_{\lambda \in T_2} 2^{-jH} \Phi_{\lambda,\alpha}(x) \eta_\lambda. \tag{18}$$

In order to define intermittency in [3] the summation was restricted to a subtree of $T_2$. But we loose the self similarity property by doing this. Therefore we restrict the summation to a forest $\mathcal{F}$ build from a family of subtrees of $T_2$. We shall choose a percolation forest.

Let $\mathcal{B}_p$ be the Bernoulli law with parameter $p$, $p \in (1/2, 1]$. Let $(B_\lambda)_{\lambda \in T_2}$ an i.i.d. family of r.v. with law $\mathcal{B}_p$. We say that an edge $\lambda \in T_2$ is open if

$B_\lambda = 1$ and closed otherwise. An infinite percolation cluster is a connected set of open edges of $T_2$. Let $A$ be such a percolation set and $r(A)$ the vertex in $A$ which is the closest to the root $*$. Then $(A, r(A))$ is a subtree of $T_2$, which is called a percolation subtree. We define the forest $\mathcal{F}_p$ as the union of all percolation subtrees.

Every permutation on $T_2$ acts on $(B_\lambda)_{\lambda \in T_2}$ and induces a permutation of the forest, denoted by $\sigma \mathcal{F}_p$. Then we know from ([8])

**Lemma 4.**
*(i) for any permutation $\sigma$ of $T_2$,*

$$Law(\mathcal{F}_p) = Law(\sigma \mathcal{F}_p)$$

*(ii) If $p > 1/2$, with probability 1 we have*

$$\mathcal{F}_p \neq \emptyset$$

*and $\mathcal{F}_p$ contains infinitely many percolation subtrees.*

Let $M$ be a symmetric $\alpha-$ stable stochastic measure on $\mathcal{R}^d$, and let us define the stochastic process $\{Y_p(x), x\}$ for $p \in (1/2, 1]$ as follows

$$Y_p(x) = \sum_{\lambda \in \mathcal{F}_p} 2^{-jH} \Phi_{\lambda,\alpha}(x) \eta_\lambda. \qquad (19)$$

**Theorem 6.** *If we suppose $M$ and $(B_\lambda)_{\lambda \in T_2}$ stochatically independent and $p > 1/2$ the process $Y_p$ is symmetric $H-$self-similar and with stationary increments.*

Proof
Let us define a stochastic measure $M_p$ by

$$M_p(dx) = \sum_{\lambda \in \mathcal{F}_p} \Psi_{\lambda,\alpha^*}(x) \eta_\lambda dx \qquad (20)$$

Then it can be checked that the conditions of theorem 2 are satisfied and consequently that $Y_p$ is the solution of

$$L_{\alpha^*} Y_p = M_p, \quad Y_p(0) = 0.$$

We give now the proofs of the properties $(SM1), \cdots, (SM4)$ for $M_p$, and we use freely the formal notation

$$< T, \varphi > = \int_{\mathbb{R}^d} \varphi(x) T'(x) \, dx$$

when $T$ is a distribution, possibly random.
proof of scaling property for $M_p$

Let us change the variable $x \in \mathbb{R}^d$ in $\mu x$, for some $\mu > 0$. We get
$$\mu^d M'_p(\mu x) = \sum_{\lambda \in \mathcal{F}_p} \mu^{d/\alpha^*} \Psi_{\lambda,\alpha^*}(\mu x) M(\mu^{d/\alpha}\Psi_{\lambda,\alpha}).$$

But $M(\mu^{d/\alpha}\Psi_{\lambda,\alpha}) = \mu^d < M'(\mu.), \mu^{d/\alpha}\Psi_{\lambda,\alpha}(\mu.) >$ and we get the following equality in law
$$M(\mu^{d/\alpha}\Psi_{\lambda,\alpha}) =_{Law} \mu^d < M'(), \mu^{d/\alpha}\Psi_{\lambda,\alpha}(\mu.) >.$$

So if we define new functions on $\mathbb{R}^d$ by
$$\Psi^{[\mu]}_{\lambda,\alpha}(x) = \mu^{d/\alpha}\Psi_{\lambda,\alpha}(\mu x), \quad \Psi^{[\mu]}_{\lambda,\alpha^*}(x) = \mu^{d/\alpha^*}\Psi_{\lambda,\alpha^*}(\mu x)$$
we get new unconditional bases of respectively $L^\alpha(\mathbb{R}^d)$ and $L^{\alpha^*}(\mathbb{R}^d)$. And we can write
$$\mu^d M'_p(\mu x) =_{Law} \sum_{\lambda \in \mathcal{F}_p} \mu^{d/\alpha} \Psi^{[\mu]}_{\lambda,\alpha^*}(x) M(\Psi^{[\mu]}_{\lambda,\alpha}(x)).$$

If we denote by $\mathcal{F}^{[\mu]}_p$ the percolation forest associated with $T_2$ encoding the dyadic cells when unit length is $\mu$, we deduce from Lemma (4) that
$$Law(\mathcal{F}^{[\mu]}_p) = Law(\mathcal{F}_p)$$
thus the following equality in law
$$\sum_{\lambda \in \mathcal{F}_p} \Psi^{[\mu]}_{\lambda,\alpha^*}(x) M(\Psi^{[\mu]}_{\lambda,\alpha}(x)) =_{Law} \sum_{\lambda \in \mathcal{F}^{[\mu]}_p} \mu^{d/\alpha} \Psi^{[\mu]}_{\lambda,\alpha^*}(x) M(\Psi_{\lambda,\alpha}).$$

Finally
$$\mu^d M'_p(\mu x) =_{Law} \mu^{d/\alpha} M'_p(x)$$
which is the scaling property we looked for.

proof of stationarity property for $M_p$

Thanks to the previous proof we have only to show that
$$M'_p(x - 1) =_{Law} M'_p(x)$$
But this is a consequence of the following equalities
$$M'_p(x-1) = \sum_{\lambda \in \mathcal{F}_p} \Psi_{\lambda,\alpha^*}(x-1) M(\Psi_{\lambda,\alpha}) \quad (21)$$
$$= \sum_{\lambda \in \mathcal{F}_p} \Psi_{\lambda,\alpha^*}(x-1) < M'(.-1), \Psi_{\lambda,\alpha}(.-1) > \quad (22)$$
$$=_{Law} \sum_{\lambda \in \mathcal{F}_p} \Psi_{\lambda,\alpha^*}(x-1) < M'(.), \Psi_{\lambda+1,\alpha} > \quad (23)$$
$$= \sum_{\lambda \in \mathcal{F}_p} \Psi_{\lambda+1,\alpha^*}(x) < M'(.), \Psi_{\lambda+1,\alpha} > \quad (24)$$
$$=_{Law} \sum_{\lambda \in \mathcal{F}^{+1}_p} \Psi_{\lambda,\alpha^*}(x) < M'(.), \Psi_{\lambda,\alpha} > \quad (25)$$

where $\mathcal{F}_p^{+1} = \{\lambda + 1, \ \lambda \in \mathcal{F}_p\}$ and of Lemma (4).

proof of unitarity property for $M_p$

This property SM2 amounts to show SM1 for $\widehat{M}$, which is proved just as above.

proof of symmetry property for $M_p$

This is a direct consequence from the symmetry of the laws of $\eta_\lambda$ and hypothesis on the percolation.

# References

1. A. Benassi and D. Roux (1999): Elliptic Self Similar stochastic processes. Preprint, Université Blaise Pascal, Clermont-Ferrand.
2. A. Benassi, S. Jaffard and D. Roux (1997): Elliptic Gaussian processes. Revista Matematica Iberoamericana 13, N.1, p. 19–90.
3. A. Benassi, S. Cohen, S. Deguy and J. Istas (to appear): Self similarity and intermittency. NSF - CBMS - Research Conference of Wavelets at the U.C.F..
4. R.L. Dobrushin (1979): Gaussian and their subordinated self-similar random generalised fields. The Annals of Probability, 7, N.1, p. 1–28.
5. I.M. Gelfand and N. Y. Vilenkin (1961): Generalised functions. Vol. 4, Moscou.
6. P.S. Kokoszka and M.S. Taqqu (1994): New classes of self-similar symmetric stable random fields. Journal of Theoretical Probability 7, N.3, p. 527–549.
7. N. Kono and M. Maejima M. (1991): Self-similar stable processes with stationary increments, Stable Processes and related topics. Boston, Birkhauser, Progress in Probability 25, p. 275–295.
8. R. Lyons (1990): Random walk and percolation on a tree. The Annals of Probability, 18, p. 937–958.
9. B.B. Mandelbrot (1977): Form, Chance and Dimension. Freeman, San Fransisco.
10. B.B. Mandelbrot and J.W. Van Ness (1968): Fractional Brownian motions, fractional noises and applications. SIAM Review 1, p. 422–437.
11. Y. Meyer (1990): Ondelettes et Opérateurs. Hermann.
12. B. Pesquet-Popescu (to appear): Statistical properties of the wavelet decomposition of certain self similar non gaussian processes. Signal Processing.
13. G. Samorodnitsky and M.S. Taqqu (1994): Stable non Gaussian random processes: stochastic models with infinite variance. Chapman - Hall, New York.
14. E.C. Titchmarsh (1937): Introduction to the theory of Fourier Integrals. Oxford, University Press.

# Wavelets for Scaling Processes

Patrick Flandrin and Patrice Abry

Ecole Normale Supérieure de Lyon
Laboratoire de Physique (UMR 5672 CNRS)
46 allée d'Italie, 69364 Lyon Cedex 07 - FRANCE
{flandrin,pabry}@physique.ens-lyon.fr

**Abstract.** Depending on the considered range of scales, different *scaling processes* may be defined, which correspond to different situations connected with self-similarity, fractality or long-range dependence. Wavelet analysis is shown to offer a unified framework for dealing with such processes and estimating the corresponding scaling parameters. Estimators are proposed and discussed on the basis of representations in the wavelet domain. Statistical (and computational) efficiency can be obtained not only for second-order processes, but also for linear fractional stable motions with infinite variance.

**Keywords.** Wavelets, scaling, self-similarity, long-range dependence, stable processes.

## 1 Introduction

Signals presenting some form of scaling behaviour can be observed in a wide variety of fields, ranging from physics (for example turbulence, hydrology, solid-state physics) or biology (DNA sequences, heart rate variability, auditory nerves spike trains, ... ) to human-operated systems (telecommunications network traffic, finance, ... ). While sharing a common property of *scale invariance*, processes used to model such observations may however differ according to the range of scales over which the invariance is effective. Indeed, relevant concepts can be either *self-similarity* (the part is, in some sense, identical to the whole), *long-range dependence* (algebraically decaying and non-integrable correlations at "large scales" result in power-law diverging spectra at low frequencies) or *irregularity of sample paths* ("small scale" scaling results in non-integer fractal dimensions). In each case, no characteristic scale exists (in a given range), the important feature being rather the existence of some invariant relation *between* scales.

Because they may correspond to non-standard situations in signal processing or time series analysis (non-stationarity, long-range dependence, ... ), *scaling processes* raise challenging problems in term of analysis, synthesis, and processing (filtering, prediction, ... ). A number of specific tools have however been developed over the years and, in a recent past, it has been realized that a natural approach was to consider scaling processes from the perspective of the multiresolution tools which had been introduced since the mid-eighties around the fruitful concept of *wavelet*. In fact, wavelets have, by construction, a built-in ability to look at a signal or a process at different scales, and to reveal potential invariant relations between scales within a proper and well-understood mathematical setting. The purpose of this paper is therefore to show which advantages can be gained from using a wavelet-based perspective when dealing with scaling processes.

More precisely, the paper is organized as follows. First, basics of wavelet analysis are briefly recalled in Section 2. Section 3 addresses the crucial issue of discussing scaling processes in the wavelet domain, showing that the proposed framework allows to consider in a unified way different types of situations (specifically, self-similar process with or without variance, as well as long-range dependent processes). Section 4 is devoted to the wavelet-based estimation of scaling parameters. Finally, a number of applications are briefly mentioned in the Conclusion, in order to support the effectiveness of the methods previously discussed.

## 2 Wavelets

Wavelet analysis [22] formalizes the idea of looking at a signal (or a process) at different *scales* or *levels of resolution*. This is achieved by decomposing any signal onto a set of elementary "building blocks" which are all deduced from a unique waveform—supposed to be reasonably well localized in both time and frequency—by means of shifts and dilations. More precisely, the *Continuous Wavelet Transform* (CWT) of a signal $X(t) \in L^2(\mathbf{R})$ is given by

$$T_X(a,t) := \frac{1}{\sqrt{a}} \int_{-\infty}^{+\infty} X(s) \, \psi_0\left(\frac{s-t}{a}\right) ds, \qquad (1)$$

with $a \in \mathbf{R}_+$, $t \in \mathbf{R}$ and where $\psi_0(.)$ stands for the *mother wavelet* of the analysis. Provided that this function is zero-mean, the wavelet transform can be inverted as

$$X(t) = C_{\psi_0} \int\!\!\int_{-\infty}^{+\infty} T_X(a,s) \, \frac{1}{\sqrt{a}} \, \psi_0\left(\frac{t-s}{a}\right) \frac{da\,ds}{a^2}. \qquad (2)$$

By varying the $t$ variable, the wavelet transform allows for a local analysis in time, whereas by varying the $a$ variable, it offers the possibility on "zooming in" on details, thus playing the role of a "mathematical microscope" and making of it a tool which is *a priori* naturally adapted to scaling processes.

Whereas the wavelet transform is highly redundant in its continuous form (1), it can be efficiently discretized on the dyadic grid $a = 2^j, t = 2^j k$ ($j$ and $k \in \mathbf{Z}$) of the time-scale plane. The framework developed for defining the corresponding *Discrete Wavelet Transform* (DWT) is referred to as *MultiResolution*

*Analysis* (MRA). Roughly speaking, the MRA of a signal space is defined by a sequence of nested subspaces $\ldots \subset V_j \subset V_{j-1} \subset \ldots$, each associated with a given level of resolution indexed by $j$ and such that passing from one subspace to the neighbouring one is obtained by a dilation of factor 2. Provided that a basis (given by integer translates of a low-pass waveform $\phi_0(.)$) exists for some subspace $V_0$ chosen as reference, the existence of a MRA guarantees that any signal can be decomposed under the form "signal = approximation + detail", with the possibility of iterating the process at coarser and coarser scales, by further decomposing successive approximations. For a decomposition of depth $J$, a signal $X(t) \in V_0$ can thus be written as

$$X(t) = \sum_{k=-\infty}^{\infty} a_X(J,k)\, \phi_{J,k}(t) + \sum_{j=1}^{J} \sum_{k=-\infty}^{\infty} d_X(j,k)\, \psi_{j,k}(t), \qquad (3)$$

where the $a_X$ and $d_X$ stand for the approximation and detail coefficients, respectively, while $\phi_{j,k}(t) := 2^{-j/2}\phi_0(2^{-j}t - k)$ and $\psi_{j,k}(t) := 2^{-j/2}\psi_0(2^{-j}t - k)$, the wavelet $\psi_0(.)$ being such that its integer translates are a basis of $W_0$, the complement of $V_0$ in $V_{-1}$. Up to the coarser approximation coefficients $a_X(J,k)$, the discrete wavelet transform is therefore given by the set of all detail coefficients $d_X(j,k)$, which measure indeed a *difference in information* between two successive approximations. Thanks to the dyadic structure of the sampling, they can be written as

$$d_X(j,k) := 2^{-j/2} \int_{-\infty}^{+\infty} X(t)\, \psi_0\left(2^{-j}t - k\right)\, dt, \qquad (4)$$

and they are therefore obtained as a projection of the analyzed signal onto the corresponding wavelet subspace $W_j$.

As in the continuous case, wavelets have to be zero-mean for being admissible, but their design can be controlled by a number of additional degrees of freedom. One important wavelet property—which will prove essential in the following—is its *number of vanishing moments*, i.e., the number $N \geq 1$ such that

$$\int_{-\infty}^{+\infty} t^k\, \psi_0(t)\, dt \equiv 0, \qquad k = 0, 1, \ldots N-1. \qquad (5)$$

Other properties may be appealing from the point of view of computational efficiency. This is especially the case of the *compact support property*, which is intimately related to the existence of FIR filterbank implementations. In fact, a key feature of the discrete wavelet transform is that the computation of the coefficients (4) can be actually achieved in a fully discrete and recursive way, leading to extremely efficient algorithms, which even outperform FFT-type algorithms.

The interested reader is referred to, e.g., [22] for a thorough presentation of wavelet transforms.

## 3 Scaling processes in the wavelet domain

In the previous Section, the wavelet transforms—either continuous or discrete—were defined for deterministic signals or functions belonging to $L^2(\mathbf{R})$. They have, however, been naturally and widely applied to the analysis of stochastic processes. In such cases, the wavelet coefficients are (continuous or discrete) random fields, raising issues concerning their existence and statistical properties. Such questions will not be addressed in detail here. Let us simply note that the wavelet coefficients will basically inherit of the properties of both the analyzed process and the mother wavelet. The statistical properties of the wavelet coefficients (existence, finiteness of moments, dependence structure) will hence depend on joint conditions on the mother wavelet and on the statistics of the analyzed process. Self-similarity or long-range dependence, which are under study here, generally involve (whenever they exist) covariance functions that are not bounded. We will assume that the mother wavelet decays at least exponentially fast in the time domain, so as to guarantee the existence of the wavelet coefficients. The interested reader is referred to, e.g., [9, 12, 16, 17, 23, 26, 27] for further details.

### 3.1 Self-similar processes with stationary increments

**Self-similarity.** A process $X = \{X(t), t \in \mathbf{R}\}$ is said to be *self-similar* with self-similarity parameter $H > 0$ (hereafter, "$H$-ss") if and only if the processes $X_1 := X$ and $X_c := \{c^{-H}X(ct), t \in \mathbf{R}\}$ have the same finite dimensional distributions for any $c > 0$. Self-similarity means that the process is statistically scale-invariant: it does not possess any characteristic scale of time or, equivalently, it is not possible to distinguish between a suitably scaled version (in time and amplitude) of the process and the process itself. Self-similarity also implies that $X$ is a nonstationary process, since it is obvious from the definition that the variance of $X$, when it exists, reads: $\mathbf{E}X^2(t) = |t|^{2H}\mathbf{E}X^2(1)$.

The (DWT) wavelet coefficients of an $H$-ss process $X$ exactly reproduce its self-similarity through the key scaling property:

**P1:** $(d_X(j,0), \ldots, d_X(j, N_j - 1)) \stackrel{d}{=} 2^{j(H+1/2)}(d_X(0,0), \ldots, d_X(0, N_j - 1))$. (6)

In the CWT framework, the same property reads (for any $c > 0$):

**P1c:** $(T_X(ca, ct_1), \ldots, T_X(ca, ct_n)) \stackrel{d}{=} c^{H+1/2}(T_X(a, t_1), \ldots, T_X(a, t_n))$. (7)

Let us emphasize that this results (non trivially) from the fact that the analyzing wavelet basis is designed from the dilation operator and is therefore, by nature, scale invariant. Such a property has been originally established in the case of the fractional Brownian motion (FBM) in [19, 23, 33], more recently in the case of the linear fractional stable motion in [16, 17, 26, 27] and more generally, for any self-similar process, in [9, 16].

The fundamental result **P1** can be given two special forms that will play key roles in the section dedicated below to the estimation of the self-similarity parameter. For processes whose wavelet coefficients have finite second-order statistics, one has:

$$\textbf{P1var:} \ \mathbf{E} d_X^2(j,k) = 2^{j(2H+1)} \mathbf{E} d_X^2(0,k), \tag{8}$$

whereas, for processes for which the quantity $\mathbf{E} \log_2 |d_X(j,k)|$ exists, one obtains:

$$\textbf{P1log:} \ \mathbf{E} \log_2 |d_X(j,k)| = j(H+1/2) + \mathbf{E} \log_2 |d_X(0,k)|. \tag{9}$$

**Stationary increments.** A process $X = \{X(t), t \in \mathbf{R}\}$ is said to have *stationary increments* (hereafter, "si") if and only if, for any $h \in \mathbf{R}$, the finite-dimensional distributions of the processes $X^{(h)} = \{X^{(h)}(t) := X(t+h) - X(t), t \in \mathbf{R}\}$ do not depend on $t$. In the case of both the DWT and the CWT, this results in a *stationarization property*, according to which

**P2:** the $\{d_X(j,k), k \in \mathbf{Z}\}$ form, at each octave $j$, a stationary sequence.

**P2c:** the $\{T_X(a,t), t \in \mathbf{R}\}$ form, at each scale $a$, a stationary process.

This is a direct consequence of the fact that the mother wavelet has at least one vanishing moment (i.e., $N \geq 1$). This has been shown in its most general form in [9, 16] and, in a more specific context, in [19, 23, 33] for the case of the FBM and in [16, 17, 26, 27] for the case of the linear fractional stable motion (LFSM).

The stationarization property **P2** can be extended to processes that do not possess stationary increments but have increments of higher order that are stationary. If we denote by $p$ the number of times one has to take increments to obtain a stationary process, the wavelet coefficients form themselves a stationary process under the condition that $N \geq p$. By stationary increments, we hereafter mean that there exists an integer $p$ such that the increments of order $p$ of $X$ are stationary and that the condition $N \geq p$ is satisfied.

Let us also note that the increments of a process $X$ can be read as a specific example of wavelet coefficients, since we have

$$X^{(ah_0)}(t) := X(t + ah_0) - X(t) \equiv \frac{1}{\sqrt{a}} T_X(a,t),$$

with $\psi_0(t) = \delta(t + h_0) - \delta(t)$, where $\delta(t)$ denotes the Dirac distribution. Such a mother wavelet has however poor regularity, possesses only one vanishing moment (i.e., $N = 1$) and cannot be constructed from a multiresolution analysis.

**Self-similarity with stationary increments.** Among all self-similar processes, there exists a subclass of particular interest over which we hereafter concentrate: the class of *self-similar processes with stationary increments* (hereafter, "$H$-sssi"). For the increments $Y(h,t) := X^{(h)}(t)$ of a $H$-sssi process, one can show that the processes $\{Y(h,t), t \in \mathbf{R}\}$ and $\{c^{-H}Y(ch,ct), t \in \mathbf{R}\}$ have the same finite-dimensional distributions, for all $c > 0$, a property that has an analagous form to—and is highly reminiscent of—the property **P1c** of the wavelet coefficients.

In the remainder of the paper, and for a sake of simplicity, we will concentrate on the coefficients of the DWT (the $d_X(j,k)$) rather than on those of the CWT (the $T_X(a,t)$), despite the fact that most results could be written in either framework.

### Finite variance and Gaussian processes

**Variance.** Let $X$ denote a zero-mean $H$-sssi process with finite variance (i.e., such that $\mathbf{E}X(t) \equiv 0$ and Var $X(t) < \infty$, for any $t \in \mathbf{R}$). From the fundamental property **P1** (eq.(6) or eq.(8)), from $\mathbf{E}d_X(j,k) \equiv 0$ and from the stationarity of the wavelet coefficients (property **P2**), it is straighforward to show that the wavelet coefficients of $X$ satisfy:

$$\text{Var } d_X(j,k) = 2^{j(2H+1)} \text{Var } d_X(0,0). \tag{10}$$

**Covariance.** Moreover, adding the usual convention that $X(0) \equiv 0$, one gets that the covariance of an $H$-sssi process reads:

$$\mathbf{E}X(t)X(s) = \frac{\sigma^2}{2}\left(|t|^{2H} + |s|^{2H} - |t-s|^{2H}\right). \tag{11}$$

From this form of the covariance of the process, the asymptotic behaviour of the covariance of the wavelet coefficients can be obtained:

**P3:** $\mathbf{E}d_X(j,k)d_X(j',k') \sim |2^{-j}k - 2^{-j'}k'|^{2(H-N)}$, $|2^{-j}k - 2^{-j'}k'| \to \infty$. (12)

This result has been established originally for the FBM in [19, 23, 33]. It shows that the range of correlation is controlled by the number $N$ of vanishing moments of the mother wavelet: the higher $N$, the shorter the range.

Finally, from the expression (11) of the covariance of $X$, it can be readily derived [1, 3, 6, 19] that:

$$\text{Var } d_X(j,k) = 2^{j(2H+1)}\sigma^2 C(\psi_0, H), \tag{13}$$

where $\sigma^2 := \mathbf{E}X^2(1)$ and $C(\psi_0, H) := \int_{-\infty}^{+\infty} |u|^{2H} \left(\int_{-\infty}^{+\infty} \psi_0(v)\psi_0(v+u)dv\right) du$.

**Gaussianity.** If ones moreover requires that the zero-mean $H$-sssi process is Gaussian, one is led to the only FBM [24]. In this case, the wavelet coefficients are Gaussian too.

### From Gaussian to stable processes

**Stable processes.** Let us suppose that we are now interested in (zero-mean) $H$-sssi processes with (possibly) infinite variance. *Stable motions* [30] offer an interesting framework to model $H$-sssi processes whose second-order statistics do not exist. By definition, such processes admit the representation

$$X(t) = \int_{-\infty}^{+\infty} f(t,u) M(du),$$

with $M(du)$ some *symmetric $\alpha$-stable* (hereafter, "S$\alpha$S") measure, with scale parameter $\sigma$, and $f(t,u)$ an integration kernel that controls the time dependence of the statistics of $X$. For instance, if $f(t,u) \equiv f(0, u-t)$, then $X(t)$ is a stationary process; if $f(ct, cu) = c^{H-1/\alpha} f(t,u)$ for any $c > 0$, then $X(t)$ is a $H$-ss process. Two examples are of particular interest, namely the *Lévy flights* and the *linear fractional stable motion* (LFSM) [30]. The first one is defined through $f(t,u) := 1$ if $u \le t$, and 0 otherwise. It is a $H$-ss process with $H = 1/\alpha$, and its increments are stationary and independent. The second one is defined through a parameter $d \le 1/2$ and the kernel $f(t,u) := (t-u)_+^d - (-u)_+^d$, where $(t)_+ = t$ if $t \ge 0$, and 0 otherwise. It is a $H$-ss process with $H = d + 1/\alpha$. Its increments are stationary but dependent, the dependence being controlled by $d$. It has been shown [9, 16, 17, 27] that, under mild joint conditions on the mother wavelet $\psi_0$ and the kernel $f(t,u)$, the wavelet coefficients of a stable motion are S$\alpha$S random variables with integral representation:

$$d_X(j,k) = \int_{-\infty}^{+\infty} h_{j,k}(u) M(du); \quad h_{j,k}(u) := \int_{-\infty}^{+\infty} f(t,u) \psi_{j,k}(t) dt.$$

**$H$-sssi stable processes.** If $X$ is a $H$-sssi stable process, then, from the stationarity of its increments and from property **P1**, one can show that the scale parameters of its wavelet coefficients satisfy the following scaling relation:

**P1infvar:** $\sigma_{j,k}^\alpha = 2^{j(H+1/2)\alpha} \sigma_{0,0}^\alpha,$  (14)

where $\sigma_{0,0}$ is a quantity that depends on both the mother wavelet $\psi_0(t)$ and the function $f(t,u)$, and therefore on $d$ for the LFSM [4, 16, 17]. This relation plays a role that is equivalent to eq.(10) in the finite variance case.

**Logarithm of $H$-sssi processes.** The property (and mainly the dependence structure) of the wavelet coefficients of $H$-sssi stable processes will not be further detailed here, and the reader is referred to [4, 5, 16, 18]. Instead, we will turn to *logarithmic transformations* of $H$-sssi processes. It is well-known [25, 30] that, if $X$ is a stable variable, then the variable $Y = \log_2 |X|$ has a finite variance, yielding the idea of considering the random variable $\log_2 |d_X(j,k)|$.

Let $X$ denote a $H$-sssi process with arbitrary (finite or infinite) variance, then it results from the fundamental properties **P1** and **P2** that [16]:

$$\mathbf{E} \log_2 |d_X(j,k)| = j(H+1/2) + \mathbf{E} \log_2 |d_X(0,0)|.$$  (15)

This equation plays a role analogous to that of eq.(10). Let us moreover note that, while eq.(10) involves the variance of $d_X(j,k)$ because of their mean which is identically zero, the equation above directly reproduce the self-similarity of $X$ through the means of $\log_2 |d_X(j,k)|$.

The covariance of $\log_2 |d_X(j,k)|$ has also been studied in the case of the LFSM. It has been shown in [16, 18] that:

$$\textbf{P3log: } |\text{Cov } \log_2 |d_x(j,k)|, \log_2 |d_x(j,k')|| \leq \tag{16}$$
$$C|k-k'|^{-(\alpha/4)(N-H)}, |k-k'| \to \infty, \tag{17}$$

evidencing again that the number of vanishing moments $N$ controls the correlation of the log-coefficients $\log_2 |d_X(j,k)|$, which can be made as small as desired by increasing $N$.

## 3.2 Long-range dependent processes

**Second-order stationarity.** Let $X$ be a second-order stationary stochastic process (supposed to be zero-mean, for a sake of simplicity), with stationary covariance function $c_X(\tau)$ and spectrum $\Gamma_X(\nu)$. The covariance of its wavelet coefficients can be expressed as [19]:

$$\mathbf{E} d_X(j,k) d_X(j',k') = \int\int_{-\infty}^{+\infty} c_X(u-v)\, \psi_{j,k}(u)\, \psi_{j',k'}(v)\, du\, dv,$$

whereas its variance reads:

$$\mathbf{E} d_X^2(j,k) = \int_{-\infty}^{+\infty} \Gamma_X(\nu)\, 2^j |\Psi_0(2^j \nu)|^2\, d\nu,$$

where $\Psi_0(\nu)$ stands for the Fourier transform of $\psi_0(t)$. The above relation may receive a standard *spectral estimation* interpretation within the time-invariant linear filtering theory, with $k$ seen as a time index and $\psi_0$ as a band-pass filter [1].

**Long-range dependence.** A second-order stationary process $X$ is said to be *long-range dependent* (LRD) if its covariance satisfies [11]:

$$c_X(\tau) \sim c_r\, \tau^{-\beta}, \tau \to +\infty,$$

with $0 < \beta < 1$. An equivalent definition amounts to saying that the spectrum of a LRD process satisfies:

$$\Gamma_X(\nu) \sim c_f\, |\nu|^{-\gamma}, \nu \to 0, \tag{18}$$

with $0 < \gamma < 1$.

From the Fourier transform, we have $\beta = 1-\gamma$ and $c_r = c_f 2\Gamma(1-\alpha)\sin(\pi\alpha/2)$, $\Gamma$ denoting (here and only here) the Gamma function [11, p. 43]. Both definitions imply that $\int_{-\infty}^{+\infty} c_X(\tau) d\tau = \infty$ or, equivalently, that $\Gamma_X(0) = \infty$, the specific marks of LRD. Note that LRD is sometimes also referred to as "long memory" or "second-order asymptotic self-similarity," for a reason made clearer in the following.

The variance of the wavelet coefficients of a LRD process reproduces the underlying power-law:

$$\text{Var } d_X(j,k) \sim 2^{j\gamma} c_f C(\psi_0,\gamma), j \to \infty, \tag{19}$$

with

$$C(\psi_0,\gamma) := \int_{-\infty}^{+\infty} |\nu|^{-\gamma} |\Psi_0(\nu)|^2 d\nu. \tag{20}$$

This relation shows that the wavelet coefficients catch the power-law behaviour of the LRD spectrum, and it is equivalent —in form, spirit and consequences— to the eqs.(10) or (13), which hold for $H$-sssi processes with finite variance. Moreover, the covariance of wavelet coefficients can be explicitly computed, yielding the following asymptotic behaviour:

**P3LRD:** $\mathbf{E} d_X(j,k) d_X(j',k') \sim |2^{-j}k - 2^{-j'}k'|^{\gamma-1-2N}$, $|2^{-j}k - 2^{-j'}k'| \to \infty$. (21)

This shows again that the number of vanishing moments $N$ controls the range of correlation and that long-range correlation within $X$ can be turned to short-range correlation within its wavelet coefficients, under the condition that $N > \gamma/2$, an inequality which is satisfied for any wavelet since, by definition, $N \geq 1$.

**Beyond long-range dependence.** In the definition (18) of LRD, one has assumed that $0 < \gamma < 1$. The extension to $\gamma < 0$ leads to a spectrum $\Gamma_X$ which still exhibits a power-law behaviour at the origin, but the corresponding process $X$ has no longer long-memory in this case. Nevertheless, the wavelet analysis described above still holds in a straightforward manner. The situation where $\gamma > 1$ is technically more difficult, since $X$ has no longer a finite variance (this situation is sometimes referred to as that of "generalized processes"). Wavelet analysis is however equally valid in this case, since the wavelet coefficients still have finite second-order statistics and eq.(19) still holds, on condition that $C(\psi_0,\gamma)$ is finite: this is satisfied as soon as $N > (\gamma-1)/2$, a condition which will be always assumed to hold in the following.

**Long range dependence and self-similarity.** Let us finally note that, despite the fact that LRD is defined independently, it has deep relations with self-similarity. It is indeed easy to check that the increments of an $H$-sssi process with finite variance are LRD processes, with $\gamma = 2H - 1$.

## 3.3 A unified framework

Let $X$ be either a $H$-sssi or a LRD process, then its wavelet coefficients exhibit the following key properties for the analysis of the underlying scaling:

- The wavelet coefficients form *stationary sequences* in time, at every scale. This is true on condition that $N > p$, $p$ being the number of times one has to take increments of a $H$-sssi process to obtain stationarity.
- The wavelet coefficients exactly *reproduce the scale invariance*, through either
$$\text{Var } d_X(j,k) = 2^{j\gamma} c_f C$$
or
$$\mathbf{E} \log_2 |d_X(j,k)| = j(H + 1/2) + C.$$

While the second of these two equations only applies to $H$-sssi processes with either finite or infinite variance, the first one gathers long-range dependence and self-similarity (with finite variance) into a single framework. In the case of self-similarity, $\gamma$ has to be read as $2H + 1$ and $c_f$ as $\sigma^2$. The constant $C$ depends on the mother wavelet and on the scaling exponent, and it has to be read according to eq.(10) or (19), accordingly. Note, moreover, that the first equality holds, gathering the various cases, on condition that $N > (\gamma - 1)/2$.
- The wavelet coefficients are *weakly correlated*—i.e., they exhibit no LRD—as soon as $N > \gamma/2$. This decorrelation is idealized into the approximation:

**ID1:** the wavelet coefficients $d_X(j,k)$ are exactly uncorrelated.

An equivalent decorrelation effect holds for $\log_2 |d_X(j,k)|$ and, again, it can be idealized into an exact decorrelation approximation:

**ID2:** the log-wavelet coefficients $\log_2 |d_X(j,k)|$ are exactly uncorrelated.

These three key properties hold regardless of the precise features of the mother wavelet, except for its number of vanishing moments that plays a role for both the stationarization property, the reproduction of the power-law and the decorrelation effect. What is noteworthy is also that they do not depend on some *a priori* assumption about the nature of the analyzed process: in any of the considered situations, the existence of a scaling behaviour is indeed evidenced by the analysis, the corresponding interpretation (in terms of either self-similarity, small scale scaling or LRD) depending on the range of scales over which it is actually observed. The following section is devoted to make use of these results for *estimating* scaling parameters evidenced by a wavelet analysis.

## 4 A wavelet framework for the estimation of scaling parameters

### 4.1 Finite variance and Gaussian processes

Focusing first on finite variance processes, a wavelet-based estimation of scaling parameters can be designed, that takes full advantage of the various properties which have been established so far. In particular, and as far as estimation is concerned, the *stationarization* property (**P2**) of the wavelet transform allows for using the quantity

$$\mu_j = \frac{1}{n_j} \sum_{k=1}^{n_j} d_X^2(j,k) \qquad (22)$$

as an estimator of the variance Var $d_X(j,k)$ at scale $j$, based on the $n_j$ coefficients available at that scale. Moreover, the coefficients involved in (22) can be considered as *almost uncorrelated* for an appropriate choice of the analyzing wavelet (in terms of its number of vanishing moments, see **P3**), making of (22) a potentially efficient estimator.

Given these attractive properties, estimating the scaling structure of a process amounts to studying the scale dependence (in $j$) of the variance estimator $\mu_j$. Since power-law behaviors are expected to occur, it proves more interesting to rather consider the quantity $y_j := \log_2 \mu_j$, thus reformulating the problem in a simple linear regression framework: together with appropriate confidence intervals about the $y_j$, the graph of $y_j$ against $j$ has been referred to as the (second order) *Logscale Diagram* [6].[1]

The Logscale Diagram is based in fact on the logarithm of the *estimated* variance $\mu_j$, but theory only guarantees that the logarithm of the *true* variance $\mathbf{E} d_X^2(j,k)$ is linear in $j$. Since, in general, $\mathbf{E} \log_2\{.\} \neq \log_2\{\mathbf{E} \,.\}$, the quantity $y_j$ cannot be expected to be unbiased, unless some correction is applied. Assuming that the analyzed process is Gaussian, and that its wavelet coefficients are uncorrelated (**ID1**), a way out [3] is to redefine $y_j$ as $y_j = \log_2 \mu_j - g_j$, with

$$g_j := \psi(n_j/2)/\log 2 - \log_2(n_j/2), \qquad (23)$$

where $\psi(.)$ is the logarithmic derivative of the Gamma function. Doing so, we readily get that the theoretical scaling relation (10) leads to

$$\mathbf{E} y_j = \gamma j + \log_2 c_f C. \qquad (24)$$

Moreover, the Gaussianity of the process carries over to the associated wavelet coefficients, so that the $y_j$ are scaled and shifted logarithms of chi-squared variables, with variance

$$\sigma_j^2 := \text{Var } y_j = \zeta(2, n_j/2)/\log^2 2, \qquad (25)$$

where $\zeta(z,\nu)$ is a generalized Riemann Zeta function.

Any kind of (weighted) linear regression of $y_j$ on $j$:

$$\hat{\gamma} = \sum_j w_j y_j, \qquad (26)$$

---

[1] It has to be observed that such an estimator shares with many other ones the common feature of characterizing straight lines in a log-log plot. Its main originality relies on the fact that wavelets allow for a well-controlled splitting of the analyzed process in a number of sub-processes at different scales, each of those being much better behaved than the original process considered as a whole.

with

$$w_j := \frac{1}{a_j} \frac{S_0 j - S_1}{S_0 S_2 - S_1^2}, \tag{27}$$

where the $a_j$ are arbitrary non-zero numbers, and $S_m := \sum_j j^m/a_j$ for $m = 0, 1, 2$, constitutes therefore an unbiased estimator of the scaling parameter $\gamma$. Among them, we will let the weights $a_j$ be precisely the variances $\sigma_j^2$ of the $y_j$, since this choice actually leads to the minimum variance unbiased estimator for the regression problem.

By construction, (26) is an unbiased estimator:

$$\mathbf{E}\hat{\gamma} = \gamma \tag{28}$$

and, if we assume—as we did—that the sequences of wavelet coefficients at different scales are uncorrelated (**ID1**), we obtain that

$$\operatorname{Var} \hat{\gamma} = \sum_j \sigma_j^2 \, w_j^2. \tag{29}$$

This is an important result, since this variance only depends on the amount of data (through the $n_j$), but not on the data itself (the $y_j$), not on the chosen wavelet (provided that its number of vanishing moments is high enough), nor on the actual (unknown) value of $\gamma$. It can be shown [1, 34] that, for data of size $n$, Var $\hat{\gamma}$ decreases as $1/n$ in the limit of large $n_j$ at each scale $j$ under consideration. Moreover, the Cramér-Rao lower bound is attained in this case, and the estimate $\hat{\gamma}$ is (asymptotically) normally distributed [1, 34]. Under the assumptions made, this permits therefore to associate confidence intervals with the points on which the linear regression is performed, thus allowing for designing tests aimed at justifying the relevance of a linear fit, as well as determining the range of scales on which such a fit makes sense [6].

Estimating $\gamma$ may not be the only issue when characterizing a scaling process. In particular, it may be most useful to also estimate a "magnitude" parameter ($\sigma^2$ in eq.(13), or $c_f$ in eq.(19)), which clearly measures in practical situations the *quantitative* importance of scaling effects in observed data. According to (24), the magnitude parameter $c_f$ is related to the intercept of the fitted straight line in the Logscale Diagram. Unfortunately, this intercept also involves the quantity $C$, which depends on both the chosen wavelet and the actual scaling parameter $\gamma$ (see eq.(20)). A two-step procedure can however be used [34], which consists first in estimating $\widehat{c_f C}$ from the intercept of the regression and, second, in estimating $\hat{c}_f$ as $\widehat{c_f C}/\hat{C}$, where $\hat{C}$ is, given $\hat{\gamma}$, an estimator of the integral $C(\gamma, \psi_0)$. The estimator constructed this way can be shown to be asymptotically unbiased, efficient and log-normally distributed [34].

## 4.2 From Gaussian to stable processes

In the case of processes with arbitrary variance (not necessarily finite), the analysis conducted so far cannot be followed. However, it has been shown previously that a number of key properties of wavelet transformed processes still apply in very general situations, irrespectively of the existence of second-order moments. In particular, the reproduction identity **P1** (eq.(6)) for $H$-ss processes guarantees that (15) holds. This allows us again to estimate $H$ by measuring a slope in a log-log plot, with the notable difference that $\mathbf{E}\log_2|d_X(j,k)|$ needs now to be estimated. If we restrict ourselves to the class of $H$-sssi processes (such as $H$-sssi S$\alpha$S processes), the stationarization property of the wavelet transform still holds, thus suggesting to make use of the quantity

$$Y_j := \frac{1}{n_j} \sum_{k=1}^{n_j} \log_2 |d_X(j,k)| \tag{30}$$

as an estimator of $\mathbf{E}\log_2|d_X(j,k)|$, with the obvious consequence that

$$\mathbf{E}Y_j = (H+1/2)j + \mathbf{E}\log_2|d_X(0,0)|. \tag{31}$$

We know [18] that, whereas S$\alpha$S processes (and, hence, the sequences of their wavelet coefficients at any scale) have infinite variance, the log-coefficients $\log_2|d_X(j,k)|$ have finite second-order statistics. Moreover, we have seen that, in the specific LFSM case, the covariance of those log-coefficients can be made arbitrarily small when using a wavelet with a high enough number of vanishing moments (see eq.(16))[2]. This results in behaviors similar to what had been established for second-order processes, the exact decorrelation idealization **ID2** leading for the variance of $Y_j$ to the closed form expression:

$$\text{Var } Y_j = \left(1 + \frac{2}{\alpha^2}\right) \frac{(\pi \log_2 e)^2}{12} \frac{1}{n_j}. \tag{32}$$

In this more general context of processes with a possibly infinite variance, stationarization and almost decorrelation are again the key ingredients for guaranteeing a relevant estimate of the scaling parameter $H$. More precisely, the estimate $\hat{H}$ follows from the linear relation (31) and can be written as:

$$\hat{H} := \sum_j w_j Y_j - \frac{1}{2}, \tag{33}$$

with the weights $w_j$ defined as in (27).

Since we have, by construction, $\sum_j w_j = 0$ and $\sum_j j w_j = 1$, it is easy to check that $\mathbf{E}\hat{H} = H$, a result which is exact regardless of the data length and of $0 < \alpha \leq 2$.

Assuming (**ID2**) an exact decorrelation between the log-coefficients, one obtains:

$$\text{Var } \hat{H} = \left(1 + \frac{2}{\alpha^2}\right) \frac{(\pi \log_2 e)^2}{12} \sum_j \frac{w_j^2}{n_j}. \tag{34}$$

---

[2] It has however to be noted that, for a given $H$, the decorrelation effect requires larger $N$'s for smaller $\alpha$'s, since $N$ has basically to behave as $1/\alpha$.

This variance is minimum for the choice $a_j = \text{Var } Y_j \sim 1/n_j$. Since $n_j$, the number of data points available at scale $j$, behaves basically as $n_j = 2^{-j}n$ for a total number $n$ of data points, we see that $\text{Var } \hat{H}$ decreases as $1/n$, regardless of the possible LRD nature of the analyzed process. Finally, numerical investigations support the claim that the wavelet-based estimate $\hat{H}$ is (asymptotically) normally distributed, thus allowing the derivation of confidence intervals from the knowledge of the variance [4].

## 4.3 Additional benefits

Besides the effectiveness of wavelet-based tools (such as the Logscale Diagram) for the analysis of "perfect" scaling processes, a number of additional benefits are offered by the proposed framework, which can prove of particular interest in more realistic situations.

**Robustness to non-Gaussianity.** Among the various assumptions which have been made for deriving analytic performance of the estimators, one concerns the Gaussianity of wavelet coefficients. Except for processes which are by themselves Gaussian, such an assumption has no reason to be relevant, but it has been shown in [6] that formulæ of the type (28) and (29) still apply in non-Gaussian cases, up to correction terms which can be explicitly included in the analysis, or even neglected in the limit of large data samples.

**Insensitivity to polynomial trends.** A common limitation of standard techniques aimed at scaling processes is that their results can be severely impaired by perturbations with respect to the ideal model of a "perfect" scaling process. This is especially the case when deterministic trends are superimposed to a process of interest, with consequences such as invalidating the stationary increments property of an actual LRD process, or mimicking LRD correlations when added to a short-range dependent process [3]. Wavelets are a nice and versatile solution to this crucial issue, since they offer the possibility of being blind to polynomial trends. As it has already been said, a wavelet needs, for being admissible, to be zero-mean (i.e., to have one, zero-th order, vanishing moment) or, in other words, to be orthogonal to constants. A natural extension of this requirement consists in imposing a higher number of vanishing moments, say $N$, so that the resulting wavelet be blind to polynomials up to orders $p \leq N - 1$.

**Computational efficiency.** The analysis of scaling processes is often faced (as it is the case with network traffic) with enormous quantities of data, thus requiring methods which are efficient from a computational point of view. Because of their multiresolution structure and their pyramidal implementation, wavelet-based methods are associated with fast algorithms overperforming FFT-based algorithms (complexity $O(n)$ vs. $O(n \log n)$, for $n$ data points).

## 5 Conclusion

Wavelet analysis has been shown to offer a natural and unified framework for the characterization of scaling processes. One of its main advantages is that it allows for a unique treatment of a large variety of processes, be they self-similar, fractal, long-range dependent, Gaussian or not, ... Further extensions can even be given to the results presented so far. For instance, it has been shown in [2, 32] how *point processes* with fractal characteristics can enter the same framework, on the basis of versatile generalizations of the Fano factor, which is of common use in this context. Other developments have been conducted for taking into account situations where *one* scaling exponent is not sufficient for a proper description and modelling. In such *multifractal* scenarii [21, 28], analyses based on higher-order statistics of wavelet coefficients are a key for efficiently computing quantities such as Legendre singularity spectra [10].

As far as applications are concerned, *telecommunications network traffic* is a domain in which wavelet-based analysis proved most useful in different respects. First, it permits to evidence in a well-controlled way the non-standard features of traffic (self-similarity, long-range dependence) which have been observed since a recent past, and to accurately estimate the corresponding scaling parameters [3, 6]. Second, its versatility allows for using, in a common framework, different aspects of traffic data (work process, interarrival delays, loss, ... ) [3], as well as for being a basis for new (wavelet-based) models [29]. Finally, its computational efficiency allows for a fast processing (with possible on-line implementations) of the very large amounts of data which are commonly encountered in this domain.

Other areas have benefited from the same tools. Among them, one can cite *turbulence*, a field in which the identification of scaling phenomena is of primary importance [20]. The equations (Navier-Stokes, ... ) that govern fluid motion are characterized by nonlinear terms that insure energy transfers from the injection to the dissipation scales, giving birth to the so-called "inertial scaling range". Scaling exponents within this range have been estimated using wavelet tools generalized to statistics different from two (see, e.g., [13]). The multifractal formalism has also been considered to model scaling within the inertial range, and refined wavelet tools have been widely used in this kind of analysis [10]. More recently, cascade models [14] have been invoked to describe a much wider variety of scaling phenomena where scaling on the time series themselves may even be barely observable. These cascades were recently rephrased in the wavelet framework [8] and used to give new insights on scaling in turbulence [7, 15].

In *biology*, scaling processes were used to model, for instance, spike trains discharges of neurons. In this context, wavelet tools such as those presented here were used to estimate the scaling parameters, directly from the observed point-processes, and to decide whether the responses are pathological or not [32].

More recently, the proposed wavelet analysis of scaling has been applied to the study of *extremal models* [31]. Such models describe in a common language a large variety of physical phenomena (wetting front motion, roughening of crack front in fracture, solid friction, fluid invasion in porous media, ... ). In all these situations, key ingredients are the competition between elastic restoring force and nonlinear pining forces, and the assumption that the dynamic of the systems is controlled, at each time step, by its *extremal* part. Wavelet analysis allowed to evidence, amongst the times series produced by extremal models, the existence of temporal statistical dependence and self-similarity, to estimate the corresponding parameters and to show the relevance of the use of LFSM models.

For any of these applications, using wavelets is in some sense "natural", in terms of a structural adequacy between the *mathematical* framework they offer (multiresolution) and the *physical* nature of the processes under study (scaling).

*Acknowledgments.* This work has been supported in part by CNRS *via* its "Programme Télécommunications", under grant # TL 97035. Thanks also to L. Delbeke and D. Veitch, with whom several of the results presented here have been derived.

# References

1. P. Abry, P. Gonçalvès and P. Flandrin (1995): Wavelets, spectrum estimation and $1/f$ processes. In *Wavelets and Statistics*, A. Antoniadis and G. Oppenheim, eds., Lectures Notes in Statistics, **105**, pp. 15–30.
2. P. Abry and P. Flandrin (1996): Point processes, long-range dependence and wavelets. In *Wavelets in Medicine and Biology*, A. Aldroubi and M. Unser, eds., pp. 413–437.
3. P. Abry and D. Veitch (1998): Wavelet analysis of long-range dependent traffic. *IEEE Trans. on Info. Theory*, **44** (1), pp. 2–15.
4. P. Abry, L. Delbeke and P. Flandrin (1999): Wavelet-based estimator for the self-similarity parameter of $\alpha$-stable processes. *IEEE-ICASSP-99*, Phoenix (AZ).
5. P. Abry, L. Delbeke, P. Flandrin and B. Pesquet-Popescu (1999): Wavelet coefficients of self-similar stable motions. Preprint.
6. P. Abry, P. Flandrin, M.S. Taqqu and D. Veitch (1999): Wavelets for the analysis, estimation and synthesis of scaling data. To appear in *Self-Similar Network Traffic and Performance Evaluation*, K. Park and W. Willinger, eds., Wiley Interscience.
7. A. Arnéodo, J.F. Muzy and S.G. Roux (1997): Experimental analysis of self-similar random cascade processes: application to fully developped turbulence. *J. Phys. II France*, **7**, pp. 363–370.
8. A. Arnéodo, E. Bacry and J.F. Muzy (1998): Random cascades on wavelet dyadic trees. *J. Math. Phys.*, **39** (8), pp. 4142–4164.
9. R. Averkamp and C. Houdré (1998): Some distributional properties of the continuous wavelet transform of random processes. *IEEE Trans. on Info. Theory*, **44** (3), pp. 1111–1124.
10. E. Bacry, J.F. Muzy, and A. Arnéodo (1994): Singularity spectrum of fractal signals from wavelet analysis: exact results. *J. Stat. Phys.*, **70**, pp. 635–674.
11. J. Beran (1994): *Statistics for Long-Memory Processes*. Chapman and Hall, New York.
12. S. Cambanis and C. Houdré (1995): On the continuous wavelet transform of second-order random processes. *IEEE Trans. on Info. Theory*, **41** (3), pp. 628–642.
13. R. Camussi and G. Guj (1997): Orthonormal wavelet decomposition of turbulent flows: intermittency and coherent structures. *J. Fluid Mech.*, **348**, pp. 177–199.

14. B. Castaing, Y. Gagne and M. Marchand (1993): Log-similarity for turbulent flows ? *Physica D*, **68**, pp. 387–400.
15. P. Chainais, P. Abry and J.F. Pinton (1999): Intermittency and coherent structures in a swirling flow: a wavelet analysis of joint pressure and velocity measurements. Preprint, submitted to *Phys. Fluids*.
16. L. Delbeke (1998): *Wavelet-Based Estimators for the Scaling Index of a Self-Similar Process with Stationary Increments*. PhD Thesis, KU Leuven.
17. L. Delbeke and P. Abry (1997): Stochastic integral representation and properties of the wavelet coefficients of linear fractional stable motion. Preprint, submitted to *Stoch. Proc. and their Appl.*.
18. L. Delbeke and J. Segers (1998): The covariance of the logarithm of jointly symmetric stable random variables. Preprint.
19. P. Flandrin (1992): Wavelet analysis and synthesis of fractional Brownian motion. *IEEE Trans. on Info. Theory*, **38** (2), pp. 910–917.
20. U. Frisch (1995): *Turbulence: The Legacy of A.N. Kolmogorov*. Cambridge Univ. Press, Cambridge.
21. J. Lévy Véhel and R. Vojak (1998): Multifractal Analysis of Choquet Capacities: Preliminary Results *Advances in Applied Mathematics*, Vol. 20, No. 1, pp. 1-43, January.
22. S. Mallat (1997): *A Wavelet Tour of Signal Processing*. Academic Press, San Diego.
23. E. Masry (1993): The wavelet transform of stochastic processes with stationary increments and its application to fractional Brownian motion. *IEEE Trans. on Info. Theory*, **39** (1), pp. 260–264.
24. B.B. Mandelbrot and J.W. van Ness (1968): Fractional Brownian motions, fractional noises and applications. *SIAM Rev.*, **10**, pp. 422–437.
25. C.L. Nikias and M. Shao (1995): *Signal Processing with Alpha-Stable Distributions and Applications*. Wiley and Sons, Inc., New York.
26. B. Pesquet-Popescu (1998): *Modélisation bidimensionnelle de processus non stationnaires et application à l'étude du fond sous-marin*. Thèse de Doctorat, Ecole Normale Supérieure de Cachan.
27. B. Pesquet-Popescu (1999): Statistical properties of the wavelet decomposition of some non-Gaussian self-similar processes. *Signal Proc.*.
28. R.H. Riedi (1995): An improved multifractal formalism and self-similar measures, *J. Math. Anal. Appl.*, **189**, pp.462–490.
29. R. Riedi, M.S. Crouse, V.J. Ribeiro and R.G. Baraniuk (1999): A multifractal wavelet model with application to network traffic. To appear *IEEE Trans. on Info. Theory*, Special Issue on Multiscale Statistical Signal Analysis and its Applications.
30. G. Samorodnitsky and M.S. Taqqu (1994): *Stable Non-Gaussian Processes: Stochastic Models with Infinite Variance*. Chapman and Hall, New York.
31. A. Tanguy, P. Abry, S. Krishnamurthy and S. Roux (1999): Extremal models and linear fractional stable motion. Preprint, submitted to *Phys. Rev. Lett.*.
32. M.C. Teich, C. Heneghan, S.B. Lowen and R.G. Turcott (1996): Estimating the fractal exponent of point processes in biological systems using wavelet- and Fourier-transform methods. In *Wavelets in Medicine and Biology*, A. Aldroubi and M. Unser, eds., pp. 383–412.
33. A.H. Tewfik and M. Kim (1992): Correlation structure of the discrete wavelet coefficients of fractional Brownian motion. *IEEE Trans. on Info. Theory*, **38** (2), pp. 904–909.

34. D. Veitch and P. Abry (1999): A wavelet-based joint estimator of the parameters of long-range dependence. To appear *IEEE Trans. on Info. Theory*, Special Issue on Multiscale Statistical Signal Analysis and its Applications.

# MULTIFRACTAL ANALYSIS

# Classification of Natural Texture Images from Shape Analysis of the Legendre Multifractal Spectrum

Piotr Stanczyk and Peter Sharpe

Intelligent Computer Systems Centre
University of the West of England
Frenchay Campus Coldharbour Lane
Bristol, BS16 1QY - U.K.
tel (44) 117 9656261
{Piotr.Stanczyk,Peter.Sharpe}@ics.uwe.ac.uk
WWW home page: http://www.ics.uwe.ac.uk/

**Abstract.** Classification of images of real textures was performed using the Legendre multifractal spectrum. Novel features have been developed from basic geometrical properties of the spectrum. Consideration was given to practical issues, in particular to the choice of measure. Classification was implemented using discriminant analysis on randomly extracted subimages. A total of 240 cases were considered. Discussion of the results is given along with areas in need of further investigation.

## 1 Introduction

Many approaches have been devised and implemented to try and characterise real textures using ideas from fractal geometry. Pentland implemented the ideas of fractional Brownian motion (fBm), [21], to devise a description of natural surfaces and consequently to build a description of them from raw image data. A generalization of fBm was introduced by L.M. Kaplan and C. -C. Jay Kuo, [9]. The Extended Self-Similarity (ESS) model relied on the generalization of the structure function, which reflects the growth of the variance of increments of a process as the scale increases, in the characterization of fBm. The resultant generalized Hurst parameter was dependent on scale and the model was no longer isotropic so that dominant directions could be considered. The main drawback of the ESS model was that the parameter values were stable only at the finest resolutions, [10]. Recent generalizations of fBm have been based on multifractal approaches, [20].

Box counting methods for estimating the fractal dimension (FD) have also been considered. Although fast, early implementations, [19], covered only a small range of the values that the fractal dimension could take. Subsequently, better methods were introduced that reduced the quantization levels which were responsible for the poor coverage of the range, [4], [11]. Chaudhuri and Sarker, [4], introduced a method of characterising texture images based mainly on the fractal dimension of the image and its transformed variants. It demonstrated the usefulness of the fractal dimension in texture analysis. However, since the fractal dimension alone is not sufficient to discriminate between visually different surfaces, [16], notions of homogeneity of the fractal dimension were considered, [11], [15].

The work presented here is concerned with the classification of images of real textures, therefore we have to apply methods capable of accommodating such structures. We extend the approach suggested by Chaudhuri and Sarker in that we choose to consider the whole of the multifractal (MF) spectrum. Sailhac and Seyler analysed ERS-1 texture images using a regional multifractal method, [22]. The results suggested that the shape of the singularity spectrum would reflect some properties of the texture. Although much of the current interest in the area is in obtaining the Large Deviation (LD) MF spectrum, [13], [1], [2], mainly using kernel methods for the density estimation, we have chosen to consider the Legendre singularity spectrum. Although it contains less information than the LD spectrum, since the shape is inherently convex, it is easier and quicker to calculate and the moments tend to smooth the data. This results in faster convergence and better stability for images of small sizes as considered in our work. Since we are interested in the classification, as opposed to segmentation, of texture images we need only concern ourselves with the global description of the image and as such there is no need to explicitly calculate the local Hölder exponents. The details of the method used are given in the next section.

The paper is organized as follows: In Sec. 2 we briefly discuss the ideas behind multifractal theory and consider the practical constraints. Sec. 3 is devoted to the development of features based on the geometry of the multifractal spectrum. We present the numerical results in Sec. 4. Conclusions and ideas for further work are included in Sec. 5.

## 2 Singularity Spectrum Estimation via the Legendre Transform

We now consider calculating the whole of the multifractal spectrum using the Legendre transform. A description of multifractal ideas and formalisms can be found in [7], [12], [14], [18]. We proceed in a similar fashion. The approach adopted here assumes that the image is viewed as a measure distributed over a compact support.

Let $\mu$ be a Borel probability measure defined on $[0,1] \times [0,1]$ and let $v_n$ be an increasing sequence of positive integers. Then define the following sub-region:

$$I_{i,j,n} = [\frac{i}{v_n}, \frac{i+1}{v_n}] \times [\frac{j}{v_n}, \frac{j+1}{v_n}] \tag{1}$$

and consider the following quantity:

$$\tau_n(q) = (\frac{-1}{\log v_n}) \log(\sum_{i,j} \mu(I_{i,j,n})^q) \tag{2}$$

Then $\mu$ is said to exhibit multifractal behavior if

$$\lim_{n\to\infty} \tau_n(q) = \tau(q) \tag{3}$$

exists for $q$ in a non-empty interval of $R$. In addition to the global measurement offered by $\tau_n$ we can consider the local structures. The point-wise Hölder exponents are defined to be:

$$\alpha(I_{i,j,n}) = \lim_{n\to\infty} \left( \frac{\log(\mu(I_{i,j,n})^q)}{-\log v_n} \right) \tag{4}$$

The Legendre multifractal spectrum, $f_l(\alpha)$, associated with $\mu$ is obtained via the Legendre transform, that is;

$$f_l(\alpha) = \inf_{q\in\Re} (q\alpha - \tau(q)) \tag{5}$$

Since we will be refering only to the Legendre multifractal spectrum from now on we can drop the $l$ subscript from further references to $f$. In practice data is collected and presented as discrete samples, images are usually stored as discrete values evaluated on a rectangular grid. If the data is subsampled by a factor of $2^n$, that is $v_n = 2^n$, then we can define the scaling ratio $r_n$ to be:

$$r_n = \frac{1}{v_n} = 2^{-n} \tag{6}$$

Consequently, the limit $n \to \infty$ can be replaced by $r_n \to 0$ in eq(3) and eq(4). For clarity we drop the $n$ subscript from $r_n$ in future references. Additionally, we have to normalize the image so that the image is defined on the unit interval $[0,1]\times[0,1]$ and $\sum_{i,j} \mu_r(i,j) = 1$.

The resultant spectrum is based on the measure that is used and a natural measure to use is the sum measure. Lévy Véhel, [12], has generalized these notions to any measures and capacities. The choice of one measure over another will reveal different aspects of the image in question. We are interested in applying multifractal methods to images of textures that may have possibly been acquired under varying lighting conditions so that the choice of the measure should reflect this. For a study of the effects of illumination on texture analysis see, for example, the work by M.J. Chantler in [3].

The choice of the sum measure is inappropriate for our purposes since it is sensitive to the mean gray level of the image. This is easily seen when considering the multiplicative cascade generated by the binomial measure. An addition of a constant level (and subsequent normalization) will result in the deformation of the spectrum.

Instead, we chose to consider the approach taken by Chaudhuri and Sarker, [5], where the differential box counting method of calculating the fractal dimension of a set is introduced. The fractal dimension of a bounded self-similar set that is the union of $N_r$ distinct copies of itself scaled up or down by a ratio $r$ can be estimated from the following relation:

$$D = \frac{\log N_r}{-\log r} \tag{7}$$

Consider an $M \times M$ image as an intensity surface occupying a 3D space. This space is then discretized so that it is covered by boxes of size $s \times s \times s$ where $M/2 > s > 1$, $s \in Z^+$. So that the scaling ratio, $r$, is given by $s/M$. Then $N_r$ is given by:

$$N_r = \sum_{i,j} n_r(i,j) \tag{8}$$

where

$$n_r = l - k + 1 \tag{9}$$

with the maximum and minimum gray level values of the image in the $(i,j)^{th}$ grid falling in the $k^{th}$ and $l^{th}$ boxes respectively. The fractal dimension is then obtained from the least-squares linear fit of $\log(N_r)$ against $\log(1/r)$. The following differential measure can be constructed

$$\mu_r(i,j) = \frac{n_r(i,j)}{N_r} \tag{10}$$

with $N_r$ and $n_r(i,j)$ defined in eq(8) and eq(9) respectively. In this formulation, it is the differences in the pixel values that are considered, as opposed to the actual pixel values as in the sum measure. This differential measure is less sensitive to the variations in the mean gray levels and should therefore be better suited for our needs.

The numerical computation of the spectrum is based on the method of moments [8], [18]. Essentially, if $\alpha = \alpha(q)$ is such that the infimum in eq(3) is satisfied then it is straightforward to show that

$$\alpha(q) = \frac{d}{dq}\tau(q) \tag{11}$$

The Legendre multifractal spectrum is obtained from eq(5) and eq(11) via the parameter $q$. Dealing with discrete data sets at finite resolution means that in practice, a least squares linear fit for the finer resolutions procedure has to be implemented.

To avoid further numerical errors that could arise from calculating the derivative of $\tau(q)$, eq(11), we follow the approach used by A. Chhabra and R.V. Jensen, [6], wherein the spectrum is calculated directly so that:

$$\alpha(q) = \lim_{r \to 0} \left\{ \frac{(\sum_{i,j} \mu_r(i,j)^q) \log \mu_r(i,j)}{\log r \sum_{i,j} \mu_r(i,j)^q} \right\} \tag{12}$$

Using eq(5) and $\alpha = \alpha(q)$ we have that:

$$f(q) = \lim_{r \to 0} \left\{ \frac{q(\sum_{i,j} \mu_r(i,j)^q) \log \mu_r(i,j) - \log \sum_{i,j} \mu_r(i,j)^q}{\log r \sum_{i,j} \mu_r(i,j)^q} \right\} \tag{13}$$

Once again we consider using least squares linear fit to obtain values for $\alpha(q)$ and $f(q)$. Since we are expecting the fine resolution values to be close to their respective limits we also perform an analysis of the residuals from the least squares to test for convergence as $r \to 0$.

# 3 Feature Development from Shape Analysis of the Spectrum

## 3.1 Robustness and Convergence Issues

Before any features based on the MF spectrum are developed we must examine whether textural images that are visually indistinguishable will result in similar spectra, i.e. the robustness of the method with respect to the particular images. We considered 36, 128×128 partially overlapping extracts from the micro-texture image Rock.00 and the macro-texture image Rock.24. From each of these extracts, the multifractal spectrum was calculated using the method described in the previous section. Each one of these curves was then evaluated at regularly spaced points so that the deviations in the calculated values of $f$ could be performed. The value of $f$ was estimated using linear interpolation. A range of $\alpha$ that was common to all 36 extracts was considered for each image. The results are shown in Fig 2, the vertical bars indicate the maximum and minimum values for the linearly interpolated $f$ over the 36 extracts for each image.

The resolution at which $f$ was evaluated was quite fine, interval length was 0.01, so that for values of $q$ near to 0 the regularly evaluated curve looked piece-wise linear. This fine resolution was needed, however, to accurately model the behaviour of $f$ for extreme values of $q$. It is clear that the majority of the deviations from the mean are present in the extremes of the calculations and that the left-hand side of the spectrum is less robust than the right-hand side. This was also the case for experiments carried out on other test images.

Additional tests were also carried out to examine the behaviour of the limiting values for $\alpha$ for the two images. Fig 1 shows the behaviour of the minimum and maximum values of $\alpha$ for the two images. It is clear that the maximum values, corresponding to the RHS of the spectrum, are more stable than the minimum, especially for the micro-texture image Rock.24.

In theory, all moments should be considered to get the whole spectrum. In practice we must find a suitable compromise so that values for $\alpha(q)$ and $f(q)$ come close to their limiting values, $q \to \pm\infty$, without numerical errors becoming substantial. We found that for all the images examined the right hand side of the spectrum converged for much smaller values of $q$ than the right-hand side. In addition, $\alpha(q)$ was quicker to converge than $f(\alpha)$. Empirically, $|q| < 50$ was a reasonable compromise for our experiments.

## 3.2 Feature Development

As suggested, the shape of the spectrum will provide information about the texture of the image. Fig 3 shows the spectrum for extracts from Rock.00 and Rock.01 images. There are some noticeable differences between the two curves that motivate the design of the features. The features are designed to capture some basic geometric characteristics. We are interested in the shape and the position of the curve. For the position, we can consider the end points and the

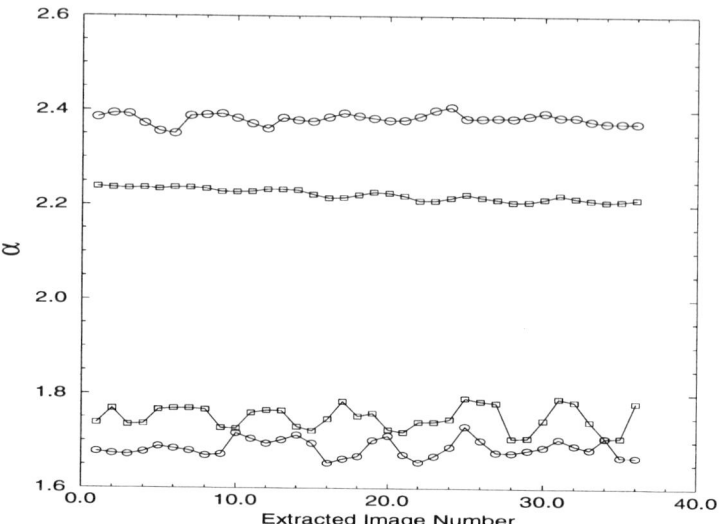

**Fig. 1.** The behaviour of the mininum and maximum values of $\alpha$ for Legendre Multifractal Spectrum for visually indistinguishable textural images. A total of 36, 128x128, extracts of Rock.00 (□-□) and Rock.24.pgm (o-o) were considered

value of $\alpha$ for which the curve attains a maximum, that is $\alpha = \alpha_0$. The end points are defined as the limiting values of $\alpha$ and $f$ as $q \to \pm\infty$ respectively, therefore it is expected that these will not be robust descriptors. However, these are included in the analysis and we expect the classifier to reject these on the basis of large within group variance, see Sec. 4 for explanation of classification procedure.

The shape of the distributions of $\alpha$ and $f(\alpha)$ (with reference to the $q$ parameter) obtained through variance, skewness and kurtosis operators could give a false description of the images since similar curves can have different parametrisations. Therefore, features based on the integrals of the left ($q > 0$) and right ($q < 0$) hand sides of the $f(\alpha)$-$\alpha$ curve have been designed. Their ratios will reveal any bias in the curve around $\alpha_0 = \alpha(0)$ indicative of the heterogeneity and anisotropy in the image, [22]. The 'total integrals' are given by:

$$F1 = \int_{\alpha_{min}}^{\alpha_0} f(\alpha)d\alpha \quad F2 = \int_{\alpha_0}^{\alpha_{max}} f(\alpha)d\alpha \quad F3 = F1 : F2$$

Since in general $f(\alpha)$ will not have the same limiting values as $q \to \pm\infty$, these integral features could also offer a biased and unrobust description, see previous section. To overcome this we have designed further integral features. Assuming, w.l.o.g., that $f(\alpha_{min}) < f(\alpha_{max})$ and letting $\alpha_1 = \min\{\alpha : f(\alpha) = f(\alpha_{max})\}$, the 'partial integrals' are given by:

$$F4 = \int_{\alpha_{min}}^{\alpha_0} f(\alpha)d\alpha - (\alpha_0 - \alpha_1)f(\alpha_{max})$$

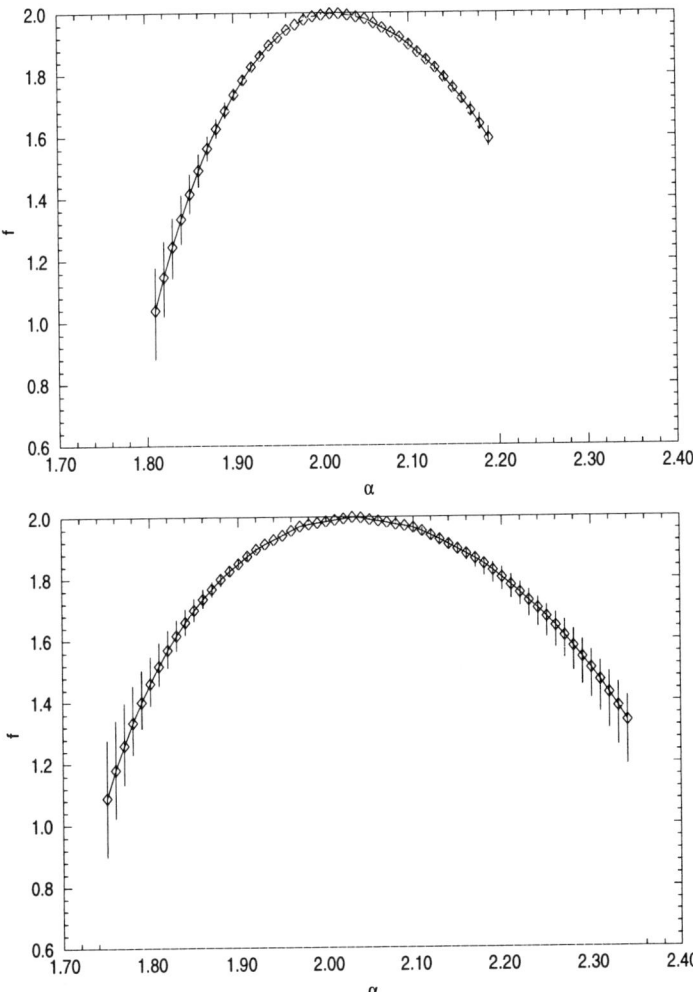

**Fig. 2.** The behaviour of the Legendre Multifractal Spectrum for visually indistinguishable textural images. A total of 36, 128×128, extracts of a)Rock.00 and b)Rock.24.pgm were considered. The spectra were sampled at regular intervals, 0.01, and linear interpolation was used to estimate $f$. The vertical bars indicate the minimum and maximum values obtained for $f$, respectively. The mean is depicted by the ◇ − ◇ line. The shown range of $\alpha$ is common to all the samples for each image

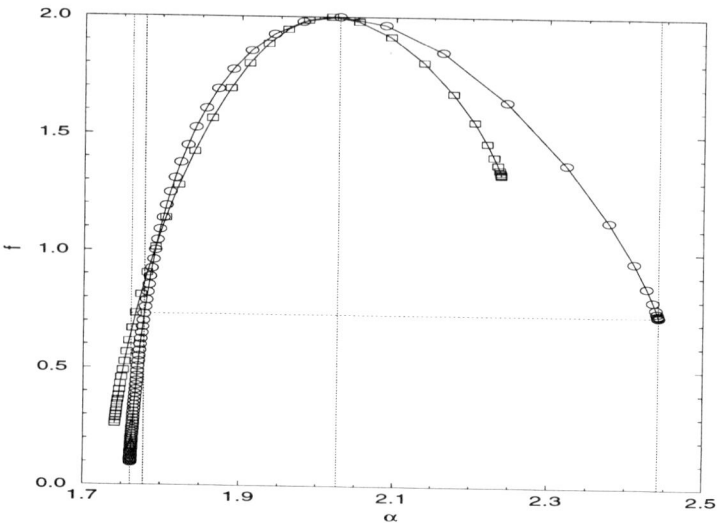

**Fig. 3.** The resultant Legendre multifractal spectrum for extracts from Rock.00 □-□ and Rock.01 o-o. The dotted lines indicate the nature of the integral features for Rock.01 case

$$F5 = \int_{\alpha_0}^{\alpha_{max}} f(\alpha)d\alpha - (\alpha_{max} - \alpha_0)f(\alpha_{max}) \quad F6 = F4 : F5$$

Note also that for most cases $f(\alpha_{min}) < f(\alpha_{max})$. Since $\alpha_{max}$ is more 'stable' than $\alpha_{min}$, see Fig 1, the partial integrals will be more robust than the total integrals. Finally, for all the integral features we consider the average integral on the $\alpha$ axis and the percentage of the total possible integral. So that for total integrals:

$$F7 = \frac{F1}{(\alpha_0 - \alpha_{min})} \quad F8 = \frac{F2}{\alpha_{max} - \alpha_0} \quad F9 = F7 : F8$$

$$F13 = \frac{F7}{(f(\alpha_0) - f(\alpha_{min}))} \quad F14 = \frac{F8}{f(\alpha_{max}) - f(\alpha_0)} \quad F15 = F13 : F14$$

and for partial integrals:

$$F10 = \frac{F4}{(\alpha_0 - \alpha_{min})} \quad F11 = \frac{F5}{\alpha_{max} - \alpha_0} \quad F12 = F10 : F11$$

$$F16 = \frac{F10}{(f(\alpha_0) - f(\alpha_{min}))} \quad F17 = \frac{F11}{f(\alpha_{max}) - f(\alpha_0)} \quad F18 = F16 : F17$$

All of the constructed features are shown in Table 1.

**Table 1.** Proposed features based on the geometry of the Legendre Multifractal Spectrum

| Feature | Description | Feature | Description |
|---|---|---|---|
| F1 | Total L.H.S. integral | F13 | Total L.H.S. integral.P |
| F2 | Total R.H.S. integral | F14 | Total R.H.S. integral.P |
| F3 | F1:F2 | F15 | F13:F14 |
| F4 | Partial L.H.S. integral | F16 | Partial L.H.S. integral.P |
| F5 | Partial R.H.S. integral | F17 | Partial R.H.S. integral.P |
| F6 | F4:F5 | F18 | F16:F17 |
| F7 | Total L.H.S. integral.S | F19 | $\alpha_0$ |
| F8 | Total R.H.S. integral.S | F20 | $\alpha_{min}=\min\{\alpha\}$ |
| F9 | F7:F8 | F21 | $\alpha_{max}=\max\{\alpha\}$ |
| F10 | Partial L.H.S. integral.S | F22 | $f(\alpha_{min})\}$ |
| F11 | Partial R.H.S. integral.S | F23 | $f(\alpha_{max})\}$ |
| F12 | F10:F11 | | |

## 4 Numerical Experiments

The work presented here has been focused on the analysis of isotropic and homogeneous natural textural images. The images were obtained from the MeasTex[1] texture data base. The image set consisted of 15, 512×512 images, illustrated in Fig 4. Each image was labelled with an arbitrary class attribute, images that were visually indistinguishable were labelled with the same class value. From each image 16, 128×128, randomly selected samples were extracted so that there was a total of 240 cases with 12 class values. For each case, features, described in the previous section, were calculated.

Classification of the cases was performed using discriminant analysis. We selected around 66% of the total cases for the discriminant analysis leaving the other 34% to test[2]. All groups were assigned equal a priori probabilities. The implementation of the discriminant analysis included the checking for within-ingroup variance for each of the features. Features F15, F16 and F18 failed this test. It is possible that these omissions are due to analysis of uncharacteristic regions of the image. These omissions are not indicative of the usefulness of these features for different image libraries, since generally different features will perform in different ways according to the images that are presented. As expected, the end points of the MF spectrum were also excluded from the analysis.

The overall correct classification rate for cases that were included in the analysis was 94.4%. The classification results for cases not included in the analysis are displayed in Table 2. 87.5% of samples were correctly classified. These are not exceptional, however, upon closer inspection we found that most of the misclassifications occurred for images that resembled each other quite closely, for example images of Rock23 (class 20) and Rock24 (class 21). Furthermore,

---

[1] MeasTex texture database is available from:
http://www.cssip.elec.uq.edu.au/g̃uy/meastex.html.

[2] All calculations were performed using SPSS, version 8.0, for Windows NT

an inspection of the values of the canonical discriminant functions for individual misclassified cases highlights that the predicted class values agree with the actual class values when $2^{nd}$ highest group probabilities are considered.

Table 2. Classification results for cases not considered in the analysis.

|   | CLASS | 01 | 02 | 05 | 06 | 10 | 11 | 13 | 14 | 16 | 18 | 20 | 21 | TOTAL |
|---|---|---|---|---|---|---|---|---|---|---|---|---|---|---|
|   | 01 | 3 | 0 | 0 | 0 | 0 | 0 | 0 | 0 | 0 | 0 | 0 | 0 | 4 |
| P | 02 | 0 | 4 | 0 | 0 | 0 | 0 | 0 | 0 | 0 | 0 | 0 | 0 | 3 |
| R | 05 | 0 | 2 | 5 | 0 | 0 | 0 | 0 | 0 | 0 | 0 | 0 | 0 | 7 |
| E | 06 | 0 | 0 | 0 | 9 | 0 | 0 | 0 | 0 | 0 | 0 | 0 | 0 | 9 |
| D | 10 | 0 | 0 | 0 | 0 | 6 | 0 | 0 | 0 | 0 | 0 | 0 | 0 | 6 |
| I | 11 | 0 | 0 | 0 | 0 | 0 | 10 | 3 | 0 | 0 | 0 | 0 | 0 | 13 |
| C | 13 | 0 | 0 | 0 | 0 | 0 | 1 | 6 | 0 | 0 | 0 | 0 | 0 | 7 |
| T | 14 | 0 | 0 | 0 | 0 | 0 | 0 | 0 | 9 | 0 | 0 | 0 | 0 | 9 |
| E | 16 | 0 | 0 | 0 | 0 | 0 | 0 | 0 | 0 | 4 | 1 | 0 | 0 | 5 |
| D | 18 | 0 | 0 | 0 | 0 | 0 | 0 | 0 | 0 | 0 | 4 | 0 | 0 | 4 |
|   | 20 | 0 | 0 | 0 | 0 | 0 | 0 | 0 | 0 | 0 | 0 | 0 | 6 | 1 |
|   | 21 | 0 | 0 | 0 | 0 | 0 | 0 | 0 | 0 | 1 | 0 | 1 | 4 | 6 |

## 5 Conclusions and Further Work

We have addressed the problem of classifying natural textures using multifractal techniques. Motivated by differences in the Legendre multifractal spectrum between different images we introduced a collection of features that captured some of the basic geometric properties of the spectrum. These features were calculated for a library of real textures consisting of 240, 128x128 gray level images. The classification of images was based on discriminant analysis. The 240 cases were divided into 'training' and 'test' sets with a ratio of approximately 2:1. Several of the features proved to have a too high within-in-group variance to be included in the analysis, but it was suggested that these might prove to be useful for other libraries of images or classification systems. The classification results warrant further investigation as to the geometrical interpretations of the features with regard to different texture images. In particular, the change in the shape and position of the spectrum with regard to textures containing different sized texels. The percentage of correct classifications for the training set was 94.4%, whilst for the cases not included in the analysis the rate dropped to 87.5%. A closer inspection of the results revealed that the misclassifications occurred mainly for images that were visually similar. This prompts us to suggest that for many applications where the associated image attributes do not form distinct classes (see [17] for example) an approach is needed that will incorporate a function of similarity between these attributes.

Current research includes generation of new features that could capture the more complex shapes found in Large Deviation multifractal spectrum as well as extending the above approach to the classification of images that may contain regions that are unsuitable for feature extraction. High magnification images of 'real' surfaces form such libraries. The regions of interest may be surrounded by areas that are out of focus due to the depth of focus problem or simply areas that are uncharacteristic of the whole image. We are investigating an approach based on the extraction of several, well chosen, subimages and classifying the image based on the frequency of classifications of the subimages.

# References

1. C. Canus, J. Lévy Véhel and C. Tricot (Oct. 1998): Continuous Large Deviation Multifractal Spectrum: Definition and Estimation. Proceedings of Fractals 98, Malta.
2. C. Canus (Apr. 1998): Robust Large Deviation Multifractal Spectrum Estimation. Proceedings of International Wavelets Conference, Tangier.
3. M.J. Chantler (Aug. 1995): Why illuminant direction is fundamental to texture analysis. IEE Proc.-Vis. Image Signal Process., Vol. 142, No. 4.
4. B.B. Chaudhuri and N. Sarker (1992): An Efficient Approach to Estimate Fractal Dimension of Textural Images. Pattern Recognition, Vol. 25, No. 9, pp. 1035–1041.
5. B.B. Chaudhuri and N. Sarker (Jan. 1995): Texture Segmentation Using Fractal Dimension. IEEE Trans. on Patt. Anal. and Mach. Intell., Vol. 17, No. 1, pp. 72–77.
6. A. Chhabra and R. V. Jensen (1989): Direct Estimation of the $f()$ singularity spectrum. Phys. Rev. Lett. 62, 1327.
7. K. Falconer (1990): Fractal Geometry: Mathematical Foundations and applications. Wiley, England.
8. T. C. Hasley, M. H. Jensen, L. P. Kadanoff, I. Procaccia and B. I. Shraiman (1986): Fractal measures and their singularities: The characterisation of strange sets. Phys. Rev. A 33, pp. 1141.
9. L.M. Kaplan and C. -C. Jay. Kuo (Dec. 1994): Extending Self-Similarity for Fractional Brownian Motion. IEEE Trans. Signal Processing, Vol. 42, No.12, pp. 3526–3530.
10. L.M. Kaplan and C. -C. Jay. Kuo (Nov. 1995): Texture Roughness Analysis and Synthesis via Extended Self-Similar (ESS) Model. IEEE Trans. on Patt. Anal. and Mach. Intell., vol.17, No. 11, pp. 1043–1056.
11. M. J. Keller, S. S. Chen and R. Crownover (1989): Texture Description and Segmentation through Fractal Geometry. Computer Vision, Graphics and Im. Proc., Vol 45, pp. 150–166.
12. J. Lévy Véhel (1996): Introduction to the Multifractal Analysis of Images. Fractal Image Encoding and Analysis. Yuval Fisher Ed., Springer.
13. J. Lévy Véhel (1996): Numerical Computation of the Large Deviation Spectrum. CFIC96 Rome.
14. J. Lévy Véhel and J.-P. Berroir (1994): Image Analysis through Multifractal Description. Fractals in the Natural and Applied Sciences (A-41), IFIP, pp. 261–274.
15. B.B. Mandlebrot (1982): Fractal Geometry of Nature. San Francisco, CA, Freeman.
16. G. Médioni and Y. Yasumoto (1984): A note on using the fractal dimension for segmentation. IEEE Computer Vision Workshop, Annapolis, MD, pp. 25–30.

17. O.K. Panagouli and A.G. Kokkalis (1998): Skid resistance and fractal structure of pavement surface. Chaos Solitons & Fractals, Vol.9, No.3, pp. 493–505.
18. H.-O. Peitgen, H. Jurgens and D. Saupe (1992): Chaos and Fractals: New Frontiers of Science. Appendix B, Springer-Verlag.
19. S. Peleg, J. Naor, R. Hartley and D. Avnir (Nov. 1984): Multiple Resolution Texture Analysis and Classification. IEEE Trans. on Patt. Anal. and Mach. Intell., vol. PAMI-6, no. 4, pp. 518–523.
20. R.F. Peltier and J. Lévy Véhel (Aug. 1995): Multifractal Brownian motion: definition and preliminary results. INRIA technical report No. 2645.
21. P. Pentland (1984): Fractal based Description of Natural Scenes. IEEE Trans. Patt. Anal. Mach. Intell. PAMI-6, No. 6, pp. 661–674.
22. P. Sailhac and F. Seyler (1997): Texture Characterisation of ERS-1 images by Regional Multifractal analysis. Fractals in Engineering, Springer, pp. 32–41.

**Fig. 4.** The collection of images used in the experiments. The illustrated images are 128x128 extracts from the centres of the original 512x512 images provided in the Meas-Tex library of images. From left to right, top to bottom the images are: Asphalt.01 (class 1), Asphalt.02 (class 2), Concrete.03 (class 5), Concrete.03 (class 6), Concrete.03 (class 6), Concrete.8 (class 10), Concrete.10 (class 11), Concrete.11 (class 11), Rock.00 (class 13), Rock.01 (class 14), Rock.02 (class 14), Rock.06 (class 16), Rock.09 (class 18), Rock.23 (class 20), Rock.24 (class 21)

# A Generalization of Multifractal Analysis Based on Polynomial Expansions of the Generating Function

Antoine Saucier[1] and Jiri Muller[2]

[1] Centre de Recherche en Calcul Appliqué
5160, boul. Décarie, bureau 400
Montréal (Québec) - CANADA H3X 2H9
antoine@cerca.umontreal.ca
[2] Institutt for energiteknikk
Box 40, N-2007 Kjeller - NORWAY
jiri@ife.no

**Abstract.** We introduce a generalization of multifractal analysis that allows to make a simple description of virtually any measure. Our representation of the generating function is constructed in such as way that it introduces a hierarchy of dimension functions $\{D_n(q), n = 0, 1, 2, ..., N\}$, instead of a single one in the case of multifractals. Our representation reduces to the multifractal description if the measure is multifractal. These dimension functions can be used to characterize the variability of irregular signals and images. Our method is demonstrated by analyzing the texture of pore space in sedimentary chalk.

## 1 Introduction

One of the objectives of textural analysis is to quantify the local variability of a signal or of an image. Such a quantification can be helpful for signals/images that typically exhibit a rather chaotic variability, e.g. digitized images of disordered materials [5], well logs [6] or remotely sensed data, which are often suitable candidates for statistical [4] or multifractal analysis [8]. In this context, a texture index will attempt to capture some aspects of the variability of a signal in a given domain. If an image has been divided into square cells on a regular grid, for instance, then the texture indices computed in each cell can be used to produce a simplified (i.e. lower resolution) representation of this image.

One way to approach the problem of texture characterization is to use the generating function $\chi_q(\delta)$. This function, commonly used in multifractal analysis, has the ability to capture multipoint correlations in a signal [7]. This ability is not restricted to pairs of points, but also includes triplets, quadruplets and more generally n-tuplets ($n$ integer). The main objective of this paper is to propose a new representation of the generating function that allows to characterize texture even if $\chi_q(\delta)$ does not take the special form $\chi_q(\delta) \sim \delta^{\tau(q)}$, i.e. the power law which is characteristic of multifractals.

For general measures the generating function does not take the form of a power law and consequently the mass exponent function $\tau(q)$ does not provide a satisfactory description. Being interested in the multipoint correlations captured by the generating function, we have been searching for simple, yet general, representations of $\chi_q(\delta)$. We propose in this paper a new representation of the generating function that can be used to describe virtually any measure. This representation is built in such as way that it introduces a hierarchy of dimension functions $\{D_n(q), n = 0, 1, 2, ..., N\}$, instead of a single one in the case of multifractals. Our representation reduces to the usual multifractal description if the measure is multifractal.

## 2  Generalization of multifractal analysis

The generating function $\chi_q(\delta)$ used for multifractal analysis is a function of two variables $q$ and $\delta$. $q$ is a real parameter and $\delta \geq 0$ is a scale ratio $\delta = \ell/L \leq 1$, where $\ell$ is a length scale and $L$ is an upper bound on $\ell$. Multifractal analysis can be done at the condition that $\chi_q(\delta)$ satisfies the power law property [2]

$$\chi_q(\delta) \sim \delta^{\tau(q)} \qquad (1)$$

in an interval $\delta \in [\delta_{min}, \delta_{max}]$ called *scaling range*. The *mass exponents* $\tau(q)$ [2] are obtained for each $q$ by using linear regressions of $\log(\chi_q(\delta))$ versus $\log(\delta)$ with the model

$$-\ln[\chi_q(\delta)] = \tau_0(q) - \tau(q) \ \ln(\delta) \qquad (2)$$

With the change of variable $x = -\ln(\delta) \geq 0 \Rightarrow \delta = \exp(-x)$ and the definition $\phi_q(x) \equiv -\ln(\chi_q(\exp(-x)))$, equation (2) can be rewritten in the equivalent form

$$\phi_q(x) = \tau_0(q) + \tau_1(q) \ x \qquad (3)$$

As emphasized in the introduction, a broad family of signals encountered in nature cannot be described satisfactorily with the simple linear model (3). If we choose to remain faithful to the idea of using the generating function to characterize texture, then our objective is to find simple and economic descriptions of the function $\phi_q(x)$, whether the measure is multifractal or not, i.e. whether (3) is satisfied or not. Our objective is therefore to generalize the model (3) to handle more general signals, while keeping a formal connection with multifractal analysis.

We propose to generalize the model (3) by expanding the function $\phi_q(x)$ on a family of orthogonal polynomials of increasing order. More precisely, we will write $\phi_q(x)$ in the form

$$\phi_q(x) = \sum_{n=0}^{N} \tau_n(q) \ P_n(x) \qquad (4)$$

where $P_n(x)$ denotes a polynomial of order $n$. The model (4) is a straightforward generalization of the simple multifractal model (3). The coefficients $\tau_n(q)$ are formal generalizations of the mass exponent function $\tau(q)$ [2]. In particular, the term $\tau_1(q)$ corresponds to the traditional $\tau(q)$ if the measure is multifractal. We assume in the following that the coordinates $x$ take discrete values $x_i, i = 1, 2, ..., N$ and that $\phi_q(x)$ is computed only for these $x_i$s (we choose equidistant $x_i$s, as is usually done for multifractal analysis). The polynomials are orthogonal with respect to a scalar product that we define for any two real functions $f$ and $g$ by

$$\langle f, g \rangle = \sum_{i=1}^{N_x} f(x_i)\, g(x_i) \tag{5}$$

where the summation extends over all the $x_i$s. We construct the polynomials $P_n(x)$ with a Gram-Schmidt orthogonalization (Appendix A), and consequently they satisfy the orthogonality relationships

$$\langle P_n, P_m \rangle = \delta_{n,m}\, \langle P_n, P_n \rangle \tag{6}$$

The coefficients $\tau_n(q)$ in the expansion (4) are simply defined by

$$\tau_n(q) = \langle P_n, \phi_q \rangle / \langle P_n, P_n \rangle \tag{7}$$

which is a consequence of the orthogonality of the $P_n$s. In this representation, the generalized mass exponent $\tau_n(q)$ is simply defined by the projection of $\phi_q(x)$ on the expansion vector $P_n(x)$.

The generating function is defined in such as way that it satisfies the property $\chi_1(\delta) = 1$ for all $\delta$, which is called the *normalization condition* (see definition (11)). It follows from this condition that $\phi_q(x)$ satisfies $\phi_1(x) = 0$ for all $x$ and therefore the $\tau_n(q)$s defined by (7) satisfy $\tau_n(1) = 0$ for each $n$. It follows that the $\tau_n(q)$s can be always written in the form

$$\tau_n(q) = (q-1) D_n(q) \tag{8}$$

and consequently the polynomial expansion (4) takes the form

$$\phi_q(x) = (q-1) \sum_{n=0}^{N} D_n(q)\, P_n(x) \tag{9}$$

where the $D_n(q)$s are formal extensions of the generalized dimensions [3]. The model (9) should be general enough to describe virtually any measure (the special case $q=0$ raises some difficulties because $\phi_0(x)$ is not always analytic, and therefore a polynomial fit is not always appropriate). Moreover, the linear component $D_1(q)$ of the fit reduces to the generalized dimension $D(q)$ if the measure is multifractal, i.e. if $N=1$.

The model (9) allows to map a function $\phi_q(x)$ of two variables $q$ and $x$ onto several functions $\{D_n(q), n = 0, 1, 2, ..., N\}$ of a single variable $q$, which simplifies the representation of $\phi_q(x)$. The number $N$ of functions required to obtain a satisfactory description of depends on the accuracy desired. For instance, the chi-square can be used to quantify the quality of fit of the expansion (9) as a function of $N$. As $N$ increases, the accuracy of the expansion improves. By contrast with multifractal analysis, where linear regressions are used to obtain mass exponents, there is no need for any fitting within this new approach. Indeed, the exponents are computed directly from the scalar product with formula (7).

## 3 The generating function

Consider a random positive scalar signal $S(\mathbf{r}_i)$ defined on a set of $D$-dimensional points $\mathbf{r}_i = (x_{i_1}, x_{i_2}, ..., x_{i_D})$ located on a regular grid. We define the *measure* of a square box $B_\mathbf{r}(\delta)$ of size $\delta$ centered on a point $\mathbf{r}$ by

$$p_\mathbf{r}(\delta) = \sum_{\mathbf{r}_i \in B_\mathbf{r}(\delta)} S(\mathbf{r}_i) \tag{10}$$

If $S(\mathbf{r}_i)$ is interpreted to be a mass localized at point $\mathbf{r}_i$, then $p_\mathbf{r}(\delta)$ is simply the total mass contained in the box $B_\mathbf{r}(\delta)$. We are interested in the moments $\langle [p_\mathbf{r}(\delta)]^q \rangle$, where $q$ is a real parameter and $\langle ... \rangle$ denotes an ensemble average, because of their ability to capture the multipoint correlations of the underlying signal $S(\mathbf{r}_i)$ [7]. These moments are not defined for non integer $q$ if $p_\mathbf{r}(\delta) = 0$, and consequently it is necessary in general to consider instead the quantity $\langle \Theta_q(p_\mathbf{r}(\delta)) \rangle$, where $\Theta_q(x) = x^q$ if $x > 0$ and $\Theta_q(x) = 0$ if $x = 0$. The two expressions $\langle \Theta_q(p_\mathbf{r}(\delta)) \rangle$ and $\langle [p_\mathbf{r}(\delta)]^q \rangle$ are equivalent if $p_\mathbf{r}(\delta) > 0$.

We always consider a collection of $M$ samples ($M \geq 2$) of equal size that are taken from the same statistical ensemble. Indeed, having several samples is necessary to estimate the uncertainties on the generating function. If a single image is available, then we split it into four disjoint quadrants of the same size that are regarded as four independent samples. Firstly, we define for each sample an *individual* generating function by $\chi_q(\delta) = \delta^{-D} \langle \Theta_q(p_\mathbf{r}(\delta)) \rangle_S$, where $D$ is the dimension of the embedding space, and angle brackets $\langle ... \rangle_S$ denote a *spatial* average. Secondly, we define the *global* generating function $\hat{\chi}_q(\delta)$ obtained from all $M$ samples by

$$\hat{\chi}_q(\delta) = \overline{\chi}_q(\delta) / [\overline{\chi}_q(\delta)]^q \tag{11}$$

where

$$\overline{\chi}_q(\delta) = \frac{1}{M} \sum_{i=1}^{M} \chi_q^{(i)}(\delta) \tag{12}$$

and $\chi_q^{(i)}(\delta)$ denotes the individual generating function of sample $i$. The definition (11) insures that $\overline{\chi}_q(\delta)$ remains proportional to the spatial average $\langle \Theta_q(p_\mathbf{r}(\delta)) \rangle_S$ estimated by combining all the $M$ samples, whereas the denominator of (11) guarantees that $\hat{\chi}_q(\delta)$ is normalized exactly, i.e. that $\hat{\chi}_1(\delta) = 1$ for all $\delta$. Our definition (11) of the generating function, based on continuous spatial averages of $p_\mathbf{r}(\delta)$, has been shown to be a natural and more flexible extension of the conventional generating function computed on a grid [6].

The uncertainty on $\hat{\chi}_1(\delta)$ can be estimated for each $\delta$ and each $q$ by taking into account the variations of $\chi_q^{(i)}(\delta)$ from sample to sample. We have shown in [7] that the uncertainty $\sigma(\phi_q(x))$ (i.e. standard deviation) on $\sigma(\phi_q(x)) \equiv -\ln(\hat{\chi}_q(\delta))$ can be estimated with the formula

$$\sigma^2(\phi_q(x)) = \frac{1}{M}\left(\left(\frac{\sigma(\chi_q)}{\langle\chi_q\rangle}\right)^2 + \left(q\frac{\sigma(\chi_1)}{\langle\chi_1\rangle}\right)^2 - 2q\left(\frac{\mathrm{cov}(\chi_q,\chi_1)}{\langle\chi_q\rangle\langle\chi_1\rangle}\right)\right) \quad (13)$$

where $\mathrm{cov}(\chi_q,\chi_1) \equiv \langle\chi_q\,\chi_1\rangle - \langle\chi_q\rangle\langle\chi_1\rangle$. In (13) the dependence of $\chi_q$ on $\delta$ has been omitted for simplicity. In practice, the expectation values appearing in (13), such as $\langle\chi_q\rangle$, are estimated by averages computed on the $M$ samples, e.g. $\langle\chi_q\rangle \approx M^{-1}\sum_{i=1}^{M}\chi_q^{(i)}(\delta)$. Once the $\sigma(\phi_q(x))$s have been estimated for each $(q,x)$ with (13), we derive the corresponding uncertainties on the $D_n(q)$s as follows. Introducing the fluctuation $\Delta\tau_n(q) \equiv \tau_n(q) - \langle\tau_n(q)\rangle$, and more generally $\Delta U \equiv U - \langle U\rangle$ for an arbitrary random variable $U$, it follows from the linearity of the scalar product in (7) that

$$\Delta\tau_n(q) = \langle P_n, \Delta\phi_q\rangle / \langle P_n, P_n\rangle \quad (14)$$

Squaring and averaging (14), and then noticing that $\sigma^2(\tau_n(q)) = \langle(\Delta\tau_n(q))^2\rangle$, we get

$$\sigma^2(\tau_n(q)) = \langle(\langle P_n, \Delta\phi_q\rangle)^2\rangle / (\langle P_n, P_n\rangle)^2 \quad (15)$$

where

$$\langle(\langle P_n, \Delta\phi_q\rangle)^2\rangle = \langle\left(\sum_{i=1}^{N_x} P_n(x_i)\,\Delta\phi_q(x_i)\right)^2\rangle$$

$$= \sum_{i=1}^{N_x}\sum_{i=1}^{N_x} P_n(x_i)P_n(x_j)\langle\Delta\phi_q(x_i)\Delta\phi_q(x_j)\rangle \quad (16)$$

If we make the assumption that the fluctuations $\Delta\phi_q(x_i)$ and $\Delta\phi_q(x_j)$ are uncorrelated when $i \neq j$, then $\langle\Delta\phi_q(x_i)\Delta\phi_q(x_j)\rangle = \delta_{ij}\langle(\Delta\phi_q(x_i))^2\rangle$ and (16) becomes

$$\langle(\langle P_n, \Delta\phi_q\rangle)^2\rangle = \sum_{i=1}^{N_x} P_n^2(x_i)\langle(\Delta\phi_q(x_i))^2\rangle = \langle P_n^2, \langle(\Delta\phi_q)^2\rangle\rangle \quad (17)$$

Noticing that $\langle(\Delta\phi_q)^2\rangle = \sigma^2(\phi_q)$ and replacing (17) in (15), we obtain the simple form

$$\sigma^2(\tau_n(q)) = \langle P_n^2, \sigma^2(\phi_q)\rangle / (\langle P_n, P_n\rangle)^2 \quad (18)$$

Finally, it follows from equation (8) that for all $q \neq 1$

$$\sigma(D_n(q)) = \sigma(\tau_n(q))/|q-1| \quad (19)$$

In this paper, the uncertainties on the dimensions $D_n(q)$s are computed with the equations (13), (18) and (19).

## 4 Example with porous rock images

We applied our method to digitized scanning electron microscope images obtained from sedimentary chalk samples (called a1, a3, a4, a10, a9 and a7) of varying porosity and permeability. For each sample, we selected 4 digitized images obtained from the same thin section. The images have a resolution of 512 x 512 pixels for a physical size of about 100 mm. A representative image from each sample is displayed in figure 1.

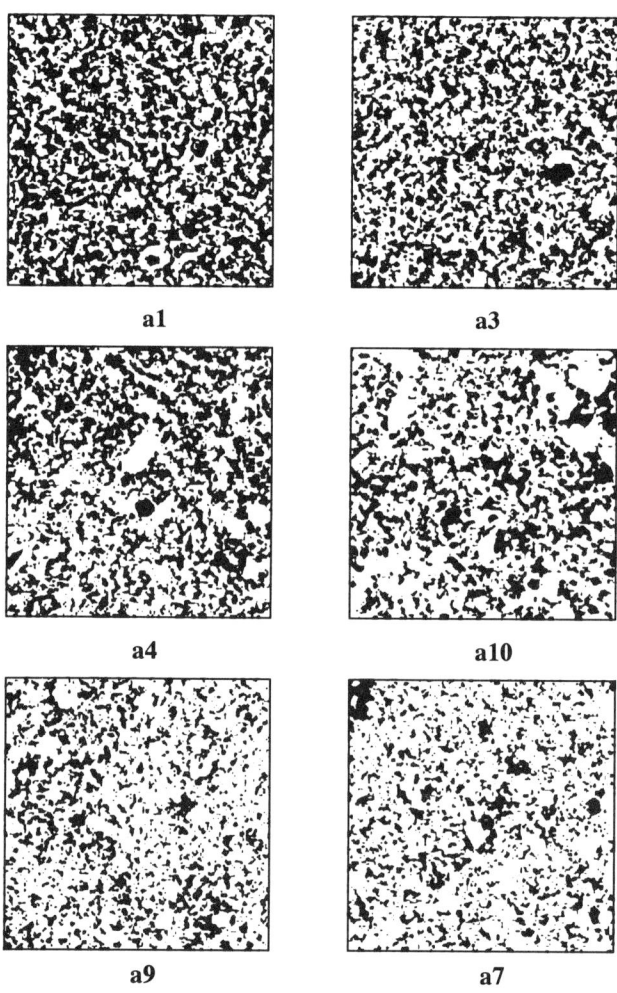

**Fig. 1.** Digitized images of chalk rock samples.

More details on these rock samples and on the digitization procedure can be found in Muller [5]. We define the signal $S(\mathbf{r})$ that determines the measure (equation (10)) by $S(\mathbf{r}) = 1$ for all rs inside the pore space (in black), and by $S(\mathbf{r}) = 0$ for all rs outside the pore space (in white) [1].

The coordinate $x$ is defined by $x = -\ln(\ell/L)$, where $L = 512$ pixels and $\ell$ is the box size measured also in pixels. We selected 48 values of $\ell$ in the range 1 pixel $\leq \ell \leq 240$ pixels, i.e. from one pixel to about half the image size, and the $\ell$s were chosen so that the values of $x$ would be distributed approximately uniformly between $x_{min} = -\ln(240/512) \approx 0.76$ and $x_{max} = -\ln(1/512) \approx 6.24$. The first polynomials constructed with the orthogonalization method (Appendix A) using these values of $x$ are shown in figure 2. It is seen that each of these polynomial is centered around zero, which is a consequence of their orthogonality to $P_0(x) = 1$.

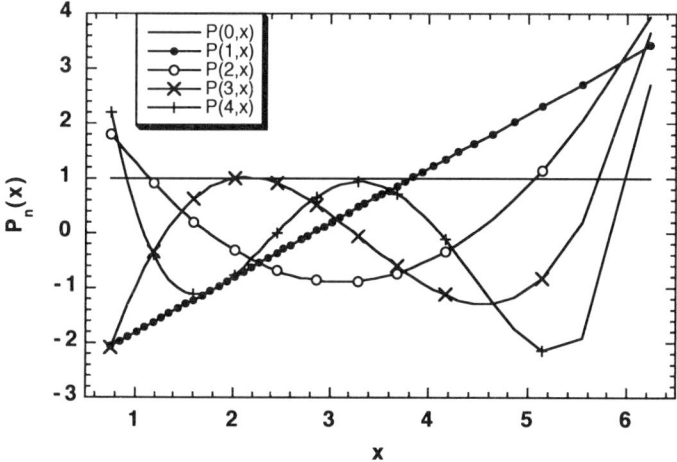

**Fig. 2.** The first four polynomials obtained with the orthogonalization method. Only the polynomial $P_1(x)$ shows (with filled round black dots) all the values of $x$ that were used.

For each rock sample, the generating function was obtained by combining all the images with the formula (11), and the uncertainty on $\phi_q$ was obtained with the formula (13). The dimensions and their uncertainties were then computed with the equations (7)-(8) and (18)-(19) respectively. For these samples, it was found that five or six terms in the polynomial expansion (9) was typically sufficient to obtain an acceptable fit of the generating function, i.e. a reduced chi-square smaller than one. The dimensions $D_1(q)$, which are the analogue of generalized dimensions $D(q)$ for multifractals, were plotted together in figure 3. We notice that the pore textures of the six chalk species are distinctly characterized by the function $D_1(q)$, as seen from the relatively small error bars. In fact we observe an interesting trend: As permeability and porosity decrease from sample a1 to sample a7, the curves $D_1(q)$ tend to have smaller values. A similar result had been obtained by Muller [5] with a conventional multifractal analysis of the same samples, where the generalized dimensions $D(q)$ had been compared.

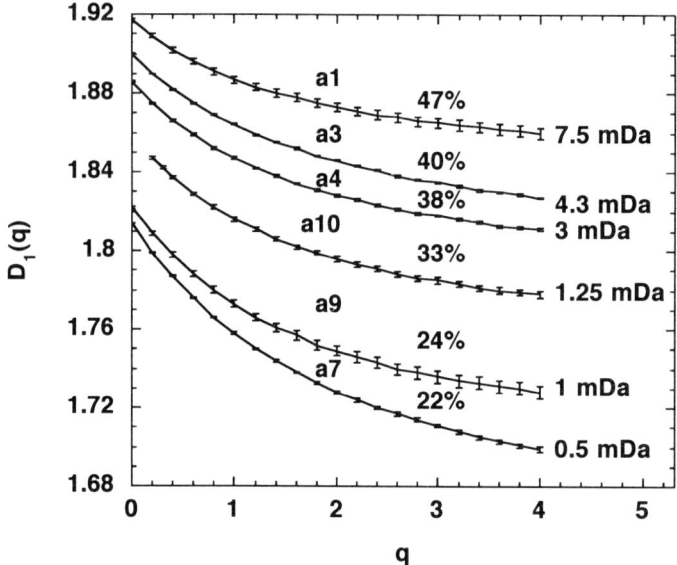

**Fig. 3.** Dimension function $D_1(q)$ for all samples. The percentages are porosities and the numbers in units of $mDa$ (milliDarcy) are permeabilities.

The dimensions $D_2(q)$ and $D_3(q)$ obtained from all rocks samples were plotted in figure 4 and 5. Again we notice that these dimensions allow to distinguish the pore space texture of the different chalk samples.

## 5 Concluding remarks

We introduced a generalized multifractal analysis that allows to make a simple description of virtually any measure. Our representation of the generating function is constructed in such as way that it introduces a hierarchy of dimension functions $\{D_1(q), D_2(q), ..., D_N(q)\}$, instead of a single one in the case of multifractals. This representation reduces to the multifractal description if the measure is multifractal. These dimension functions can be used to characterize the variability, or texture, of irregular signals and images. Our method was illustrated by analyzing the texture of pore space in sedimentary chalk. In the future, we plan to apply this method to different types of signals/images for which the characterization of texture can be of practical importance.

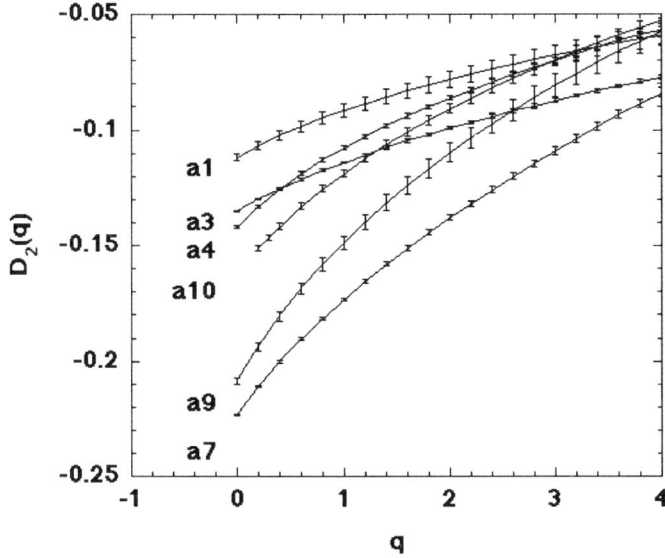

**Fig. 4.** Dimension function $D_2(q)$ for all samples. The samples a1 and a3, which are the highest permeability samples, are somewhat more horizontal than the other curves.

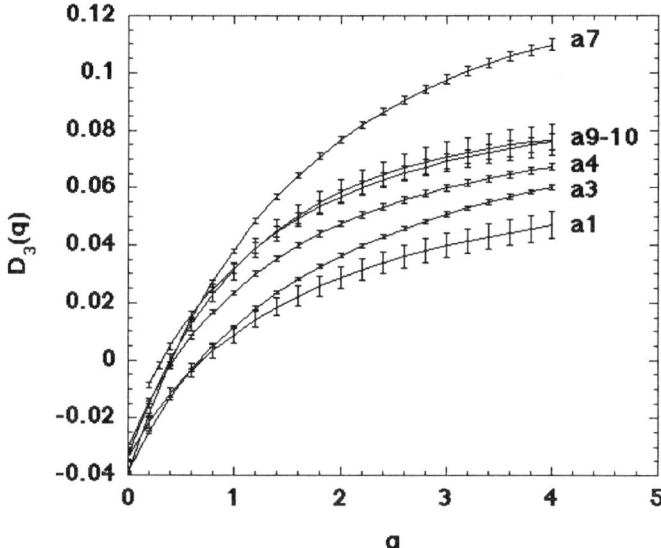

**Fig. 5.** Dimension function $D_3(q)$ for all samples.

**Acknowledgments**: Antoine Saucier would like to thank Jacques Richer for valuable comments and insights during the elaboration of this work.

## References

1. Saucier A. and J. Muller (1993): Remarks on some properties of geometrical multifractals. *Physica A*, 199:350–362.
2. C. J. G. Evertsz and B. B. Mandelbrot (1992): Multifractal measures. In H. O. Peitegen, H. Jurgens, and D. Saupe, editors, *Chaos and Fractals*. Springer-Verlag, New York, Appendix B.
3. H. G. E. Hentschel and I. Procaccia (1983): The infinite number of generalized dimensions of fractals and strange attractors. *Physica 8D*, pages 435–444.
4. F. P. Miranda, L. E. N. Fonseca, J. R. Carr, and J. V. Taranik (1996): Analysis of jers-1 (fuyo-1) sar data for vegetation discrimination in northwestern brazil using the semivariogram textural classifier (stc). *int. j. remote sensing*, 17(17):3523–3529.
5. Jiri Muller (1994): Characterization of the north sea chalk by multifractal analysis. *Journal of geophysical research*, 99:7275–7280.
6. A. Saucier, O. K. Huseby, and J. Muller (1997): Electrical texture characterization of dipmeter microresistivity signals using multifractal analysis. *Journal of geophysical research*, 102:10327–10337.
7. A. Saucier and J. Muller (October 25–28, 1998): Multifractal approach to textural analysis. In Miroslav M. Novak, editor, *Fractals and beyond, complexities in the sciences*, pages 161–171. World Scientific Publishing Co., Singapore, London, New Jersey, Hong Kong, ISBN 981-02-3593-3.
8. D. Schertzer and S. Lovejoy (1987): Physical modeling and analysis of rain and clouds by anisotropic scaling multiplicative processes. *J. Geophys. Res.*, 92:9693–9714.

## Appendix A: Construction of the orthogonal polynomials

In this appendix we show how the orthogonal polynomials are constructed. We choose to set the polynomial of order zero to unity, i.e. we define

$$P_0(x) \equiv 1 \tag{20}$$

Let us suppose that we already have constructed the orthogonal polynomials $\{P_0(x), P_1(x), ..., P_{n-1}(x)\}$, and let us ask ourselves how the next one, i.e. $P_n(x)$ with $n \geq 1$, can be obtained. We look for a $P_n(x)$ of the form

$$P_n(x) = x^n - \{\beta_n(0) + \beta_n(1)P_1(x) + \beta_n(2)P_2(x) + ... + \beta_n(n-1)P_{n-1}(x)\} \tag{21}$$

where the constants $\beta_n(k)$ are to be determined. The orthogonality conditions $\langle P_k, P_n \rangle = 0$ with $k = 0, 1, 2, ..., n-1$ yield immediately the unknown constants, which are given by

$$\beta_n(k) = \langle P_k(x), x^n \rangle / \langle P_k, P_k \rangle \tag{22}$$

for $k = 0, 1, 2, ..., n-1$. The recursive method (20)-(21)-(22) is well adapted to programming. The only input needed for the construction is the set of $x_i$s and the scalar product (5). Once the orthogonal polynomials are constructed, we normalize them with the transformation $P_n(x) \to P_n(x)/\sqrt{\langle P_n, P_n \rangle}$, applied for all $n \geq 2$ ($P_0(x)$ and $P_1(x)$ are left untouched). It is stressed that we chose *not* to normalize $P_1(x)$, i.e. we leave it in the form $P_1(x) = x + \texttt{constant}$, so that the coefficient $\tau_1(q)$ remains identical to $\tau(q)$ if the measure is multifractal.

It should be noted that high order polynomials can be obtained accurately only if the spacing between the $x_i$s is small enough. Indeed, if the density of the $x_i$s is too low to capture the oscillations of high order polynomials, then the resulting scalar products are not accurate enough and it becomes impossible to satisfy accurately the orthogonality conditions.

# Local Effective Hölder Exponent Estimation on the Wavelet Transform Maxima Tree

Zbigniew R. Struzik

Centre for Mathematics and Computer Science (CWI)
Kruislaan 413, 1098 SJ Amsterdam - THE NETHERLANDS
Zbigniew.Struzik@cwi.nl

**Abstract.** We present a robust method of estimating an effective Hölder exponent locally at an arbitrary resolution. The method is motivated by the multiplicative cascade paradigm, and implemented on the hierarchy of singularities revealed with the wavelet transform modulus maxima tree. In addition, we illustrate the possibility of the direct estimation of the scaling spectrum of the effective Hölder exponent, and we link it to the established partition functions based multifractal formalism. We motivate both the local and the global multifractal analysis by showing examples of computer generated and real life time series.

**keywords**
multifractal analysis, wavelet transform, Hölder exponent

## 1 Introduction

The application of the wavelet transform (modulus maxima) representation of a signal to multi-fractal analysis has almost reached the status of a standard. The formalism developed by Arneodo et al in the early nineties [1,2] has been extensively used to test many natural phenomena and has contributed to substantial progress in each domain in which it has been applied [3–5]. Nevertheless the respective methodology is intrinsically statistical in nature and provides only global estimates of scaling (of the moments of relevant quantity). While this is often a required property, there are cases when local information about scaling provides more relevant information than the global spectrum. This is particularly true for time series where scaling properties are non-stationary, whether it be due to intrinsic changes in the signal scaling characteristics or even boundary effects.

We, therefore, address the problem of estimation of the *local* scaling exponent through the paradigm of the multiplicative cascade. We reveal the hierarchy of the scaling branches of the cascade with the wavelet transform modulus maxima tree, which has proven to be an excellent tool for the purpose [2,6]. Contrary to the intrinsically instable local slope of the maxima lines, this estimate is robust and provides a stable, effective Hölder exponent, local in scale and position. From this an attempt can be made to derive the multifractal spectra directly from log-histogram scaling evaluation, linking the local analysis with the global multifractal spectra approach. Not as stable as the global scaling estimates from the partition functions method, the direct histogram of the effective Hölder exponent provides considerably more information about the relative density of local scaling exponents, and with some added on stabilisation, may prove to be an interesting alternative in multifractal spectra estimation.

The structure of the paper is as follows. In section 2, we focus on the relevant aspects of the wavelet transformation, in particular the ability to characterise scale-free behaviour through the Hölder exponent. Together with the hierarchical scale-wise decomposition provided by the wavelet transform, it will enable us to reveal the scaling properties of the tree of the multiplicative cascading process. In section 3, we introduce a technical model enabling us to estimate the scale-free characteristic (the effective Hölder exponent) for the branches of such a process. In section 4, we use the derived effective Hölder exponent for the local temporal description of the time series characteristics at a given resolution (scale). This is followed by an analysis of distributions of local h and the (scaling) evolution of the log-histogram and its relation to the standard partition functions based multifractal formalism. We motivate both the local and multifractal analysis by showing examples of generated and real life time series. Section 5 closes the paper with conclusions and suggestions for future developments.

## 2 Continuous Wavelet Transform and its Maxima Used to Reveal the Structure of the Time Series

The recently introduced Wavelet Transformation (WT), see e.g. Ref. [7,8], provides a way of analysing the local behaviour of functions. In this, it fundamentally differs from global transforms like the Fourier Transformation. In addition to locality, it possesses the often very desirable ability of filtering the polynomial behaviour to some predefined degree. Therefore, correct characterisation of time series is possible, in particular in the presence of *non-stationarities* like global or local trends or biases. One of the main aspects of the WT which is of great advantage for our purpose is the ability to reveal the *hierarchy* of (singular) features, including the scaling behaviour. [2]

Conceptually, the wavelet transformation is a convolution product of the time series with the scaled and translated kernel - the wavelet $\psi(x)$, usually a $n-th$ derivative of a smoothing kernel $\theta(x)$. Usually, in the absence of other criteria, the preferred choice is the kernel, which is well localised both in frequency and position. In this report, we chose the Gaussian $\theta(x) = \exp(-x^2/2)$ as the smoothing kernel, which has optimal localisation in both domains.

The scaling and translation actions are performed by two parameters; the scale parameter $s$ 'adapts' the width of the wavelet kernel to the *microscopic resolution* required, thus changing its frequency contents, and the location of the analysing wavelet is determined by the parameter $b$:

$$Wf(s,b) = \frac{1}{s} \int_{-\infty}^{\infty} dx\, f(x)\, \psi(\frac{x-b}{s}) \,, \tag{1}$$

where $s, b \in \mathbf{R}$ and $s > 0$ for the continuous version (CWT).

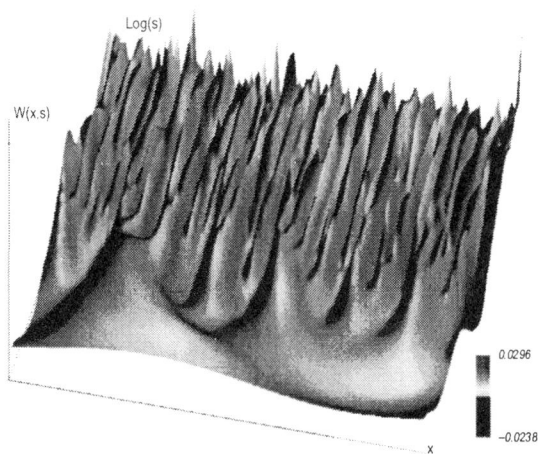

**Fig. 1.** Continuous Wavelet Transform representation of the random walk (Brownian process) time series. The wavelet used is the Mexican hat - the second derivative of the Gaussian kernel. The coordinate axes are: position $x$, scale in logarithm $\log(s)$, and the value of the transform $W(s, x)$.

In figure 1, we show the wavelet transform of a random walk sample decomposed with the Mexican hat wavelet - the second derivative of the Gaussian kernel. From the definition, the transform retains all of the temporal locality properties - the position axis is in the forefront of the 3D plot. The standard way of presenting the CWT is using the logarithmic scale, therefore the scale axis pointing 'in depth' of the plot is $\log(s)$. The third vertical axis denotes the magnitude of the transform $W(s, b)$.

The 3D plot shows how the wavelet transform reveals more and more detail while going towards smaller scales, i.e. towards smaller $\log(s)$ values. Therefore, the wavelet transform is sometimes referred to as the 'mathematical microscope', due to its ability to focus on weak transients and singularities in the time series. The wavelet used determines the optics of the microscope; its magnification varies with the scale factor $s$.

## 2.1 Assessing Singular Behaviour with the Wavelet Transformation

Quite frequently it is the singularities, the rapid changes, discontinuities and frequency transients, and not the smooth, regular behaviour which are interesting in the time series. Let us, therefore, demonstrate the wavelet's excellent suitability to address singular aspects of the analysed time series in a *local* fashion. The singularity strength is often characterised by the Hölder exponent.

If there exists a polynomial $P_n$ of degree $n < h$, such that

$$|f(x) - P_n(x - x_0)| \leq C|x - x_0|^h , \tag{2}$$

the supremum of all $h$ such that the above relation holds, is termed the Hölder exponent $h(x_0) \in (n, n+1)$ of the singularity at $x_0$. $P_n$ can often be associated with the Taylor expansion of $f$ around $x_0$, but Eq. 2 is valid even if such expansion does not exist [11]. The Hölder exponent is therefore a function defined for each point of $f$, and it describes the local regularity of the function (or distribution) $f$.

Let us take the wavelet transform $W^{(n)}f$ of the function $f$ in $x = x_0$ with the wavelet of at least $n$ vanishing moments, i.e. orthogonal to polynomials up to degree $n$:

$$\int_{-\infty}^{+\infty} x^m \psi(x)\, dx = 0 \quad \forall m,\ 0 \leq m < n .$$

For the sake of illustration, let us assume that the function $f$ can be characterised by Hölder exponent $h(x_0)$ in $x_0$, and $f$ can be locally described as:

$$f(x)_{x_0} = c_0 + c_1(x - x_0) + \cdots + c_n(x - x_0)^n + C|x - x_0|^{h(x_0)} .$$

Its wavelet transform $W^{(n)}f$ with the wavelet with at least $n$ vanishing moments now becomes:

$$W^{(n)}f(s, x_0) = \frac{1}{s} \int C|x - x_0|^{h(x_0)} \psi(\frac{x - x_0}{s}) dx$$
$$= C|s|^{h(x_0)} \int |x'|^{h(x_0)} \psi(x')\, dx' .$$

Therefore, we have the following power law proportionality for the wavelet transform of the (Hölder) singularity of $f(x_0)$:

$$W^{(n)}f(s, x_0) \sim |s|^{h(x_0)} .$$

Note: One should bear in mind that the above relation is an approximate case for which exact theorems exist [9]. In particular, we will restrict the scope of this paper to Hölder singularities for which the local and pointwise Hölder exponents are equal [10]. Thus we will not take into consideration the 'oscillating singularities' (e.g $x^\alpha sin(1/x^\beta)$) requiring two exponents [11, 12]. Nevertheless, it is sufficient for our purpose to state that the continuous wavelet transform can be used for characterising the Hölder singularities in the time series even if masked by the polynomial bias.

It can be shown [13] that for Hölder singularities, the location of the singularity can be detected, and the related exponent can be recovered from the scaling of the Wavelet Transform, along the so-called *maxima line*, converging towards the singularity. This is a line where the wavelet transform reaches local maximum (with respect to the position coordinate). Connecting such local maxima within the continuous wavelet transform 'landscape' gives rise to the entire tree of maxima lines. As we show in the following subsection, it appears that restricting oneself to the collection of such maxima lines provides a particularly useful representation of the entire CWT.

Let us consider the following set of examples of simple singular structures, see figure 2 left; a single Dirac pulse at $D(1024)$, the saw tooth consisting of an integrated Heaviside step function at $I(2048)$, and the Heaviside step function for $S(3072^+)$, where + denotes the right-handed limit. The Hölder exponent of a Dirac pulse is $-1$ by definition. For Hölder singularities, the process of integration and differentiation respectively adds and subtracts one from the exponent. We, therefore, have $h = 0$ for the right-sided step function $S(3072^+)$ and $h = 1$ for the integrated step $I(2048)$.

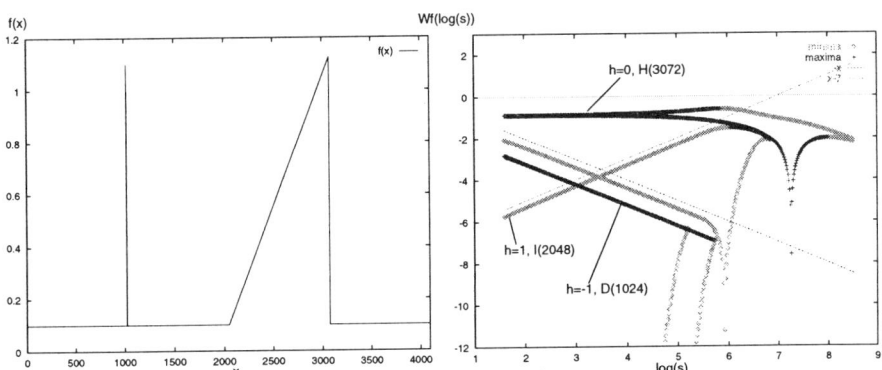

**Fig. 2.** Left: the test signal consisting of the Dirac pulse $D(1024)$, the change in slope - integrated Heaviside step $I(2048)$, and the Heaviside step $H(3072)$. Right: the log-log plot of the maxima, together with their respective logarithmic derivatives, corresponding to all three singularities: $D(1024)$, $I(2048)$ and $H(3072)$. Lines of theoretical slope are also indicated; these are $-x$ for $D(1024)$, $x$ for $I(2048)$ and a constant for $H(3072)$. The wavelet used is the Mexican hat.

These values can also be verified in the scaling of the corresponding maxima lines. We obtain the (logarithmic) slopes of the maxima values very closely following the correct values of these exponents, see figure 2 right. This, of course, suggests the possibility of the estimation of the Hölder exponent of (Hölder) singularities from the slope of the maxima lines approaching these singularities. An important limitation is, however, the requirement for the singularities to be *isolated* for this procedure to work. Note that the scaling of the maxima lines becomes stable in the log-log plot in figure 2 right only below some critical scale $s_{crit}$, below which the singularities effectively become isolated for the analysing wavelet. Indeed, the distance between the singular features in the test time series in figure 2 left equals 1024, which is in the order of three standard deviations of the analysing wavelet at $(\log(s_{crit}) = 5.83 = \log(1024/3)$. This example largely simplifies the issue since the singular structures are of the same size, resulting in one characteristic scale at which they appear in the wavelet transform. Also, generally, the scaling of the maxima lines for other than the presented simple examples will not follow a stright line even for isolated singularities. Still, the rate of decrease of (the supremum of) the related wavelet transform maximum will be consistent, thus allowing estimation of $h$.

## 2.2 Wavelet Transform Modulus Maxima Representation

The continuous wavelet transform described in Eq. 1 is an extremely redundant representation, much too costly for most practical applications. This is the reason why other, less redundant representations, are frequently used. Of course, in going from high redundancy to low redundancy (or even orthogonality), certain (additional) design criteria are necessary. For our purpose of analysis of the local features of time series, one critical requirement is the translation shift invariance of the representation; nothing other than the boundary coefficients of the representation should change, if the time series is translated by some $\Delta x$.

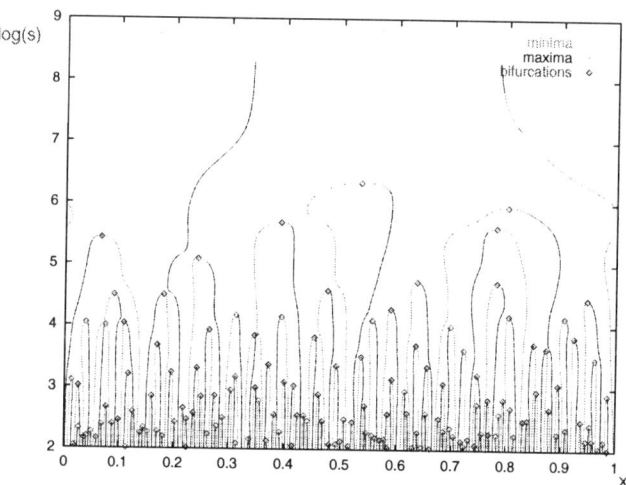

**Fig. 3.** WTMM representation of the time series and the bifurcations of the WTMM tree. Mexican hat wavelet.

A useful representation satisfying this requirement and of much less redundancy than the CWT is the representation making use of the local maxima of the WT as suggested in the previous section. Such maxima interconnected along scales form the so-called Wavelet Transform Modulus Maxima (WTMM) representation, first introduced by Mallat [14]. In addition to translation invariance, the WTMM also possesses the ability to characterise fully the local singular behaviour of time series, as illustrated in the previous subsection.

Moreover, the wavelet transform and its WTMM representation can also be shown to be invariant with respect to the rescaling/renormalisation operation [6, 2, 15, 12]. This property makes it an ideal tool for revealing the renormalisation structure of the (hypothetical) multiplicative process underlying the analysed time series.

Suppose we have the time series $f$ invariant with respect to some renormalisation operation $\mathcal{R}$:

$$f = \mathcal{R}f \ .$$

The wavelet transform of $f$ will, for a certain class of $\mathcal{R}$, in particular for multiplicative cascades, show the invariance with respect to an operator $\mathcal{T} \leftrightarrow \mathcal{R}^{-1}$. This can be recovered from the invariance of the wavelet transform of $F$:

$$W(f) = \mathcal{T}\, W(f)$$

and in particular from the invariance of (the hierarchy of) the WTMM tree [15, 6].

The aforementioned properties of the maxima lines representation make it particularly useful for our purpose.

## 3 Estimation of the Local, Effective Hölder Exponent Using the Multiplicative Cascade Model

We have shown in the previous section that the wavelet transform and in particular its maxima lines can be used in evaluating the Hölder exponent in isolated singularities. In most real life situations, however, the singularities in the time series are not isolated but densely packed. The logarithmic rate of increase or decay of the corresponding wavelet transform maximum line is usually not stable but fluctuates wildly, often making estimation impossible due to divergence problems when the value of the WT along the maximum line approaches zero.

As a remedy for the estimation problems, we will use the characterisation with the model based approximation of the local scaling exponent, which we will refer to as an *effective* Hölder exponent of the singularity.

In order to estimate this exponent in real life time series with dense singular behaviour, we need to approach the problem of diverging maxima values in log-log plots and the problem of slope fluctuations. We used the procedure of bounding the local Hölder exponent as described in [16] to pre-process the maxima. The crux of the method lies in the explicit calculation of the bounds for the (positive and negative) slope locally in scale. The parts of the maxima lines for which the slope exceeds the bounds imposed are simply not considered in calculations. (For the technical details of local slope calculation and bounding, we refer the reader to [16].) E.g. compare in figure 2 the example log-log slopes above the critical scale $s_{crit}$, where the singularities can no longer be considered as isolated. In particular, note that the local slope near scale $\log(s) = 6$ and $\log(s) = 7$ reaches $\pm\infty$. Such diverging slopes are thresholded and removed by applying the bounding procedure. In this example and throughout this paper we use $|\check{h}| < 2$ bound on the local slope $\check{h}$ of each maximum. The output of this procedure is, therefore, the set of non-diverging values of the maxima lines corresponding to the singularities in the time series.

In figure 4 below, we show the effect of the procedure on the distribution of the maxima values for a fixed scale $s = 5$, for a fractional Brownian motion (fBm) record with H=0.6. The unbounded distribution has a Gaussian shape as expected, which shows as a parabola in the logarithmic plot in 4 left. Bounding local slopes to $|\check{h}| < 2$ results in a rapid decay of small values of maxima towards the limit of 0 value, see the filled histogram, thus making negative moments well defined.[1] In figure 4 right, we verify in log-log coordinates that the decay of the small values follows a power law.

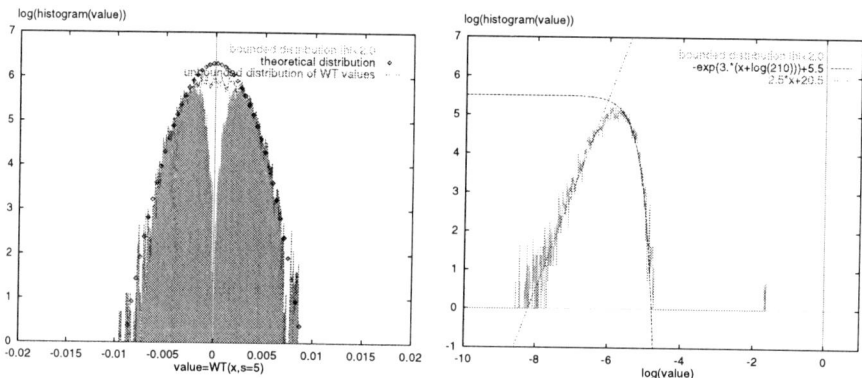

**Fig. 4.** Left: distribution of WT values of fBm of H=0.6 with and without local slope bounding in log-normal coordinates. Right: bounded distribution in log-log scale. Bound $|\check{h}| < 2$, scale $s = 5$. Sample length 16386.

Even though instead of fluctuating wildly between $+\infty$ and $-\infty$, the WT values are now more tempered, they still fluctuate, with the local slope changing

---

[1] One can argue about which is better: removing the apparently relevant information from the distribution of the maxima values or lacking the negative moments... Since we do not have proof that the information removed is redundant, we do not provide a definitive answer here.

from point to point (within bounds). Of course, this is why it is not possible to evaluate the Hölder exponent by a linear fit in log-log plot, something we can do for isolated cases giving a stable maximum value decay/increase. Therefore, we resort to the second assumption, in which we model the singularities as created in some kind of a collective process of a very generic class. For the estimation of the local Hölder exponent in such time series, we will use a multiplicative cascade model. This will allow us to construct a stable estimate of a local $h(x_0)$ exponent. The multiplicative cascade model is a generalisation of a binomial multiplicative process, otherwise known as the Besicovitch binomial process.

## 3.1 Multiplicative Cascade Model

Let us take the well known example of the Besicovitch measure on the Cantor set, see e.g. [17]. The set of transformations $B_i$, $i \in \{1,2\}$ describing the Besicovitch construction can be expressed as:

$$B_i\, f(x) = p_i\, f\left(\frac{x + b_i}{c_i}\right) ;$$

with the normalisation requirement:

$$p_1 + p_2 = 1 . \qquad (3)$$

Additionally, we put conditions ensuring non-overlapping of the transformations:

$$\frac{1 + b_1}{c_1} < \frac{0 + b_2}{c_2}$$

while all the respective values $b_1/c_1, b_2/c_2, c_1^{-1}, c_2^{-1}$ are from the interval $(0,1)$.

For equal ratios, $p_1 = p_2 = 1/2$ and $c_1 = c_2 = 3$ with $b_1 = 0$ and $b_2 = 2$, we recover the middle-third, homogeneous distribution of measure on the Cantor set. We have the Besicovitch measure for non-equal $p_i$, with the other settings above retained. Finally, for non-equal $p_i$, regardless of normalisation Eq 3 and with $c_1 = c_2 = 2$ with $b_1 = 0$ and $b_2 = 1$, we have the multiplicative cascade on $(0..1)$ interval.

Each point of this cascade is uniquely characterised by the sequence of weights $(s_1...s_n)$ taking values from the (binary) set $\{1,2\}$, and acting successively along a unique process branch leading to this point. Suppose that we denote the density of the cascade at the generation level $F_i$ ($i$ running from 0 to $max$) by $\kappa(F_i)$, we then have

$$\kappa(F_{max}) = p_{s_1} ... p_{s_n}\, \kappa(F_0) = P_{F_0}^{F_{max}}\, \kappa(F_0)$$

and the local exponent is related to the product $P_{F_0}^{F_{max}}$ of these weights:

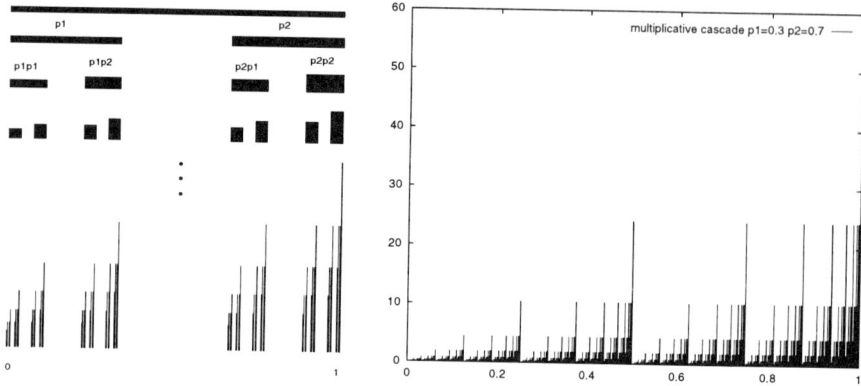

**Fig. 5.** Left: the Besicovitch measure on the Cantor set, generations $F_0$ through $F_3$ and the generation $F_6$. The distribution of weights is $p_1 = 0.4$ and $p_2 = 0.6$. The standard middle third Cantor division is retained. Right: similar construction but on 0..1 support instead of the Cantor set, leading to multiplicative cascade. $p_1 = 0.3$ and $p_2 = 0.7$, generation $F_{13}$.

$$h^{F_0}_{F_{max}} = \frac{\log(P^{F_{max}}_{F_0})}{\log((1/2)^{max}) - \log((1/2)^0)} \ .$$

In any experimental situation, the weights $p_i$ are not known and $h_i$ has to be estimated. This can be simply done using the fact that for the multiplicative cascade process of the kind just described, the effective product of the weighting factors is reflected in the difference of logarithmic values of the densities at $F_0$ and $F_{max}$ along the process branch:

$$h^{F_0}_{F_{max}} = \frac{\log(\kappa(F_{max})) - \log(\kappa(F_0))}{\log((1/2)^{max}) - \log((1/2)^0)} \ .$$

The densities along the process branch can be estimated with the wavelet transform, using its remarkable ability to reveal the entire process tree of a multiplicative process [6]. It can be shown that the densities $\kappa(F_i)$ can be estimated from the value of the wavelet transform along the maxima lines corresponding to the given process branch. The estimate of the effective Hölder exponent becomes:

$$\hat{h}^{s_{hi}}_{s_{lo}} = \frac{\log(Wf\omega_{pb}(s_{lo})) - \log(Wf\omega_{pb}(s_{hi}))}{\log(s_{lo}) - \log(s_{hi})} \ ,$$

where $Wf\omega_{pb}(s)$ is the value of the wavelet transform at the scale $s$, along the maximum line $\omega_{pb}$ corresponding to the given process branch. Scale $s_{lo}$ corresponds with generation $F_{max}$, while $s_{hi}$ corresponds with generation $F_0$.

For the estimation of $h$, we need $s_{hi}$ and $Wf\omega_{pb}(s_{hi})$. We can, of course, pick any of the roots of the sub-trees of the entire maxima tree in order to evaluate exponents of the partial process or sub-cascade. But for the entire sample available we must use the entire tree and for this purpose, we can only do as well as taking the sample length to correspond with $s_{hi}$, i.e.:

$$s_{hi} \equiv s_{SL} = \log(SampleLength).$$

Unfortunately, the wavelet transform coefficients at this scale are heavily distorted by finite size effects. This is why we estimate the value of $Wf\omega_{pb}(s_{hi})$ using the mean $h$ exponent.

### 3.2 Estimation of the Mean Hölder Exponent

For a multiplicative cascade process, a mean value of the cascade at the scale $s$ can be defined as:

$$\mathcal{M}(s) = \frac{\mathcal{Z}(s,1)}{\mathcal{Z}(s,0)}, \qquad (4)$$

where the $\mathcal{Z}(s,q)$ is the partition function of the $q$-th moment of the measure distributed over the wavelet transform maxima at the scale $s$ considered:

$$\mathcal{Z}(s,q) = \sum_{\Omega(s)} (Wf\omega_i(s))^q, \qquad (5)$$

where $\Omega(s) = \{\omega_i(s)\}$ is the set of all maxima $\omega_i(s)$ at the scale $s$, satisfying the constraint on their local logarithmic derivative in scale [16]. This mean gives the direct possibility of estimating the mean value of the local Hölder exponent as a linear fit to $\mathcal{M}$:

$$\log(\mathcal{M}(s)) = \bar{h} \log s + C. \qquad (6)$$

We will not, however, use the definition 4 since we want the Hölder exponent to be the local version of the Hurst exponent. This compatibility is easily achieved when we take the second moment in the partition function to define the mean $\bar{h}'$:

$$\mathcal{M}'(s) = \sqrt{\frac{\mathcal{Z}(s,2)}{\mathcal{Z}(s,0)}}.$$

Therefore, we estimate our mean Hölder exponent $\bar{h}'$ from 6 substituting $\mathcal{M}$ with $\mathcal{M}'$. The estimate of the local Hölder exponent, from now on to be denoted $\hat{h}(x_0,s)$ or just $\hat{h}$, now becomes:

$$\hat{h}_{s_{lo}}^{s_{SL}} \cong \frac{\log(Wf(s_{lo})) - (\bar{h}' \log s + C)}{\log(s_{lo}) - \log(s_{SL})}.$$

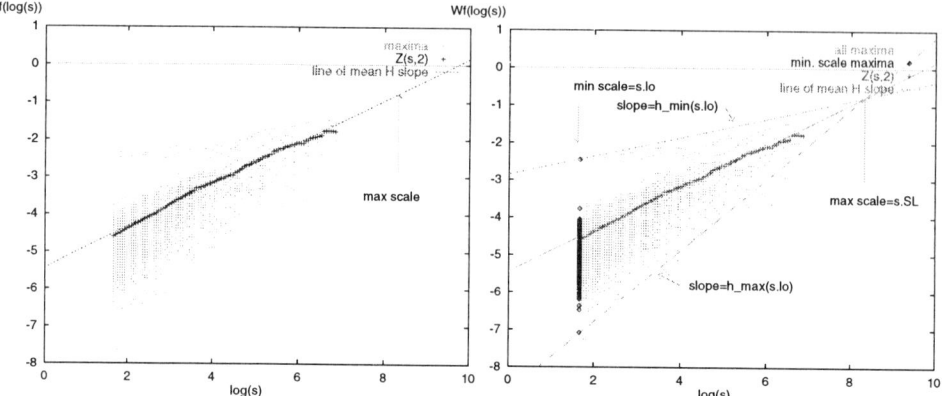

**Fig. 6.** Left: the projection of the maxima lines of the WT along time. The mean value of the Hölder exponent can be estimated from the log-log slope of the line shown. Also, the beginning of the cascade at the maximum scale $s_{hi}$ is indicated. Right: the maxima at the smallest scale considered are shown in the projection along time. The effective Hölder exponent can be evaluated for each point of the maximum line at $s_{lo}$ scale. Two extremal exponent values are indicated, for minimum and maximum slope.

## 4 Employing the Effective Hölder Exponent in Local and Global Spectra Estimation

Such an estimated local $\hat{h}(x_0, s)$ can be depicted in the temporal fashion, for example with a background colour, as we have done in figure 7. The first example time series is a computer generated sample of fractional Brownian motion with $H = 0.6$. It shows almost monochromatic behaviour, centred at $H = 0.6$ the colour green is dominant. There are however several instances of darker green and light blue indicating locally smooth components.

The second example time series is a record of the $S\&P500$ index from time period [1984-1988]. There are significant fluctuations in colour in this picture, with the green colour centred at $H = 0.5$, indicating both smoother and rougher components. In particular, one can observe an extremal red value at the crash '87 coordinate, followed by very rough behaviour (a rather obvious fact, but to the best of our knowledge not reported to date in the rapidly growing coverage of this time series record), see e.g. [18].

The third example is a real life biological time series and comes from aphids. This is the temporal record of electrical resistance, a 'penetratiogram', reflecting the penetration of the tongue of the aphid through the plant cell wall. We attempt to characterise the different regions of the time series, visible as a number of hierarchical 'pits' of certain depths within the signal. With green focused at the mean Hurst exponent equal $H = 0.5$, the result quite convincingly shows the patchy difference in characterisation of the pits at (two) different levels of pit hierarchy. Stripes of a different colour spectrum indicate a high level of non-stationarity of $\hat{h}$ distribution. Note that the obvious amplitude difference does not influence the colour in the plot due the fact that constant offset is filtered out by the wavelet used - the colour is due to a genuine difference in the local scaling exponent.

The last example shows a record of heartbeat intervals recorded from a healthy human heart. Contrary to the two previous examples which show a high degree of localisation (or non-stationarity) of the exponent strength, this plot shows an intricate structure of interwoven singularities at various strengths. This behaviour has been recently reported [19] to correspond with the multifractal behaviour of the heartbeat. The green is centred at $\hat{h} = 0.1$.

Note that these examples are only meant for illustration purposes. A detailed discussion of the implications of the local $\hat{h}$ analysis applied will appear elsewhere.

## 4.1 Scale-wise Evolution of the Effective Hölder Exponent

In addition to one scale plot showing the colour spectrum of singular behaviour, we can also see the scale position locations where the effective Hölder exponent is near a particular value. We show an example *band* of $\hat{h}_\epsilon(s)$ of width $\epsilon = 0.02$, by selecting $\hat{h} = -0.5 \pm 0.01$ in figure 8 for the record of white noise. The number of locations that fall within the band range visibly grows with scale and this growth determines the dimension $D(h)$ which can be associated with the particular $\hat{h}$, at the band resolution $\epsilon$.

Such $D(h)$ can be estimated for the entire range of $h$, resulting in the so-called *spectrum of singularities*. It is a standard way of visualising the distribution of singularities - it gives the (fractal) dimension $D(h)$ of the supporting set of singularities for each exponent value $h$ in the time series.

$$D(\hat{h}) = dim(\{x_0\} \ : \ Tf(x-x_0) \sim |x-x_0|^{h(x_0)}) \sim \lim_{\epsilon \to 0} \lim_{s_{lo} \to 0} \frac{\log(\mu_\epsilon(\hat{h}(s_{lo})))}{\log(s_{lo})} \ ,$$

where $\mu_\epsilon$ is the measure of the total number of locations (selected maxima) that fall within the band of size $\epsilon$ at a particular scale location $s_{lo}$.

Due to the fact that it relies on selecting a very narrow band of exponents, this procedure is, however, inherently sensitive to the choice of parameters such as the band width and the density of sampling of the scale axis. Therefore, it provides considerably less stable scaling estimates than the commonly used *partition functions, or Legendre transform*, method, see Ref [2]. The partition functions method is actually at the other extreme, taking *all* the maxima as the support for the measure of which the moments are calculated. This is done through the partition function $\mathcal{Z}$, Eq. 5. The $D(h)_m$ spectrum is then obtained from the scaling of the partition function using the Legendre transformation: $D_m(h) \leftrightarrow \tau(q)$, if $\mathcal{Z}(s,q) \sim s^{\tau(q)}$. In this construction, the moment parameter $q$ has the purpose of 'selecting' an adequate range of Hölder exponents from the global quantity $\mathcal{Z}$. As a result, the partition function method provides only rough, 'outline' information about the $D_m(h)$ spectrum.

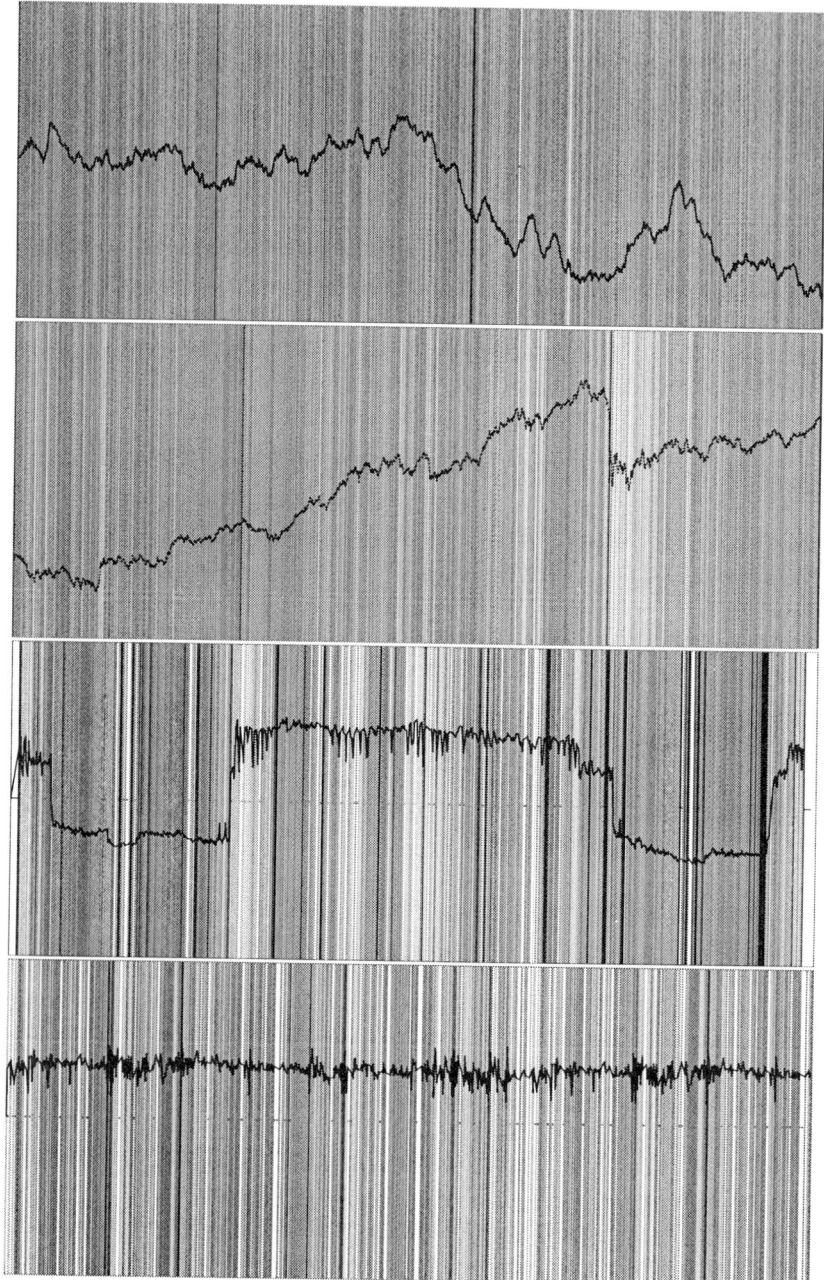

**Fig. 7.** Example time series with local Hurst exponent indicated in colour. From top to bottom: fBm with $H = 0.6$, a record of S&P500 index, penetratiogram of aphids, and the last is the record of healthy heart interbeat intervals. The background colour indicates the Hölder exponent locally, centred at the Hurst exponent at green, colour goes towards blue for higher $\hat{h}$ and towards red for lower $\hat{h}$. Local slope bounds for all the plots $|\check{h}| \leq 2$.

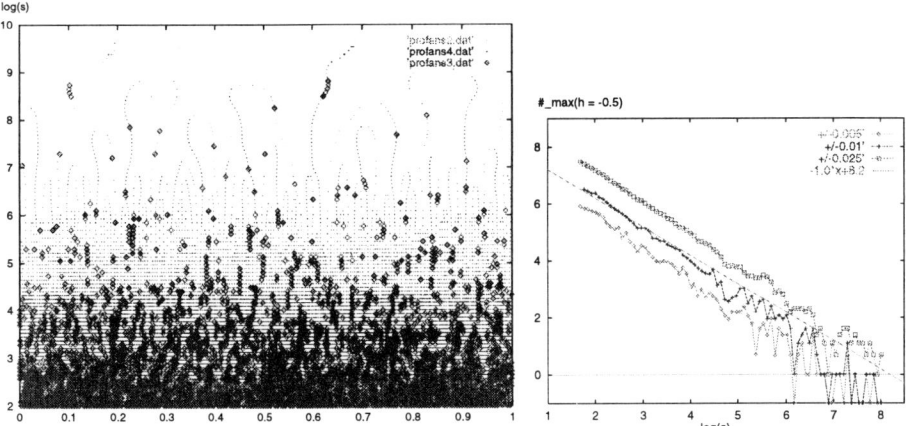

**Fig. 8.** Left: WTMM representation of a sample of white noise. The maxima are highlighted where the effective Hölder exponent reaches a particular value of $\hat{h} = -0.5 \pm 0.01$, i.e. $\epsilon = 0.02$. Right: for three values of the bin size $\epsilon = 0.01, \epsilon = 0.2, \epsilon = 0.05$, the logarithm of the sum of the highlighted maxima is shown for each scale with respect to the $\log(s)$ axis. Consistent scaling of $-1$ rate is shown for *supremum* of $\epsilon = 0.02$ plot.

It seems possible to take a middle path in order to calculate more stable scaling estimates of the $D(h)$ in the direct way from the scaling of 'selected' maxima parts. This can be done by weighted selection, replacing the histogram box centred at $\hat{h}$ and of $\epsilon$ width, with a smooth, say Gaussian, kernel of $\epsilon$ standard deviation, centred at $\hat{h}$.

While we will not pursue this approach further here, leaving it to a separate treatment, it seems that a similar idea has successfully been applied to large deviation multifractal spectra estimation on dyadic partitions as reported in [20]. For a description of this multifractal spectrum, see [21, 22]

### 4.2 Log-histograms of the Effective Hölder Exponent

Let us for now, instead of selecting one $\hat{h}_\epsilon(s)$ value band across scales and analysing its scaling, group the estimated local scale-wise $\hat{h}(x_0, s)$ into histograms for each scale value, using bin size $\epsilon$ centred at $h$.

We will analyse histograms of $\hat{h}$, taking the logarithm of the measure in each histogram bin. This conserves the monotonicity of the original histogram, but allows us to compare the log-histograms with the spectrum of singularities $D(h)$. There is a direct correspondence between our log-histograms and the $D(h)$ through the scaling of the measure $\mu_\epsilon(s)$ in the bin of size $\epsilon$ of the histogram. Estimation of the rate of growth of this measure, would in fact be an identical procedure to the scaling estimate for each $\epsilon$ wide band of $\hat{h}$, as discussed in the previous subsection.

There is, however, a substantial amount of information in the log-histogram at a particular scale which can be analysed without performing the scaling analysis. The log-histogram shows the relative probability density of the Hölder exponent per scale. Assuming that the scaling of log-histogram is linear in log-log scale, the shape of the log-histogram remains invariant across scales and converges towards the shape of the $D(h)$, except for 'normalisation' of the maximum of $P(\hat{h})$, which corresponds with the scaling of the zeroth moment in the partition functions method. (Note that the scaling assumption can be verified both by analysing the scaling of the moments and by the scaling of the $\hat{h}$ bin contents.) This is why in figure 9, we show the non-normalised histograms for different scale values, and compare them to the $D_m(h)$ spectrum obtained with the partition functions method.

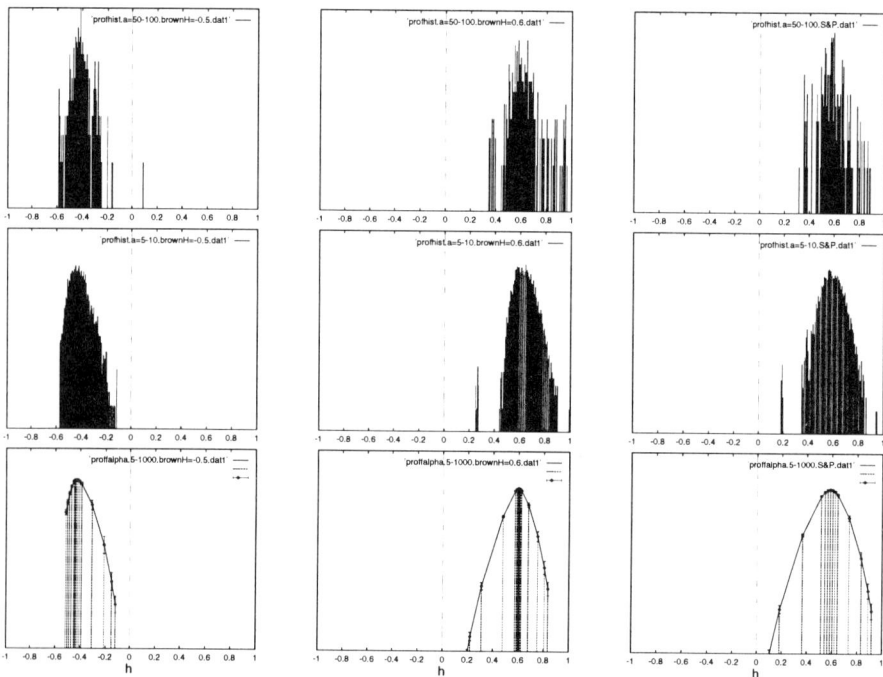

**Fig. 9.** Two sets of $\hat{h}$ histograms for respective scales $\log(s) \simeq 3.9$, $\log(s) \simeq 1.6$, in top first and centre row. Below, in the bottom row, the $D_m(h)$ spectrum obtained with the partition functions method (moments $-5 \leq q \leq 5$). Left column: for 4096 samples of white noise. Centre: 4096 samples of fractional Brownian motion with $H = 0.6$. Right: S&P 500 index, first 4096 samples from figure 7. Local slope bounds for all the plots $|\hat{h}| \leq 2$.

For three example time series, we show in figure 9 log-histograms of the exponent $\hat{h}$ at different scales. The time series considered are a white noise sample, a fractional Brownian motion with $H = 0.6$, and a record of the S&P index. Starting at the top, the row of histograms is made for the scale $\log(s) \simeq 3.9$, followed by histograms for $\log(s) \simeq 1.6$. The upper histograms show considerable fragmentation. Several modes become visible and for all the scales above this scale, the fluctuations will dominate the distribution and consistent statistical behaviour will become dispersed. On the contrary, while going down with the scale, the bulk of consistent behaviour converges to one distribution. The consistent statistical behaviour is also captured in the $D_m(h)$ spectrum obtained with the partition functions method shown in the bottom row of figure 9.

Several aspects of the $D_m(h)$ versus $P(\hat{h})$ already discussed in [23] are visible in the plots. The $P(\hat{h})$ evidently contains more information than $D_m(h)$, in particular, $D_m(h)$ is a convex hull over the $P(\hat{h})$. This is particularly visible in the second (and third) sample, where a pair of extremal $\hat{h}$ values, disconnected from the distribution, corrupts the $D_m(h)$ spectrum. It can be verified through analysing the corresponding maxima that these values are the end of the sample artifacts and thus do not belong to the distribution. Such maxima can, of course, be removed prior to $D_m(h)$ evaluation, but we kept them for the purpose of illustration: the $P(\hat{h})$ evidently shows that these values disconnect from the bulk of the distribution while $D_m(h)$ is inherently unable to do so.

Both the distributions $P(\hat{h})$ and the $D_m(h)$ spectra are evaluated using the local slope bound of $|\check{h}| \leq 2$, on a relatively short samples of 4096 data points. In figure 10, we show both the distribution $P(\hat{h})$ and the $D_m(h)$ spectrum evaluated with maximally tight and relaxed bounds and a relatively long length of $2^{15}$ samples of white noise. We can verify that for the maximally tight bounds, both the distribution and the $D_m(h)$ spectrum are well clustered around the $\hat{h} = H$ value, indicating that the bounding procedure does not remove any information relevant to recovering this part of the distribution/spectrum. Also, the width of the spectrum is slightly reduced when compared to the results from 8 times shorter data such as used in figure 9. The spectrum remains within the theoretical predictions $1 - H - h$ indicated with the line plot. The same data set processed with maximally relaxed bounds gives the $D_m(h)$ spectrum with its part for the negative moments exceeding the theoretical bounds. The histogram, however, fills the entire space granted by the theoretical bounds.

## 5 Conclusions

We have presented a method of estimating an effective Hölder exponent locally for an arbitrary resolution. The method is motivated by the multiplicative cascade paradigm, and implemented on the hierarchy of the wavelet transform modulus maxima tree. Contrary to the intrinsically unstable local slope of the maxima lines, this estimate is robust and provides a stable, effective Hölder exponent, local in scale and position.

We have presented a number of real life examples using this local exponent estimate. The colour exponent panels included show the intricate scale-free structure of the time series. The exact implications of this structure, of course depend on the application.

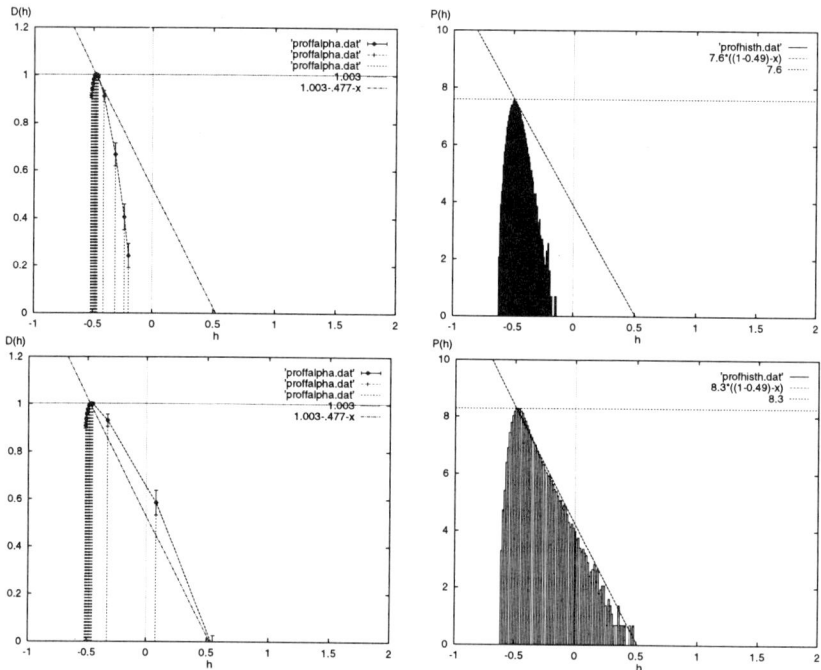

**Fig. 10.** White noise sample, length 32.768 data points. Left: the $D_m(h)$ spectrum obtained with the partition functions method (moments $-5 \leq q \leq 5$). Right: the corresponding distribution $P(\hat{h})$ for $\log(s) = 1.61$. Top row is obtained using the local slope bound of $|\check{h}| \leq 2$, lower row with relaxed local slope bound. Theoretical predictions $C(1 - H - h)$ are indicated with the line plot.

In addition, we have illustrated the possibility of direct estimation of scaling spectrum of the effective Hölder exponent and linked it to the established partition functions multifractal formalism. Not as stable as the global scaling estimates from the partition functions method, the direct histogram of the effective Hölder exponent provides considerably more information about the relative density of local scaling exponents, and with some added stabilisation through the appropriate choice of a smoothing kernel, may prove to be an interesting alternative in multifractal spectra estimation.

## Acknowledgments

The author would like to thank Maria Haase and Jacques Lévy Véhel for discussions during Fractal '98 on the behaviour of negative moments in the multifractal spectra. Thanks also to Hans van den Berg for providing the aphids data (originally produced by Dr. Freddie Tjallingii from the Department of Entomology at Wageningen Agricultural University) and for related discussions. Plamen Ivanov, thanks for providing the heartbeat data [24] and for discussions. Shlomo Havlin deserves special thanks for suggesting we include a colour plot of the $h$ exponent in [19], which was the starting point of this work. Arno Siebes, thanks for supporting this work. This work has been carried out with the financial support from the Impact project.

## References

1. A. Arneodo, E. Bacry and J.F. Muzy (1991): Wavelets and Multifractal Formalism for Singular Signals: Application to Turbulence Data. *PRL*, **67**, No 25, 3515–3518.
2. A. Arneodo, E. Bacry and J.F. Muzy (1995): The Thermodynamics of Fractals Revisited with Wavelets. *Physica A*, **213**, 232–275.
   J.F. Muzy, E. Bacry and A. Arneodo (1994): The Multifractal Formalism Revisited with Wavelets. *International Journal of Bifurcation and Chaos* **4**, No 2, 245–302.
3. A. Arneodo, A. Argoul, J.F. Muzy, M. Tabard and E. Bacry (1995): Beyond Classical Multifractal Analysis using Wavelets: Uncovering a Multiplicative Process Hidden in the Geometrical Complexity of Diffusion Limited Aggregates. *Fractals* **1**, 629–646.
4. A. Arneodo, E. Bacry, P.V. Graves and J.F. Muzy (1995): Characterizing Long-Range Correlations in DNA Sequences from Wavelet Analysis. *PRL*, **74**, No 16, 3293–3296.
5. P.Ch. Ivanov, M.G. Rosenblum, C.-K. Peng, J. Mietus, S. Havlin, H.E. Stanley and A.L. Goldberger (1996): Scaling Behaviour of Heartbeat Intervals Obtained by Wavelet-based Time-series Analysis. *Nature*, **383**, 323–327.
6. Z.R. Struzik (1995): The Wavelet Transform in the Solution to the Inverse Fractal Problem. *Fractals* **3** No.2, 329–350.
   Z.R. Struzik (1996): From Coastline Length to Inverse Fractal Problem: The Concept of Fractal Metrology. *Thesis*, University of Amsterdam.
   Z.R. Struzik (1997): Fractals under the Microscope or Reaching Beyond the Dimensional Formalism of Fractals with the Wavelet Transform. *CWI Quarterly*, **10**, No 2, 109–151.

7. I. Daubechies (1992): *Ten Lectures on Wavelets.* S.I.A.M..
8. M. Holschneider (1995): *Wavelets - An Analysis Tool.* Oxford Science Publications.
9. S. Jaffard and Y. Meyer (1996): Wavelet methods for pointwise regularity and local oscillations of functions. *Memoirs of AMS*, 123(587).
10. B. Guiheneuf and J. Lévy Véhel (1998): 2-Microlocal Analysis and Application in Signal Processing, *In proceedings of International Wavelets Conference, Tangier.*
11. S. Jaffard (1997): *Multifractal Formalism for Functions: I. Results Valid for all Functions, II. Self-Similar Functions.* SIAM J. Math. Anal., 28(4): 944–998.
12. A. Arneodo, E. Bacry and J.F. Muzy (1995): Oscillating Singularities in Locally Self-Similar Functions. *PRL*, **74**, No 24, 4823–4826.
13. S.G. Mallat and W.L. Hwang (1992): Singularity Detection and Processing with Wavelets. *IEEE Trans. on Information Theory* **38**, 617–643.
14. S.G. Mallat and S. Zhong (1992): Complete Signal Representation with Multiscale Edges. *IEEE Trans. PAMI* **14**, 710–732.
15. A. Arneodo, E. Bacry and J.F. Muzy(1994): Solving the Inverse Fractal Problem from Wavelet Analysis. *Europhysics Letters*, **25**, No 7, 479–484.
16. Z.R. Struzik (1998): Removing Divergences in the Negative Moments of the Multi-Fractal Partition Function with the Wavelet Transformation. *CWI Report*, **INS-R9803**. Also in 'Fractals and Beyond - Complexities in the Sciences', M.M. Novak, Ed., World Scientific, 351–352.
17. K. Falconer (1990): *Fractal Geometry - Mathematical Foundations and Applications*, John Wiley.
18. Y. Liu, P. Cizeau, P. Gopikrishnan, M. Meyer, C.-K. Peng and H.E. Stanley (1998): Volatility Studies of the S&P 500 Index. *Preprint.*
    P. Cizeau, Y. Liu, M. Meyer, C.-K. Peng and H.E. Stanley (1997): Volatility Distribution in the S&P 500 Stock Index. *Physica A*, **245**, 441–445.
19. P.Ch. Ivanov, M.G. Rosenblum, L.A. Nunes Amaral, Z.R. Struzik, S. Havlin, A.L. Goldberger and H.E. Stanley (1998): Multifractality in Human Heartbeat Dynamics. *Preprint.*
20. Ch. Canus, J. Lévy Véhel and C. Tricot (1998): Continuous Large Deviation Multifractal Spectrum: Definitions and Estimation. 'Fractals and Beyond - Complexities in the Sciences', M.M. Novak, Ed., World Scientific, 117–128.
21. J. Lévy Véhel and R.Vojak (1998): Multifractal Analysis of Choquet Capacities: Preliminary Results *In Advances in Applied Mathematics*, Vol. 20, No. 1, pp. 1-43, January.
22. J. Lévy Véhel (1996): Numerical Computation of the Large Deviation Multifractal Spectrum, *In CFIC, Rome.*
23. J. Lévy Véhel and R. H. Riedi (1997): Fractional Brownian Motion and Data Traffic Modeling: The Other End of the Spectrum. *Fractals in Engineering*, J. Lévy Véhel, E. Lutton, C. Tricot, Eds., Springer Verlag.
    R. H. Riedi and J. Lévy Véhel (1997): TCP Traffic is Multifractal: a Numerical Study. *INRIA Research Report*, No. 3129.
24. *Heart Failure Database* (Beth Israel Deaconess Medical Center, Boston, MA).

# Easy and Natural Generation of Multifractals: Multiplying Harmonics of Periodic Functions

Marc-Olivier Coppens[1] and Benoit B. Mandelbrot[2]

[1] Delft Department of Chemical Technology
Delft University of Technology
Julianalaan 136, 2628 BL Delft - THE NETHERLANDS
M.O.Coppens@stm.tudelft.nl
[2] Department of Mathematics
Yale University, New Haven, CT 06520-8283 - U.S.A.

**Abstract.** Simple multifractal measures are constructed by multiplying a periodically extended function with copies of itself. The frequencies of the copies form a geometric series, $(1, b, b^2, \ldots, b^n, \ldots)$, where $b$ is a *real* number larger than 1. This deterministic construction leads to measures that are similar to random multifractal measures, yet are easier to build. At the same time, they do not have the unphysical disadvantages of other deterministic multifractals, such as the multinomial measures. The effect of random phase shifts is also considered.

## 1 Introduction

Multifractals are encountered in many fields [22]. Examples are the energy dissipation in a turbulent fluid [14, 15, 24, 25, 22], the growth rate along a DLA-cluster [23, 4, 6], the reaction rate along a fractal catalyst surface [8, 2], the current distribution in a percolation cluster [28], and time in a model for price variation [21]. This variety of multifractal measures calls for a better, more intuitive understanding of how multifractals originate, and for alternative simple constructions. Some constructions, such as the popular multinomial multiplicative cascade, are recursive and subdivide space into boxes, using an integer base $b$. In general, $b$ is nonphysical, so that it would be desirable to minimize its role. The new method presented in this paper does just that.

A self-similar fractal set is identical at all scales; in a (multiplicative) multifractal measure $m$, the distribution of a quantity to be called mass is similar at different resolutions. The space over which mass is distributed is divided into fractal subsets, each with the same value for $m$. The fractal dimensions of these subsets can be plotted as a function of the Hölder exponent $\alpha$ defined by $m \sim \delta^\alpha$; when the resolution $\delta$ approaches zero, the plot yields the multifractal spectrum $f(\alpha)$. While this formalism may not be the most general one (see, *e.g.*, [16, 19, 20]), it suffices for most applications. For a general introduction to multifractals, see [17, 18, 5] and the recently published book [22].

Much work on multifractals, especially after their popularization by Frisch and Parisi [7] and Halsey *et al.* [10], assumes implicitly that a multifractal measure is close to being multinomial or even simply binomial. The construction of

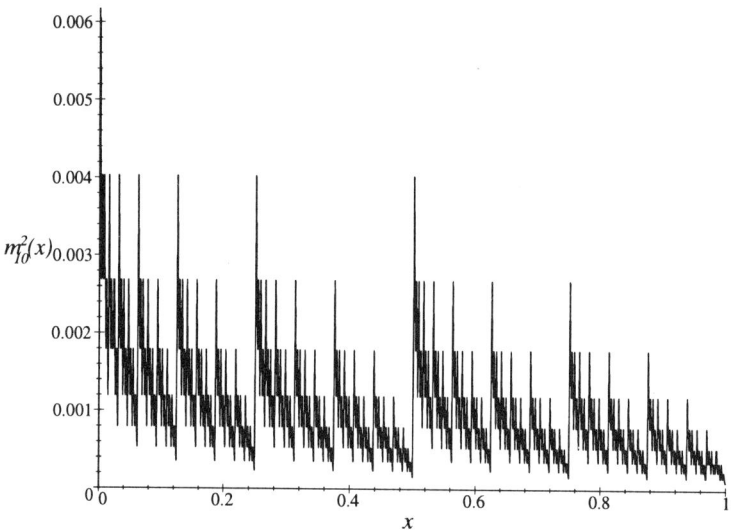

**Fig. 1.** Binomial measure, after 10 generations, with $p_1 = 0.6$ and $p_2 = 0.4$.

a multinomial measure on a line segment [0,1) starts by dividing the segment in $b$ pieces of length $1/b$, and associating a different weight or probability $p_i$ to each of these pieces. Each of the $b$ shorter segments is replaced by a $b$ times smaller copy of this generator, in which mass or probability is redistributed in the same way as in the generator; therefore, the $j$-th piece in the $i$-th segment of the first iteration is given the weight $p_i p_j$. The number of segments $N$ with the same weight $p = p_i p_j$ follows the multinomial distribution, and this *multiplicative cascade* continued *ad infinitum* creates a multinomial multifractal measure. For $b = 2$, the measure is binomial. An example with $p_1 = 0.6$ and $p_2 = 0.4$ is shown in Fig. 1. When the resolution $\delta = b^{-n}$ approaches zero in this cascade, sets with the same Hölder exponent $\alpha = \lim_{\delta \to 0} \log p / \log \delta$ are fractal, and they have a fractal similarity dimension $f(\alpha) = -\lim_{\delta \to 0} \log N / \log \delta$. The link between $f(\alpha)$ and the moment generating function involves a Legendre transform (the thermodynamic formalism), and the explicit formulas for $\alpha$ and $f(\alpha)$ are well-known and will not be repeated here (see [10, 16–18]).

The multinomial measures are a fine first example, yet inadequate for most practical purposes. In the first place, why the arbitrary division in segments, based on some integer $b$ ? Why should this multiplicative cascade proceed with the same set of probabilities $\{p_1, \ldots, p_b\}$ ? Moreover, the measures obtained in turbulence [14, 15] and growth phenomena [23, 4, 6] are definitely *not* binomial or multinomial. Even the best known signature, the $f(\alpha)$ spectrum, need not have the familiar ∩-shape, unless special conditions are satisfied as Coppens discusses in the practical example of catalysis [2].

The random multiplicative multifractals introduced by Mandelbrot [14, 15, 22] can lead to $f(\alpha)$ spectra that have different shapes, and do not require an arbitrary choice of a base $b$. Ingenious mathematical examples, such as the Minkowski measure [9, 20], were studied to show how the spectrum can become left-sided (*i.e.*, the maximum value of $f(\alpha)$ is obtained for $\alpha \to \infty$) [27], or how $\alpha$ or $f(\alpha)$ can be negative. Moreover, there can be several maxima in the $f(\alpha)$ curve [26].

The experimental evidence shows that these spectra are not just esoteric creations. But they are complicated and make it very desirable to have some simple method to generate various spectra other than the inadequate multiplicative scheme behind the multinomial measures. The method introduced in this paper responds to this strong desire and proves particularly powerful, despite its simplicity. The method merely involves the repeated multiplication of a periodic function $w(t)$ (the *generator*) with rescaled copies $w(b^k x)$ of itself. We will show how this product of harmonics leads to a multifractal measure.

The main advantage of our method is threefold. First of all, a great variety of measures can be constructed through a multiplication of functions that is not much harder than the generation of multinomial measures, but much more general. Secondly, as opposed to the multinomial measure, in which the "base" $b$ is an integer larger than 1, the base is now any *real* number larger than 1. Thirdly, we will prove how the essential characteristics of measures constructed using this completely *deterministic* method are the same as those of Mandelbrot's *random* multifractal measures, when the multipliers are chosen in a way to be discussed in this paper.

This is a preliminary announcement of the method and some of the general features of the *multifractal product of function (MPF)*. Detailed mathematical proofs will appear soon [3].

## 2 Methodology of the *MPF*

Define a function $w(x)$ with period 1, *i.e.*:

$$w(x+k) = w(x), \qquad \forall k \in \mathbf{N}. \tag{1}$$

This "generator" and a base $b > 1$ define the functions $w(b^{i-1}x), i = 1, 2, 3, \ldots$ These functions' frequency is $b^{i-1}$ times higher than for $w(x)$. Now, let:

$$m_n^b(x) = \prod_{i=1}^{n} w(b^{i-1}x). \tag{2}$$

Like $w(x)$, the functions $m_n^b(x)$ have period 1, if $b$ is an integer.

When $w(x) = p_j$ for $(j-1)/b \leq x < j/b$, with $\sum_j p_j = 1$, the $b$-nomial measure is recovered. Note the resemblance to a Fourier series, where a *sum* of periodic functions $w_i$ (cosines, sines) with a geometric series of frequencies converges to the limit Weierstrass function. Here, the sum is replaced by a product, and, if a nondegenerate limit $m(x) \equiv m_\infty^b(x)$ exists (in the sense of measures rather than functions), it can be shown to be a multifractal measure that we call a *multifractal product of functions* (*MPF*).

The multifractal spectrum $f(\alpha)$ is defined as follows. First, the coarse-grained Hölder exponents of the $b^n$ intervals $[ib^{-n}, (i+1)b^{-n}) = [i\epsilon_n, (i+1)\epsilon_n)$ are evaluated for all $i = 0, \ldots, b^n - 1$:

$$\alpha(i) = -\frac{1}{n} \log_b \int_{ib^{-n}}^{(i+1)b^{-n}} m_n^b(x) dx = \frac{\log \int_{i\epsilon_n}^{(i+1)\epsilon_n} m_n^b(x) dx}{\log \epsilon_n}. \tag{3}$$

The histogram $N(\alpha)$ is then constructed, by distributing the Hölder exponents in bins $[\alpha, \alpha + \Delta\alpha)$. The normalized logarithm of this histogram is the premultifractal spectrum:

$$f_n(\alpha) = \frac{1}{n} \log_b N(\alpha) = -\frac{\log N(\alpha)}{\log \epsilon_n}. \tag{4}$$

The limit multifractal spectrum is:

$$f(\alpha) = \lim_{n \to \infty} f_n(\alpha). \tag{5}$$

Another way to derive $f(\alpha)$ is from the cumulative distribution $N(A > \alpha)$. It should be noted that $b$ can be any positive real number larger than one and does not have to be an integer in the *MPF*, as opposed to the multiplicative cascades leading to multinomial multifractals.

The multifractals constructed in this way are entirely deterministic, but can be randomized by introducing a random phase shift with every iteration, *i.e.*, we let

$$m_n^b(x) = \prod_{i=1}^{n} w[b^{i-1}(x + \phi_i)], \tag{6}$$

in which $\phi_i$ is a random deviate uniformly distributed between 0 and 1.

## 3 Examples

An unlimited variety of measures is generated through this *MPF* by changing the generator $w(x)$. A few examples are presented below.

- A multinomial measure is constructed when $w(x)$ is constant within each of $b$ intervals of length $1/b$.
- The potential well or "fjord", $w(x) = 1 - 0.25[1 - \cos(2\pi x)]^2$, shown in Fig. 2 generates the measure shown in Fig. 3 after respectively 5 (above) and 10 (below) generations. This could serve as a crude model for the accessibility measure along a rough surface. Even visually, the measures in this and the previous example are quite different from the binomial measure in Fig. 1.

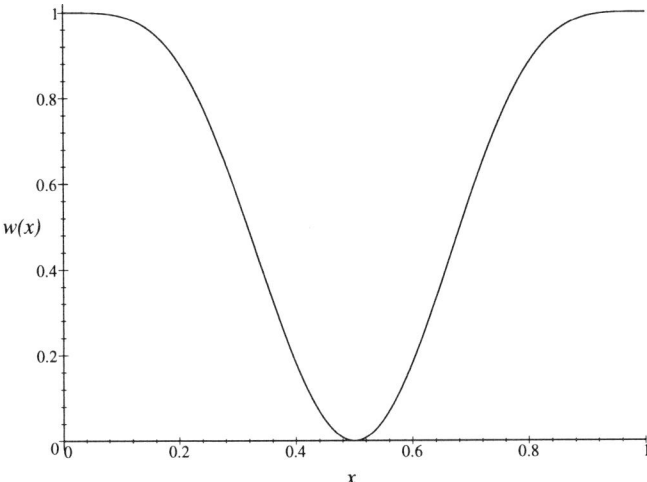

**Fig. 2.** Potential well or "fjord", $w(x) = 1 - 0.25[1 - \cos(2\pi x)]^2$.

- For $w(x) = \sqrt{x(1-x)}$ (a semi-circle), the tenth generation $w_{10}^2(x)$ looks as in Fig. 4. The pre-multifractal spectrum corresponding to this "circle measure" is shown as well. It is clearly different from the familiar symmetrical spectrum of a binomial measure. For higher pre-multifractal generations, similar spectra were found with similar $\alpha_{\min}$ and $\alpha_{\max}$, yet the fluctuations become smaller and occur at other places. They appear to be a result of the typical relatively slow overall convergence and the way in which the spectrum was generated through binning. This could be avoided through a procedure similar to the one described in [1].
- The following class of base functions is particularly interesting, because their shape can be qualitatively changed by modifying the value of the parameters:

$$w(x) = \alpha + \beta x^a + \gamma(1-x)^b, \qquad 0 < x < 1. \tag{7}$$

Depending on the values of the 5 parameters, a wide range of *MPFs* can be constructed. This example will be discussed elsewhere.

## 4 Link with random multiplicative multifractals

The deterministic measures generated using our new method are interesting by themselves, but also because of their close relation to random canonical multiplicative multifractals ($CM^2$) [15, 22].

The construction of the random $CM^2$ measures relies on an integer base $b$. It starts with a uniform measure of density 1 and begins by dividing $[0, 1]$ into $b$ equal $b$-adic parts. After the first construction step, the density is constant on each $b$-adic part and its $b$ values $w_i$ are independent and identically dis-

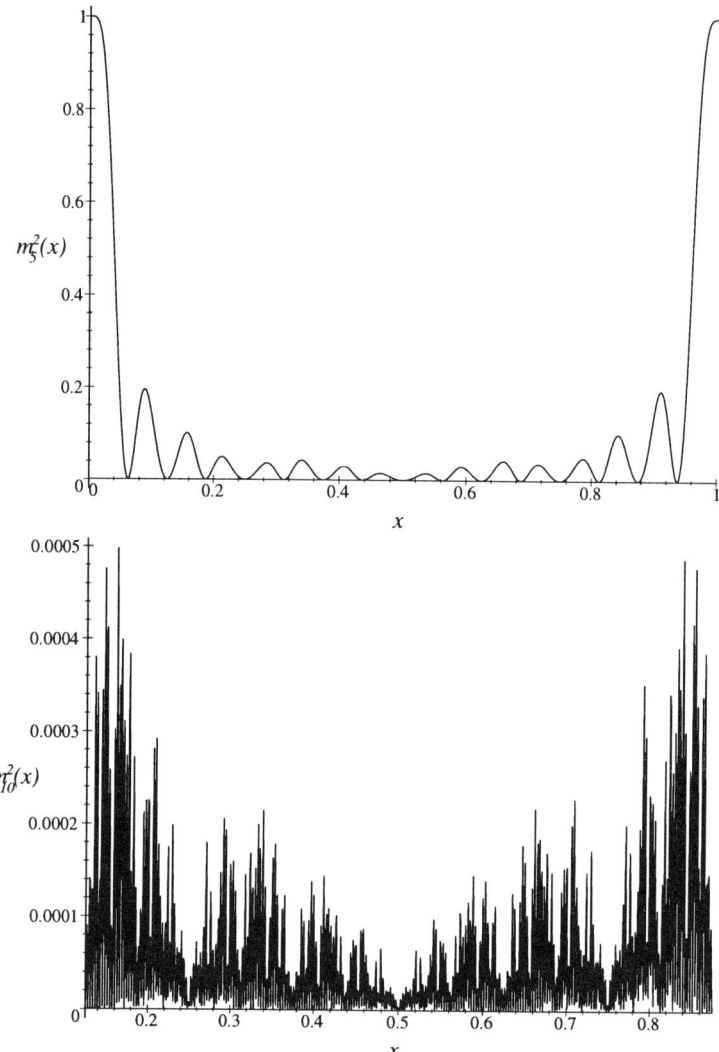

**Fig. 3.** Fifth (above) and tenth (below) generation of the potential well *MPF* shown in the previous Figure, formed by a binary multiplicative cascade of $w(x) = 1 - 0.25[1 - \cos(2\pi x)]^2$. The interval [1/8, 7/8] is shown for the tenth generation, because a plot of $m_{10}^2(x)$ over the interval [0,1] would be completely dominate by the peaks $m_n^b(0) = m_n^b(1)$. This is typical for a multifractal measure.

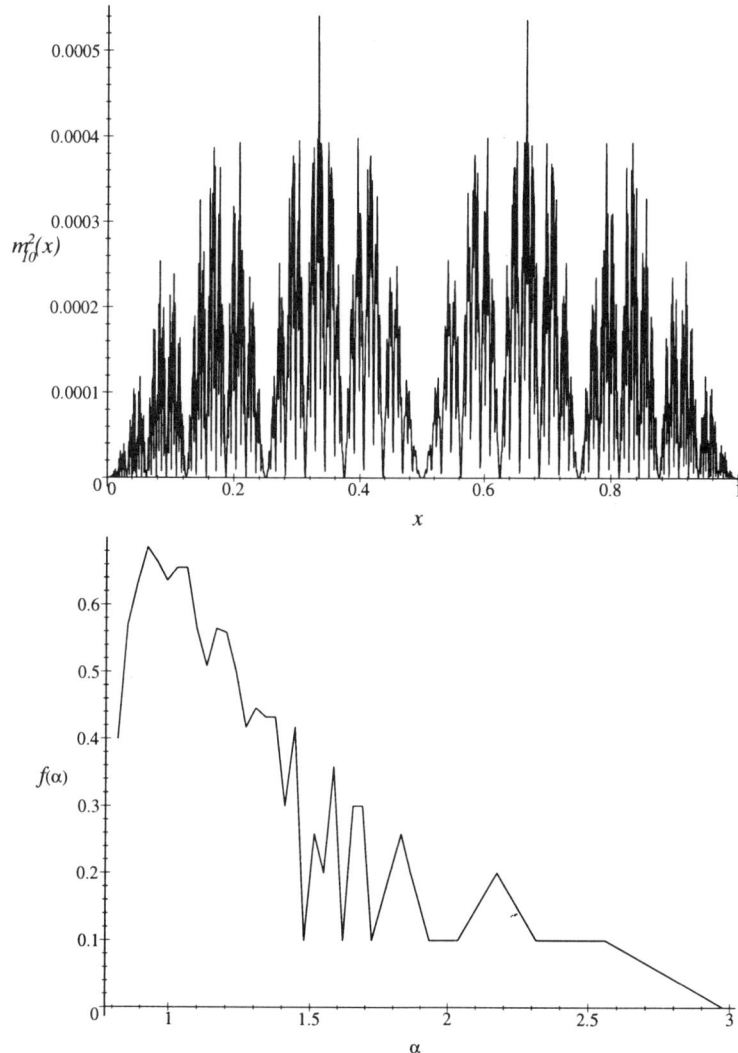

**Fig. 4.** Above: tenth generation of the "circle measure", formed by a binary multiplicative cascade of a semi-circle. Below: the corresponding pre-multifractal spectrum.

tributed random variables with cumulative distribution function $F(w) = $ Probability $\{W < w\}$ and average $\langle w \rangle = 1$. The notation $w$ is preserved for reasons that will transpire momentarily, but $w$ no longer denotes a periodic and non-random function. The second step of the construction introduces $b^2$ equal $b$-adic parts. The corresponding multipliers are independent and identically distributed random variables $w_{ij}$ and the corresponding densities at the end of the second step take the form $w_i w_{ij}$. When this multiplicative procedure is iterated, the random multifractal $CM^2$ ensues.

The cumulative distribution function $F(w)$ is of course non-decreasing and such that $F(0) = 0$ and $F(\infty) = 1$. If necessary, the graph of $F(w)$ can be processed by filling in each jump by a vertical straight interval. Exchanging the coordinate axes for this filled-in graph yields the filled-in graph of an inverse function $F^{-1}(x)$, which in turn can be made into a left-continuous inverse function. Using integration by parts, $\langle w \rangle = \int_0^1 w dF(w) = \int_0^1 F^{-1}(x) dx$. This result begins to justify using the same letter $w$ for different purposes. Indeed, the same formal condition of conservation on the average applies to $CM^2$, in the form $\langle w \rangle = 1$, and also to the MPF, in the form $\int_0^1 w(x) dx = 1$. When $b$ is an integer, the inverse $F^{-1}(x)$ of the cumulative probability distribution used to construct the random $CM^2$ is the function $w(x)$ used to construct the MPF.

In the $CM^2$ construction, only an interval $[0, 1]$ is considered; in the MPF construction, the function $w(x)$ is periodically extended beyond this interval. For almost all $x$, the infinite series of multipliers $w(b^{i-1}x)$ used to construct the MPF samples the interval $[0, 1)$ in a uniformly dense way, as is the case for the $CM^2$, because the $\beta$-map:

$$(x, \{bx\}, \{b^2 x\}, \ldots),  \qquad (8)$$

where $\{y\} = y \bmod 1$ is almost surely uniformly dense in $[0, 1)$. The latter was shown by Weyl, Hardy and Littlewood [29, 11, 12] and in a more general case by Kuipers and Niederreiter [13].

To conclude, a left-continuous function $w(x)$ that satisfies conservation on the average can be used in two distinct ways: when $b$ is an integer, to construct a $CM^2$ measure; for all $b$, to construct a MPF measure. There is clearly a close relationship between the previously studied multifractal properties of the $CM^2$ and the simpler to construct MPF, which is also multifractal.

This raises the question of whether or not the multifractal formalism that [15, 22] developed to apply to $CM^2$ also applies to the MPF. For integer $b$, this is already clear from the relation discussed above. The answers provided in [3] for general $b$ are to the affirmative. The criterion of non-degeneracy is:

$$\int_0^1 w(x) \log_b w(x) dx < 1 \qquad (9)$$

and the $\tau(q)$ and $f(\alpha)$ functions are those familiar from the study of $CM^2$.

To be more precise, the preceding affirmative answers apply directly to the randomized MPF in which $w(b^{i-1}x)$ is replaced by $w[b^{i-1}(x+\phi_i)]$, with a random phase $\phi_i$, as in Eq. (6). For irrational $b$, the same affirmative answer applies to the non-random product, Eq. (2), for $n \to \infty$, taken over the interval $[X, X+1]$, where $X$ is asymptotically large.

An interesting corrolary of this link between the random measures and the limit *MPF* is that a large number of transformations on $w(x)$ do not change the spectrum either. Such transformations include translations, rotations or reflections of parts of $w(x)$ about vertical axes. However, note in passing that finite generations of the *MPF* do not sample the interval $[0,1)$ uniformly for all $x$, so that the identity of measures and spectra is not valid for finite generations.

## 5 Conclusions

This paper introduces a new, simple way to generate a great variety of multifractal measures. We call these measures *Multifractal Products of Functions (MPF)*. They are constructed by repeatedly multiplying a periodically extended function with copies of itself, each copy having a frequency $b$ times higher than the previous one in the series, where $b$ is a real number larger than 1. The generation is therefore similar to the iterative construction of certain basic fractals, or the cascade leading to a multinomial measure, but is clearly much more general than the latter.

Apart from simplicity, the most important advantage of the *MPF* is the link between them and random multiplicative measures. This opens the way to many applications, because of the ease with which the *MPF* operates. Despite the fact that the *MPF* is either completely *deterministic* or slightly randomized (by using Eq. (6) with random phases), the multifractal spectrum is the same as that of a *random* multiplicative measure.

We sketched the methodology and gave some examples; other papers will refine the mathematical background and discuss the generated multifractal spectra in more detail. The method can be used in various applications in which multiplicative processes are present, such as in the description of turbulence, growth processes and the study of the accessibility distribution over rough interfaces.

## References

1. C. Canus, J. Lévy Véhel and C. Tricot (1998) *In proceedings of Fractals 98*, Malta, October 1998.
2. M.-O. Coppens (1997): *Fractals in Engineering*. Eds. J. Lévy-Véhel, E. Lutton and C. Tricot (Springer, Berlin), 336.
3. M.-O. Coppens and B.B. Mandelbrot (1999). *Submitted for publication*.
4. C.J.G. Evertsz and B.B. Mandelbrot (1992). *J. Phys. A: Math. Gen.* **25**, 1781.
5. C.J.G. Evertsz and B.B. Mandelbrot (1992): *Chaos and Fractals*. Eds. H.-O. Peitgen, H. Jürgens and D. Saupe (Springer, New York), 921.
6. C.J.G. Evertsz and B.B. Mandelbrot (1992). *Physica A* **185**, 77.
7. U. Frisch and G. Parisi (1985): *Turbulence and Predictability in Geophysical Fluid Dynamics and Climate Dynamics*. International School of Physics "Enrico Fermi", Course 88. Eds. M. Ghil *et al.* (North-Holland, Amsterdam), 84. Excerpt in [18]

8. R. Gutfraind, M. Sheintuch and D. Avnir (1991). *J. Chem. Phys.* **95**, 6100.
9. M.C. Gutzwiller and B.B. Mandelbrot (1988). *Phys. Rev. Lett.* **60**, 673.
10. T.C. Halsey, M.H. Jensen, L.P. Kadanoff, I. Procaccia and B.I. Shraiman (1986). *Phys. Rev. A* **33**, 1141.
11. Hardy, G.H., Littlewood, J.E. (1914). *Acta Mathematica* **37**, 155.
12. Hardy, G.H., Wright, E.M. (1938). *An Introduction to the Theory of Numbers* (Clarendon Press, Oxford), p. 381.
13. Kuipers, L., Niederreiter, H. (1974) *Uniform Distribution of Sequences* (Wiley, New York).
14. B.B. Mandelbrot (1972): *Statistical Models and Turbulence. Proc. Symp. Univ. Calif. San Diego (La Jolla), July 15-21, 1971*. Eds. M. Rosenblatt and C. Van Atta (Springer, Berlin). Reproduced in [18]
15. B.B. Mandelbrot (1974). *J. Fluid Mech.* **62**, 331. Reproduced in [18].
16. B.B. Mandelbrot (1989): *Fractals' Physical Origins and Properties. Proceedings of the Erice Meeting, 1988*. Ed. L. Pietronero (Plenum, New York), 3.
17. B.B. Mandelbrot (1989). *Pure Appl. Geophys.*, **131**, 5.
18. B.B. Mandelbrot (1989): *The Fractal Approach to Heterogeneous Chemistry*. Ed. D. Avnir (Wiley, Chichester), 45.
19. B.B. Mandelbrot (1990). *Physica A* **168**, 95.
20. B.B. Mandelbrot (1993): *Chaos in Australia*. Ed. G. Brown and A. Opie (World Scientific, Singapore), 1.
21. B.B. Mandelbrot (1997): *Fractals and Scaling in Finance. Discontinuity, Concentration, Risk.* Springer, New York.
22. B.B. Mandelbrot (1999): *Multifractals and 1/f Noise: Wild Self-Affinity in Physics.* Springer, New York.
23. B.B. Mandelbrot and C.J.G. Evertsz (1990): *Nature* **348**, 143.
24. C. Meneveau and K.R. Sreenivasan (1987). *Phys. Rev. Lett.* **59**, 1424.
25. R.R. Prasad, C. Meneveau and K.R. Sreenivasan (1988). *Phys. Rev. Lett.* **61**, 74.
26. R. Riedi (1995). *J. Math. Anal. Appl.* **189**, 462.
27. R.H. Riedi and B.B. Mandelbrot (1995). *Adv. Appl. Math.* **16**, 132.
28. D. Stauffer and A. Aharony (1994): *Introduction to Percolation Theory*. 2nd Ed. (Taylor and Francis, London).
29. Weyl, H. (1916). *Math. Annalen* **77**, 313.


*Acknowledgements*— Discussions with Prof. Michael Frame are gratefully acknowledged. Funding for MOC during the first stages of this project was provided by a Fellowship from the Belgian American Educational Foundation, and a F.W.O.–N.S.F. grant.

# IFS

# IFS-Type Operators on Integral Transforms

Bruno Forte[1,2], Franklin Mendivil[2,3], and Edward R. Vrscay[2]

[1] Facoltà di Scienze MM. FF. e NN. a Cà Vignal,
Università Degli Studi di Verona, Strada Le Grazie, 37134 Verona, ITALY
forte@biotech.sci.univr.it
[2] Department of Applied Mathematics, Faculty of Mathematics,
University of Waterloo, Waterloo, Ontario - CANADA N2L 3G1
ervrscay@links.uwaterloo.ca
[3] School of Mathematics, Georgia Institute of Technology,
Atlanta, Georgia 30332-0160 - U.S.A.
mendivil@math.gatech.edu

**Abstract.** Most standard fractal image compression techniques rely on using an IFS operator directly on the image function. Sometimes, however, it is more convenient to work on a faithful representation of the image which, in certain applications, may be a transformed version of the image. For example, if an MRI image is scanned in as frequency data it may be more natural to work on the Fourier transform of the image rather than on the image itself. After a brief introduction to fractal transforms and classical fractal image compression, we discuss some generalities of IFS operators on transform spaces. We then illustrate with examples from Fourier, wavelet and Lebesgue transforms.
We emphasize that the operations can be done completely in the transform domain. In some applications, e.g. measures, we may not even need or desire to return to the spatial domain.

## 1 Introduction

In this paper we consider the construction of fractal, IFS-type (for Iterated Function Systems) operators on transforms of functions. This represents a continuation of earlier work on *generalized fractal transforms* acting on complete metric spaces, e.g. probability measures [10], moments of probability measures [6], fuzzy sets [2], $L^p$ spaces [5], distributions [7], Fourier transforms [8] and discrete wavelet transforms [12,16]. The "spirit of IFS" permeates this work as in the past. We also note a related work on Fourier and Laplace transforms of fractal sets and multifractal measures [9].

First, we begin with an appropriate complete metric space, $(\mathcal{F}, d_\mathcal{F})$, the elements of which may represent "images" defined over a compact metric space $(X, d)$, the *base* or pixel space. For an element $y \in \mathcal{F}$, construct a set of *fractal components* of $y$, denoted by $y_i$, which are distorted copies of $y$ supported on subsets $X_i = w_i(X)$, where $w_i : X \to X$ denote IFS contraction maps. Then define an associated *fractal transform* $T : \mathcal{F} \to \mathcal{F}$ that combines the fractal components $y_i$, $1 \leq i \leq N$, in a suitable manner so that $Ty \in \mathcal{F}$. Under certain (not too restrictive) conditions, $T$ will be contractive on $(\mathcal{F}, d_\mathcal{F})$, implying that there exists a unique $\bar{y} \in \mathcal{F}$ such that $T\bar{y} = \bar{y}$.

Given an element $y \in \mathcal{F}$, the *inverse problem of fractal approximation* involves finding an operator $\mathcal{F}$ whose fixed point $\bar{y}$ approximates $y$ to some prescribed accuracy, i.e. given an $\varepsilon > 0$, find a contractive map $T_\varepsilon$ such that $d_\mathcal{F}(\bar{y}_\varepsilon, y) < \varepsilon$ where $T_\varepsilon \bar{y}_\varepsilon = \bar{y}_\varepsilon$. In practical applications, e.g. fractal image compression [4, 11], it is the operator $T_\varepsilon$ that is stored in computer memory. Given any $y_0 \in \mathcal{F}$ (for example, a blank screen), Banach's Fixed Point Theorem implies that the sequence $y_n = T_\varepsilon y_{n-1}$ converges to $\bar{y}_\varepsilon$, the suitable approximation to $y$.

In most practical applications, in particular image compression, fractal transforms are applied directly to the image. The space $(\mathcal{F}, d_\mathcal{F})$ is an appropriate function space, typically $L^2([0,1]^2, m)$. However, as is done in other approximation schemes or image compression methods, one may work with faithful representations of images in a kind of "dual space" $(\mathcal{G}, d_\mathcal{G})$. Examples are moments of probability measures [6], Fourier transforms of $L^2$ functions [8], and discrete wavelet expansions of $L^2$ functions [3, 12, 15, 16]. In these cases, the IFS operator $T$ on $(\mathcal{F}, d_\mathcal{F})$ induces an affine operator $M$ on $(\mathcal{G}, d_\mathcal{G})$ as indicated in the following diagram:

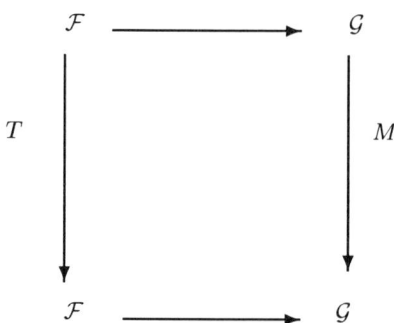

Inverse problems of fractal approximation in $(\mathcal{F}, d_\mathcal{F})$ are then transformed to inverse problems in $(\mathcal{G}, d_\mathcal{G})$. Under the conditions that $T$, hence $M$, is contractive, solutions to the inverse problem in $(\mathcal{G}, d_\mathcal{G})$ using the Collage Theorem may then be formulated. Indeed, this was the procedure followed in Refs. [6, 8, 16].

We also mention that once function approximation/image compression is achieved in the space $\mathcal{G}$, it may not be necessary to return to the space $\mathcal{F}$. In fact, in some cases, e.g. moments of measures/images, a return may not be practically possible, nor would it be of interest to do so. In such cases, one works exclusively in the transform domain. We emphasize, however, that when working in such "dual spaces," it is important to establish a number of properties, including

1. Completeness of the dual space $(\mathcal{G}, d_\mathcal{G})$.
2. The "faithfulness" of $\mathcal{G}$ (i.e. is it an isomorphism of $\mathcal{F}$?) as well as the operator $M$ (i.e. does $M$ map $\mathcal{G}$ to itself?)

Assuming that $T$ is contractive, the above properties are necessary to ensure the existence of a unique fixed point $\bar{g} \in \mathcal{G}$, i.e. $\bar{g} = M\bar{g}$, by Banach's Fixed Point Theorem.

A natural question that arises in the study of induced operators is, "How does operating on $\mathcal{G}$ instead of on $\mathcal{F}$ relate to operating directly on $\mathcal{F}$?" For example, what is "self-similarity" in the transform domain if, in fact, this is a meaningful question?

Although it appears very natural to consider the induced operator $M$ on the space $\mathcal{G}$, we are not constrained to use it. As we shall see below, it may be advantageous to use another operator, depending upon the application. Nevertheless, the conditions of contractivity as well as completeness listed above must still be established.

In this paper $(\mathcal{G}, d_\mathcal{G})$ will be an appropriate space of *function transforms*, in particular *integral transforms*. There are two major motivations for this approach:

1. In many cases, the data which we seek to represent or compress is the result of an integral transform on some function space, e.g. MRI data, blurred images.
2. It may be more convenient to work in certain spaces of integral transforms. For example, as we show below, Lebesgue transforms of normalized nonnegative $L^1$ functions are nondecreasing and continuous functions. These latter functions may be easier to work with, especially in the sense of approximability.

In Section 2, standard IFS-fractal transforms on functions are very briefly reviewed, mostly for purposes of notation. In Section 3, we consider the integral transform, with kernel $K$, of a "fractally transformed" function, i.e. $Tf$, and relate it to the integral transform of $f$. This equation simplifies if $K$ satisfies a general functional equation. It is then of interest to examine whether the kernel $K$ itself can satisfy an IFS-type equation, for which it is necessary to examine the general space of kernels. A special class of solutions for this functional equation are considered. Finally, we present some examples.

For ease of notation and clarity of discussion, the following discussion is retricted to the one-dimensional case. However, the extension to two (or more) dimensions is straightforward.

## 2 Iterated Function Systems with Maps (IFSM)

Briefly, standard fractal image compression schemes are variations of affine IFSM with associated fractal transform operators [5, 7]. Let $(X, d)$ denote the base or pixel space and $m$ be a measure on $X$ (usually the unit square with Lebesgue measure for the computer screen). Let $w_i : X \to X$, $1 \leq i \leq N$ be contraction maps and denote $X_i = \{w_i(x)|x \in X\}$. (In most practical applications, $w_i : D_i \to R_i$, where $D_i, R_i \in X$ denote, respectively, *domain* and *range* blocks so that $X_i = R_i$.) For simplicity, as is done in practice, the $w_i$ are assumed to be affine contractions. In our one-dimensional treatment, $X = [0,1]$ and we denote:

$$w_i(x) = c_i x + a_i, \quad c_i, a_i \in \mathbf{R}, \quad 0 \le c_i < 1, \quad 1 \le i \le N. \tag{1}$$

Associated with each IFS map $w_i$ is an affine grey-level map:

$$\phi_i(t) = \alpha_i t + \beta_i, \quad \alpha_i, \beta_i \in \mathbf{R}. \tag{2}$$

The $N$-map IFSM, denoted in vector notation as $(\mathbf{w}, \Phi)$, defines a *fractal transform operator* $T : \mathcal{F} \to \mathcal{F}$. The action of $T$ on an image function $u \in L^p(X, m)$ is given by [5, 7]

$$(Tu)(x) = \sum_{i=1}^{N} u_i(x), \tag{3}$$

where the $u_i(x)$ denote the *fractal components* of $X$:

$$u_i(x) = \begin{cases} \phi_i(u(w_i^{-1}(x))), & x \in X_i, \\ 0, & x \notin X_i. \end{cases} \tag{4}$$

Because of the additivity of integrals, the natural combination of fractal components in $L^p$ spaces is addition. A straightforward calculation yields the bound

$$\| Tu - Tv \|_p \le C_p \| u - v \|_p, \quad \forall u, v \in L^p(X, m), \tag{5}$$

where

$$C_p = \sum_{i=1}^{N} |c_i|^{1/p} |\alpha_i|. \tag{6}$$

The scalars $c_i$ and $\alpha_i$ are seen to determine whether or not the operator $T$ is contractive in $L^p(X, m)$. If $T$ is contractive, then there exists a unique fixed point $\bar{u} = T\bar{u}$. Note that if all coefficients $\beta_i = 0$, then $T$ is linear in $L^p(X, m)$, implying that $\bar{u} = 0$ is a fixed point.

## 3 Fractal Transforms of Integral Transforms

### 3.1 Derivation of a Functional Equation for the Kernel

In this section we let $\mathcal{S} : \mathcal{F} \to \mathcal{G}$ denote an integral transform with kernel $K : X \times \mathbf{R} \to \mathbf{R}$,

$$\widehat{f}(s) = (\mathcal{S}f)(s) = \int_X K(t, s) f(t) \, dt. \tag{7}$$

We shall also write this transform in inner product form as $\mathcal{S}f = <K, f>$.

Let $T$ be an affine IFSM operator as defined in Eq. (3). For an $f \in L^p(X)$, let $g = Tf$. Then the transform $\widehat{g} = \mathcal{S}(g)$ is given by

$$\widehat{g}(s) = \int_X K(t,s) \sum_{i=1}^N [\alpha_i f(w_i^{-1}(t)) + \beta_i] I_{X_i}(t) \, dt$$

$$= \sum_{i=1}^N \alpha_i \int_{X_i} K(t,s) f(w_i^{-1}(t)) \, dt + \sum_{i=1}^N \beta_i \int_{X_i} K(s,t) dt$$

$$= \sum_{i=1}^N \alpha_i c_i \int_X K(c_i u + a_i, s) f(u) \, du + \widehat{\beta}(s), \qquad (8)$$

where

$$\widehat{\beta}(s) = \sum_{i=1}^N \beta_i \widehat{I_{X_i}}(s). \qquad (9)$$

(Note that $\widehat{\beta}(s)$ depends only on the $\beta_i$ - and, of course, the $X_i$ - but not on $f$.)
Eq. (8) may be written in the form

$$<K, Tf> = <T^\dagger K, f> + L(s), \qquad f \in \mathcal{F}, \qquad (10)$$

where the operator $T^\dagger$ may be interpreted as a kind of "adjoint" fractal operator on the kernel $K$,

$$(T^\dagger K)(t,s) = \sum_{i=1}^N \alpha_i c_i K(c_i t + a_i, s), \qquad (11)$$

and $L$ as a kind of condensation function. However, the dilations in the spatial variable produced by $T^\dagger$ in Eq. (11) represent *expansions*. In contrast to IFSM fractal transforms on functions, the transform $K$ is tiled with expanded copies of itself.

We now focus on Eq. (8) and attempt to rewrite the integrals involving $K$ as *bona fide* integral transforms of $f$. First, we write

$$\int_X K(c_i u + a_i, s) f(u) du = \int_X \frac{K(c_i u + a_i, s)}{K(u, \zeta_i(c_i, a_i, s))} K(u, \zeta_i(c_i, a_i, s)) f(u) du, \qquad (12)$$

where the $\zeta_i$ functions perform a renormalization or scaling of the transform variable $s$. It is desirable that the quotient in the integrand on the right be independent of the integration variable $u$, i.e. constant with respect to $u$. Most generally, this implies that

$$K(c_i u + a_i, s) = C_i(c_i, a_i, s) K(u, \zeta_i(c_i, a_i, s)), \qquad i = 1, 2, \ldots, N. \qquad (13)$$

However, allowing each scaling relation to possess its own functions $C_i$ and $\zeta_i$ may be too general since, for example, no "self-similarity" property is imposed on $K$. Therefore, we postulate the following functional relation to be satisfied by the kernel $K$ and the functions $C$ and $\zeta$:

$$K(c_i u + a_i, s) = C(c_i, a_i, s) K(u, \zeta(c_i, a_i, s)), \tag{14}$$
$$\forall u \in X, \quad i = 1, 2, \ldots, N.$$

Eq. (14) may be considered in several ways, including:

1. as a functional relation between the kernel $K$, the constant $C$ and scaling function $\zeta$,
2. as a functional equation in the unknown functions $K$ and $\zeta$, given $C$,
3. as a functional equation in the unknown functions $C$ and $\zeta$, given $K$.

As in the case of differential equations, the solution of functional equations requires "initial conditions." In addition, however, an admissible space of functions in which solutions are sought must also be specified. This is the subject of future research. A few simple results are presented in Section 3.3.

## 3.2 Induced Fractal Operators on the Fractal Transforms

If the functional equation in (14) is satisfied by the kernel $K$, then the integrals in the first sum of Eq. (8) simplify to

$$\sum_{i=1}^{N} \alpha_i c_i \int_X C(c_i, a_i, s) K(u, \zeta(c_i, a_i, s)) f(u) \, du = \sum_{i=1}^{N} \alpha_i c_i C(c_i, a_i, s) \widehat{f}(\zeta(c_i, a_i, s)).$$

The net result is the relation

$$\widehat{g}(s) = (M\widehat{f})(s)$$
$$= \sum_{i}^{N} \alpha_i c_i C(c_i, a_i, s) \widehat{f}(\zeta(c_i, a_i, s)) + \widehat{\beta}(s), \tag{15}$$

a kind of self-similarity equation defining the action of operator $M$ in the figure of Section 1.

If we now assume that $T$ is contractive in $L^p(X)$ with fixed point $\bar{f}$, then $\widehat{\bar{f}} = \mathcal{S}(\bar{f})$ satisfies the fixed-point equation $\widehat{\bar{f}} = M\widehat{\bar{f}}$ or

$$\widehat{\bar{f}}(s) = \sum_{i=1}^{N} \alpha_i c_i C(c_i, a_i, s) \widehat{\bar{f}}(\zeta(c_i, a_i, s)) + \widehat{\beta}(s). \tag{16}$$

However, there remains the question whether the operator $M$ is *contractive* in the space of transforms $\mathcal{G}$ and with respect to what metric. Assuming that $\mathcal{G}$ is a Banach space with norm denoted $\|\cdot\|_\mathcal{G}$, we define the metric $d_\mathcal{G}(u,v) = \| u-v \|_\mathcal{G}$ for $u, v \in \mathcal{G}$. If $M$ is contractive in this metric then, by Banach's Fixed Point Theorem, $\widehat{\bar{f}}(s)$ may be generated by standard iteration: Start with any function $v_0 \in \mathcal{G}$ and define $v_{n+1} = Mv_n$. Then $v_n \to \widehat{\bar{f}}$ as $n \to \infty$ in the $d_\mathcal{G}$ metric.

In many practical examples, including Fourier and wavelet transforms, the coefficients $C(c_i, a_i, s)$ in Eq. (15) may be bounded with respect to $s$. In such cases, a straightforward calculation yields

$$d_{\mathcal{G}}(Mu, Mv) \leq D d_{\mathcal{G}}(u, v), \qquad u, v \in \mathcal{G}, \tag{17}$$

where

$$D = \sum_{i=1}^{N} c_i |\alpha_i \bar{C}_i(c_i, a_i, s)| |J_i|^{-1} \tag{18}$$

where $\bar{C}_i = \max_s C(c_i, a_i, s)$ and $|J_i|$ denotes the (maximum of the) Jacobian of the transformation $s \to \zeta(c_i, a_i, s)$. Contractivity of $M$ is guaranteed if $D < 1$.

## 3.3 Some Remarks on the Functional Equation for the Kernel

It is natural to inquire about the actual meaning of the functional equation in (14). Suppose that $K$ is a solution. Furthermore, consider the particular case in which the sets $X_i = w_i(X)$ (or the range blocks $R_i$) $1 \leq i \leq N$, form a partition of $X$, herewith to be referred to as an *IFS partition* of $X$, the case normally employed in practical fractal image and signal compression. In this nonoverlapping case, each point $x \in X$ has only one fractal component (neglecting boundary points in the continuous case). As a result, we may "invert" Eq. (14) to give

$$K(t, s) = C(c_i, a_i, s) K(w_i^{-1}(t), \zeta(c_i, a_i, s)), \quad t \in X_i, \quad i = 1, 2, \ldots, N. \tag{19}$$

The nonoverlapping nature of the $X_i$ allows us to express this result as follows,

$$K(t, s) = (\mathcal{M}K)(t, s)$$
$$= \sum_{i=1}^{N} C(c_i, a_i, s) K(w_i^{-1}(t), \zeta(c_i, a_i, s)). \tag{20}$$

Thus, as in the case of the IFSM fractal transform $T$, cf. Eq. (3), $K$ is now written as a linear combination of its own *fractal components* under the action of the IFS maps $w_i$. In other words, $K$ is the fixed point of a fractal transform $\mathcal{M}$ that operates on kernels. Note that there is no restriction on the IFS partition of $X$, implying that $K$ satisfies a kind of *universal self-similarity*. Clearly, this is a special property.

The following proposition shows that the functional equation in (14) is equivalent to this type of universal self-similarity.

**Proposition 1.** *The function $K(t, s)$ satisfies the functional equation Eq. (14) for fixed functions $C(c, a, s)$ and $\zeta(c, a, s)$ if and only if for every IFS partition of $X$ with IFS maps of the form $w_i(x) = c_i x + a_i$ there are functions $C_i(s)$ and $\zeta_i(s)$ so that $K$ is the fixed point of the fractal transform operator*

$$(\mathcal{M}K)(t, s) = \sum_{i=1}^{N} C_i(s) K(w_i^{-1}(t), \zeta_i(s)). \tag{21}$$

*Proof.* By the comments immediately preceding the statement of the proposition, we know that if $K$ satisfies the functional equation, then for any IFSM partition of $X$, $K$ is the fixed point of the IFSM (21) where $C_i(s) = C(c_i, a_i, s)$ and $\zeta_i(s) = \zeta(c_i, a_i, s)$.

Conversely suppose that for any IFSM partition of $X$ there are functions $C_i(s)$ and $\zeta_i(s)$ such that $K$ is the fixed point of the induced IFSM operator (21). In order to show that $K$ is a solution to the functional equation, we must define the functions $C(c, a, s)$ and $\zeta(c, a, s)$.

To this end, let $c$ and $a$ be fixed such that $w_1(x) = cx + a$ defines a contractive map from $X$ to itself. Choose $w_2, w_3, \ldots, w_n$ to be affine maps such that the IFS $\{w_1, w_2, \ldots, w_n\}$ is an IFS partition of $X$. Then by hypothesis we know that there are functions $C_i(s)$ and $\zeta_i(s)$ so that $K$ is the fixed point of the induced IFSM given by Eq. (21). Define

$$C(c, a, s) = C_1(s)$$

and

$$\zeta(c, a, s) = \zeta_1(s).$$

Then for all $s$ and $t$ and for this specific choice of $c$ and $a$ we have

$$K(ct + a, s) = C(c, a, s) K(t, \zeta(c, a, s))$$

and so $K$ satisfies the functional equation Eq. (14) for this specific choice of $c$ and $a$.

Clearly, since $c$ and $a$ were arbitrary, the above procedure can be performed for all $c$ and $a$, thus constructing functions $C(c, a, s)$ and $\zeta(c, a, s)$ so that $K$ satisfies the functional equation. □

To repeat, the functional equation is equivalent to the property of universal self-similarity. The solution $K$ is the fixed point of an IFSM-type operator on kernel functions. Note, however, that it is *not* guaranteed that the coefficients $C_i(s)$ are contractive. An additional complication arises from the fact that the operator $\mathcal{M}$ in Eq. (20) is linear in $K$. In order to avoid the trivial solution $K(t, s) = 0$, it may be necessary to restrict the solution space of the functional equation to appropriate "shells," as is done, for example, in the case of IFSP Markov operators and probability measures. These are open questions for further research. We now examine the functional equation for some very special cases.

**Proposition 2.** *Suppose that a kernel $K$ satisfies the functional equation in Eq. (14) for $\zeta(c_i, a_i, s) = s$. Then $K$ is independent of $t$, i.e. $K(t, s) = K(s)$.*

*Proof.* For simplicity of notation, let us drop the subscripts $i$. Choose fixed values of $c$ and $a$. Then for $K(t, s) \neq 0$ we have

$$C(c, a, s) = \frac{K(ct + a, s)}{K(t, s)}.$$

Since $t \in X$ and $s \in \mathbf{R}$ are independent variables, it follows that both sides of the equation are independent of $t$ (since the left-hand side is $t$-independent).

Now, choose the value $t^* = a/(1-c)$ so that $ct^* + a = t^*$. (The existence of such a $t^* \in X$ is guaranteed by the assumption on the IFS maps that $w_i : X \to X$.) Inserting this value of $t$ into the above equation yields

$$C(c, a, s) = \frac{K(t^*, s)}{K(t^*, s)} = 1.$$

This result is true for all values of $c, a, s$. Therefore, the functional equation reduces to

$$K(ct + a, s) = K(t, s),$$

the only solution of which is $K(t, s) = f(s)$, a function of $s$ only. □

The following result is obtained in a very similar fashion.

**Proposition 3.** *Suppose that the kernel $K$ satisfying the functional equation in Eq. (14) is independent of $s$, i.e. $K(t, s) = K(t)$. Then $K$ is a constant.*

These two simple results illustrate the importance of "mixing" between the spatial variable $t \in X$ and the transform variable $s \in \mathbf{R}$. In the following, the particular consequences of separability of the kernel $K$ are examined.

**Proposition 4.** *Suppose that the kernel $K$ satisfying the functional equation in Eq. (14) is separable, i.e. $K(t, s) = K_1(t)K_2(s)$. Then $K_1$ is constant on $X$ and $K_2$ satisfies the relation*

$$K_2(s) = C(c_i, a_i, s) K_2(\zeta(c_i, a_i, s)). \tag{22}$$

*Proof.* Once again, for simplicity of notation, we omit the subscripts $i$ and choose fixed values of $c$ and $a$. Then, assuming separability (as well as $K(t,s) \neq 0$), a rearrangement of Eq. (14) yields

$$\frac{K_1(ct + a)}{K_1(t)} = C(c, a, s) \frac{K_2(\zeta(c, a, s))}{K_2(s)} = A, \tag{23}$$

where $A$ is a real constant, since $t \in X$ and $s \in \mathbf{R}$ are independent. For the particular value $t = t^* = a/(1-c)$, we find that $A = 1$, which must hold true for all values of $c, a, s$. Therefore $K_1(cu + a) = K_1(u)$, implying that $K_1$ is constant on $X$. The functional relation (22) for $K_2$ then follows immediately. □

**Proposition 5.** *Let $T : L^p(X) \to L^p(X)$ be the fractal transform operator associated with an $N$-map affine IFSM, as defined in Eq. (3) of Section 2. For an $f \in L^p(X)$, let $g = Tf$. Let $\widehat{f}$ and $\widehat{g}$ denote the integral transforms of $f$ and $g$ respectively, assuming that the kernel $K$ satisfies the functional equation in Eq. (14) and is separable, i.e. $K(t,s) = K_1(t)K_2(s)$. Then*

$$\widehat{g}(s) = \left[\sum_{i=1}^{N} \alpha_i c_i\right] \widehat{f}(s) + \widehat{\beta}(s), \tag{24}$$

*where $\widehat{\beta}(s)$ is defined in Eq. (9).*

*Proof.* From the previous proposition, it follows that $K_1(x) = B$, a constant, on $X$. Therefore,

$$\widehat{f}(s) = BK_2(s) \int_X f(t)dt. \tag{25}$$

Substitution into Eq. (15) yields

$$\widehat{g}(s) = \sum_{i=1}^{N} \alpha_i c_i B K_2(s) \int_X f(t) dt + \widehat{\beta}(s), \tag{26}$$

which, when rearranged, gives the desired result. □

The reader may compare Eq. (24) with Eq. (15). When the kernel $K$ is separable, the resulting operator $M$ relating $\widehat{g}$ to $\widehat{f}$ is rather simple in form, involving no dilations in the transform variable $s$. (This is a consequence of the fact that $K$ is constant with respect to variations in the spatial variable $t \in X$.) If we further assume that the IFSM operator $T$ is contractive with fixed point $\bar{f}$, then, from Eq. (24), with $f = g = \bar{f}$, it follows that

$$\widehat{\bar{f}}(s) = \widehat{\beta}(s) \left[1 - \sum_{i=1}^{N} \alpha_i c_i\right]^{-1}. \tag{27}$$

## 4 Examples

### 4.1 Fourier Transform

The kernel is $K(t, \omega) = e^{i\omega t}$, so that

$$K(c_i u + a_i, \omega) = e^{i(c_i u + a_i)\omega} = e^{ia_i\omega} e^{iuc_i\omega} = e^{ia_i\omega} K(u, c_i\omega). \tag{28}$$

Thus, $C(c_i, a_i, \omega) = e^{ia_i\omega}$ and $\zeta(c_i, a_i, \omega) = c_i\omega$. If $g = Tf$, then Eq. (16) becomes

$$\widehat{g}(\omega) = \sum_i \alpha_i c_i e^{ia_i\omega} \widehat{f}(c_i\omega) + \widehat{\beta}(\omega). \tag{29}$$

Notice that if $T$ has contractive spatial maps (the $w_i$'s) then the induced operator will have expansive spatial maps. Intuitively, this happens because large frequencies correspond to small scales and small frequencies correspond to large scales. Computationally this happens because the kernel is of the form $K(t, s) = M(st)$. For a kernel of the form $K(t, s) = M(t/s)$, one would have a direct relationship between frequency and scale.

It is well known that $f, g \in L^2(X)$ implies that $\widehat{f}, \widehat{g} \in L^2(\mathbf{R})$. Thus it is convenient to use the usual $L^2$ metric in $\mathcal{G}$. Following the calculation of Eq. (18), we find

$$\| M\widehat{u} - M\widehat{v} \|_2 \leq \sum_{i=1}^{N} c_i^{1/2} |\alpha_i| \, \| \widehat{u} - \widehat{v} \|_2, \quad \widehat{u}, \widehat{v} \in L^2(\mathbf{R}). \tag{30}$$

From Eq. (6), with $p = 2$, contractivity of the IFSM operator $T$ implies contractivity of $M$.

In the case of measures, i.e. $f, g \in \mathcal{M}(X)$, the set of Borel probability measures on $X$, some care must be taken in the construction of a suitable metric on the space of transforms $\mathcal{G}$ of measures [7]. It can then be shown that contractivity of $T$ implies contractivity of $M$. We refer the reader to [7] for details.

Finally, from a historical viewpoint, we recall Zygmund's [17] analysis of the Fourier transform of (uniform) Cantor-Lebesgue measure on the classical Cantor set. Not surprisingly, his analysis, which exploited the self-similarity of the problem, was quite analogous to the fractal transform method.

### 4.2 Wavelet Transform

In this case, the kernel is given by $K(t, s, b) = \psi\left(\frac{t-b}{s}\right)$, where $\psi(x)$ denotes a mother wavelet function. There are two transform variables, $s$ and $b$, corresponding to scaling and translation, respectively.

$$K(c_i u + a_i, s, b) = \psi\left(\frac{c_i u + a_i - b}{s}\right)$$
$$= \psi\left(\frac{u - c_i^{-1}(b - a_i)}{sc_i^{-1}}\right) \quad (31)$$

so that the functional equation satisfied by $K$ is

$$K(c_i u + a_i, s, b) = K(u, sc_i^{-1}, c_i^{-1}(b - a_i)). \quad (32)$$

Here, $C(c_i, a_i, s, b) = 1$ and the scaling function for the parameter $s$ is

$$\zeta(c_i, a_i, s, b) = sc_i^{-1}.$$

Thus, for $g = Tf$, we obtain

$$\widehat{g}(s, b) = \sum_i \alpha_i c_i \widehat{f}\left(sc_i^{-1}, c_i^{-1}(b - a_i)\right). \quad (33)$$

Numerous authors have studied the possibilities of mixing IFS with wavelets (see, for example, [3, 8, 12, 15, 16] and references therein) with good results. The multiresolution structure of the wavelet transform makes it an ideal candidate for fractal analysis.

### 4.3 Lebesgue Transform

$$K(t, s) = \begin{cases} 1, & \text{if } 0 \leq t \leq s \\ 0, & \text{if } s \leq t \leq 1. \end{cases}$$

This kernel satisfies the functional equation with

$$K(c_i u + a_i, t) = K(u, \frac{t - a_i}{c_i}),$$

where we recall that $c_i > 0$. Another (perhaps more useful) way to write the Lebesgue transform of $f$ is as

$$\widehat{f}(s) = \int_0^s f(t)\, dt. \tag{34}$$

If we restrict $f$ to be positive (as is the case for image functions) then it may be viewed as a density function on $X = [0,1]$. Then $\widehat{f}(s)$ will be the cumulative distribution function (CDF) for $f$. For $f \in L^1$, $\widehat{f}(s)$ is nondecreasing and continuous, with $(\mathcal{S}f)(0) = 0$. If we assume $f$ to be normalized, i.e. $\|f\|_1 = 1$, then $\widehat{f}(1) = 1$.

If we relax the condition that the Lebesgue transform $\widehat{f}(s)$ be continuous in $s$, then the space $\mathcal{G}$ is given by

$$\mathcal{G} = \{F : [0,1] \to [0,1] : F(0) = 0, F(1) = 1, F \text{ nondecreasing }\},$$

which is the set of CDFs for probability measures on [0,1]. A suitable choice of fractal-type transforms $M$ on this space is as follows [13, 14]. We again assume that the sets $w_i(X)$ overlap only at their endpoints. Then for an $f \in \mathcal{G}$:

$$(Mf)(x) = \alpha_i f(w_i^{-1}(x)) + \beta_i, \quad x \in X_i, \tag{35}$$

where

$$\beta_1 = 0 \quad \text{and} \quad \alpha_i + \beta_i \leq \beta_{i+1} \leq 1 \quad \text{and} \quad \alpha_N + \beta_N = 1.$$

(The reader may verify that the above conditions guarantee that $M$ preserves the nondecreasing property. Technically, the above equation is not valid at any points of intersection of the $w_i(X)$. At those points, one would choose either the maximum or minimum of the two "fractal components" in order to preserve right or left continuity, respectively.) The papers [13, 14] contain an extended discussion of this example with some very nice applications to image representation.

This definition of $M$ on $\mathcal{G}$ is slightly more general than the operator induced by the IFSM map $T : \mathcal{F} \to \mathcal{F}$ as it permits the introduction of point masses at the boundaries of the intervals $X_i = w_i(X)$. It essentially represents a "grand unification" of all IFS-type schemes, since $\mathcal{G}$ may now include the Lebesgue transforms of measures, functions and distributions.

### 4.4 Moments of Measures

This final example is perhaps more of a "nonexample," in two aspects. First, the domain of the transform is the space of Borel probability measures on $[0,1]$. Second, the transform space is a sequence space – the space of moments of measures on $[0,1]$ (as discussed in [6]).

Let $\mathcal{M}(X)$ be the collection of probability measures on $X = [0,1]$ and $\mu \in \mathcal{M}(X)$. We define the moment sequence by

$$g_k(\mu) = \int_X x^k \, d\mu(x) \quad \text{for } k = 0, 1, 2, \ldots \tag{36}$$

so the "kernel" for this transform is the function $K(x,n) = x^n$. Clearly, this kernel does not satisfy equation (14).

Note that $\mu_0 = 1$ since $\mu$ is a probability measure. Furthermore, $g_{k+1} \leq g_k$ since $x^{k+1} \leq x^k$ for $x \in [0,1]$. Thus, $(g_k) \in l^\infty$. We define the space

$$\bar{l}_0^2 = \{\mathbf{c} = (c_0, c_1, \ldots) \in l^\infty | c_0 = 1\}$$

with the weighted inner product

$$\langle \mathbf{c}, \mathbf{d} \rangle = 1 + \sum_{k=1}^{\infty} \frac{c_k d_k}{k^2}.$$

Then $\bar{l}_0^2$ is a complete metric space and the operator on $\bar{l}_0^2$ induced from the Markov operator on $\mathcal{M}(X)$ is a contraction (see [6]). However, this operator is not of IFS type, since the kernel does not satisfy the functional equation (14).

## 5 Acknowledgements

This research was supported in part by the Natural Sciences and Engineering Research Council of Canada (NSERC) in the form of an Operating Grant (ERV) as well as a Collaborative Projects Grant (ERV and C. Tricot) which are gratefully acknowledged.

## References

1. M.F. Barnsley (1988): *Fractals Everywhere*. Academic Press, New York.
2. C.A. Cabrelli, B. Forte, U.M. Molter and E.R. Vrscay (1992): Iterated fuzzy set systems: a new approach to the inverse problem for fractals and other sets. J. Math. Anal. Appl. **171**, 79–100.
3. G. Davis (1998): A wavelet-based analysis of fractal image compression. IEEE Trans. Image Proc. **7**, No. 2, 141–154.
4. Y. Fisher (1995): *Fractal Image Compression, Theory and Application*. Springer-Verlag, New York.
5. B. Forte and E.R. Vrscay (1995): Solving the inverse problem for function and image approximation using Iterated Function Systems. Dyn. Cont. Disc. Imp. Syst. **1**, 177–231.
6. B. Forte and E.R. Vrscay (1995): Solving the Inverse Problem for Measures using Iterated Function Systems: A New Approach. Adv. Appl. Prob. **27**, 800–820.
7. B. Forte and E.R. Vrscay (1998): Theory of Generalized Fractal Transforms. *Fractal Image Encoding and Analysis*, NATO ASI Series F, Vol. 159, Y. Fisher Ed., Springer Verlag, Berlin Heidelberg.
8. B. Forte and E.R. Vrscay: Inverse Problem Methods for Generalized Fractal Transforms. *Fractal Image Encoding and Analysis*, ibid.
9. M. Giona (1997): Analytic expressions for the structure factor and for the moment-generating function of fractal sets and multifractal measures. J. Phys. **A 30**, 4293–4312.

10. J. Hutchinson (1981): Fractals and self-similarity. Indiana Univ. J. Math. **30**, 713–747.
11. N. Lu (1997): *Fractal Imaging*. Academic Press, New York.
12. F. Mendivil and E.R. Vrscay (1997): Correspondence between fractal-wavelet transforms and Iterated Function Systems with grey level map. *Fractals in Engineering*, J. Lévy Véhel, E. Lutton and C. Tricot, Editors, Springer-Verlag, pp. 54–64.
13. A. Perazzolo (1997): IDFS, Un nuovo approccio alla compressione e decompressione dei dati immagine. Tesi di Laurea in Scienze dell'informazione, Facoltà di Scienze, M.F.N., Università degli Studi di Verona, Italy.
14. D. Romani (1997): Transformata Frattale IDFS su uno schema finito. Tesi di Laurea in Scienze dell'informazione, Facoltà di Scienze, M.F.N., Università degli Studi di Verona, Italy.
15. A. van de Walle (1995): *Relating fractal compression to transform methods*. Master of Mathematics Thesis, Department of Applied Mathematics, University of Waterloo.
16. E.R. Vrscay (1998): A generalized class of fractal-wavelet transforms for image representation and compression. Can. J. Elec. Comp. Eng. **23**, Nos. 1-2, 69–83.
17. A. Zygmund (1968): *Trigonometric Series*. Vol. 1, Cambridge University Press, pp. 194–197.

# Comparison of Dimensions of a Self-Similar Attractor

Serge Dubuc and Jun Li

Département de mathématiques et de statistique,
Université de Montréal
C.P. 6128, Succ. Centre-ville, Montréal (Québec) - CANADA H3C 3J7
{dubucs,li}@dms.umontreal.ca

**Abstract.** The fractal dimension of a self-similar attractor in the plane is always overestimated by its similarity dimension. We illustrate the case where the fractal dimension and the similarity dimension coincide and the case where they differ from one another. Finally, we will discuss a special fractal curve whose fractal dimension is determined by the largest eigenvalue of a square matrix of order 5.

**Keywords:** dimension, fractal, attractor, self-similarity.

## 1 Introduction

In fractal geometry, it is possible to define many dimensions for the same category of geometric objects. For example, for any self-similar attractor in the plane we have the topological dimension, the Hausdorff dimension, the similarity dimension, and the fractal dimension (which is sometimes called the box-counting dimension). A natural question to ask is when do two of these dimensions differ from one another. Our discussion will deal especially with the comparison between the similarity dimension and the fractal dimension. The numerical computation of the similarity dimension is always easy, but this is not the case for the fractal dimension when it differs from the similarity dimension. For this reason, we will particularly discuss the numerical evaluation of the fractal dimension of a certain fractal curve. This curve is due to the first author [4] and is called the forest curve. This self-similar curve has similarity dimension 2, but its fractal dimension is 1.656614. We reproduce its graph in Figure 1.

## 2 Self-similar attractors

We describe, first of all, the class of self-similar attractors that was proposed by Mandelbrot [10] and by Hutchinson [8]. Given a finite set $\mathcal{S} = \{S_1, S_2, \ldots, S_N\}$ of $N$ maps of the plane, an *attractor* with respect to $\mathcal{S}$ is a set $K$ in the plane such that
$$K = \cup_{i=1}^{N} S_i(K).$$

Fig. 1. The forest curve

When every map $S_i$ of the finite set $\mathcal{S}$ is a similitude, we say that $K$ is a *self-similar attractor*. The key theorem is given by Hutchinson.

**Theorem 1 (Hutchinson [8]).** *Let $\mathcal{S} = \{S_1, \ldots, S_N\}$ be a finite set of contraction maps of the plane. Then there exists a unique closed bounded set $K$ in the plane such that $K = \cup_{i=1}^{N} S_i(K)$. For arbitrary set $A$ in the plane, let $\mathcal{S}(A) = \cup_{i=1}^{N} S_i(A)$, $\mathcal{S}^p(A) = \mathcal{S}(\mathcal{S}^{p-1}(A))$. Then for any closed bounded set $A$, $\mathcal{S}^p(A)$ converges to the attractor $K$ in the Hausdorff metric.*

Barnsley [1] gives details of the proof of this result. There is one situation where the Hutchinson's theorem works very well. This situation, first conceived by Lévy [9], is described by Mandelbrot [10] and is thoroughly investigated by Tricot [12]. Given $P_0, P_1, \ldots, P_N$, $N+1$ points in the plane, we denote by $S_i$ the similitude of the plane which sends the segment $P_0 P_N$ onto the segment $P_{i-1} P_i$; we will suppose that all similarity factors, $r_i = ||P_{i-1}P_i||/||P_0P_N||$, are less than 1. Then the attractor with respect to the set of similitudes $\{S_1, \ldots, S_N\}$ is the limit curve of the sequence of polygonal lines $\mathcal{S}^p(P_0 P_N)$. In Mandelbrot's terminology, the polygonal line $P_0 P_1 \ldots P_N$ is the generator of the attractor. Tricot gives a natural parametrisation of this curve and says that the curve has a self-similar structure.

In a more general way, if we can find two polygonal lines, $P_0 P_1 \ldots P_N$ (the generator) and $Q_0 Q_1 \ldots Q_M$ (the initiator), it is possible to construct a curve as the limit of a sequence of polygonal lines: $Q^n$, $n = 0, 1, 2, \ldots$ where $Q^0$ is the initial line. We obtain $Q^{n+1}$ from $Q^n$ by replacing every segment of $Q^n$ by $N$ segments; if $A_j = [Q_{j-1}^n, Q_j^n]$ is the $j$-th segment to replace and $T_j$ is the simililitude of the plane which sends the segment $A_j$ to the segment $[P_0, P_N]$, then the $N$ replacement segments are $T_j^{-1}(S_i(T_j(A_j)))$ for $i = 1, 2, \ldots, N$.

## 3  Dimensions of a self-similar attractor

We first recall some definitions of dimension. Let $K$ be a compact set in the plane and let $\mathcal{N}(\epsilon)$ denote the minimum number of disks of radius $\epsilon$ needed to cover $K$. We define the *upper fractal dimension* of $K$ as $\limsup_{\epsilon \to 0} \ln \mathcal{N}(\epsilon)/\ln(1/\epsilon)$ and the *lower fractal dimension* of $K$ as $\liminf_{\epsilon \to 0} \ln \mathcal{N}(\epsilon)/\ln(1/\epsilon)$. We denote the lower fractal dimension by $\delta(K)$ and the upper fractal dimension by $\Delta(K)$. If $\delta(K) = \Delta(K)$, this number gives the *fractal dimension* of $K$. The fractal dimension was first studied by Bouligand [3] and is sometimes called the *box-counting dimension* [6, 12].

Let $\mathcal{S} = \{S_1, \ldots, S_N\}$ be a finite set of similitudes and $r_i$ the ratio of similarity of $S_i$. We suppose that $\forall i\ r_i < 1$. The *similarity exponent* of $\mathcal{S}$ is by definition the unique positive solution $x$ of the equation $\sum_{i=1}^{N} r_i^x = 1$. We will call this number by $e(\mathcal{S})$. If $K$ is the attractor associated with the finite set $\mathcal{S}$, we call the similarity exponent of $\mathcal{S}$ the *similarity dimension* of $K$. We may find this definition of similarity dimension in many references after being defined by Mandelbrot [10], (cf. [5, 8, 12]).

We present here the principal relation between the similarity dimension and the fractal dimension.

**Theorem 2 (Barnsley, p. 184 [1], adaptation).** *If $K$ is a self-similar attractor in the plane with respect to a finite set of contractive similitudes $\mathcal{S} = S_{i\,i=1}^{N}$, then $\Delta(K) \leq e(\mathcal{S})$.*

*Proof.* Barnsley [1], p. 184, states this result without proof and there is none in the references that he mentions. We prove it here. We set $e = e(\mathcal{S})$ and we distinguish two cases.

Case 1: All the scaling factors of similarity $r_i$ are positive.
Consider at first a disk $D$ of radius $R$ with center $C$ covering the attractor $K$. Given $\epsilon \in (0, R)$, we say a sequence $i_1, i_2, \ldots, i_k$ of index in $\{1, 2, \ldots, N\}$ is acceptable at level $\epsilon$ if the following double inequality is satisfied:

$$R \times r_{i_1} \times r_{i_2} \ldots \times r_{i_k} < \epsilon \leq R \times r_{i_1} \times r_{i_2} \ldots \times r_{i_{k-1}}. \tag{1}$$

Write $\mathcal{I}$ for the set of sequences of index acceptable at level $\epsilon$. The union of disks $S_{i_k} \circ \ldots S_{i_2} \circ S_{i_1}(D)$ with $i_1, i_2, \ldots, i_k \in \mathcal{I}$ will cover $K$. Let $\tau = \min\{r_i\}$; then we have $\tau > 0$. If $i_1, i_2, \ldots, i_k$ is acceptable, then from the right part of the double inequality (1) we can deduce that $\epsilon \tau \leq R \times r_{i_1} \ldots \times r_{i_k}$. Raising each member of the last inequality to the power $e$ and summing over $\mathcal{I}$, we obtain

$$\#\mathcal{I} \epsilon^e \tau^e \leq R^e \sum_{i_1 \ldots i_k \in \mathcal{I}} (r_{i_1} \ldots r_{i_k})^e.$$

It can be shown that the sum over $\mathcal{I}$ of the terms $(r_{i_1} \ldots r_{i_k})^e$ is 1, so the cardinality of $\mathcal{I}$ is bounded above by $(R/(\tau \epsilon))^e$. In addition, from the left part of the double inequality (1), we find that the union of the disks centered at $S_{i_k} \circ \ldots S_{i_2} \circ S_{i_1}(C)$ with radius $\epsilon$ and $i_1, i_2, \ldots, i_k \in \mathcal{I}$ will cover $K$. Thus the minimum number of disks of radius $\epsilon$ needed to cover $K$ will not be larger than $(R/(\tau\epsilon))^e$. The upper fractal dimension is then overestimated by $e$.

Case 2: One of the scaling factors of similarity $r_i$ is 0.

We consider the finite set $\mathcal{S}'$ of non-degenerated similitudes $S_i$ for which $r_i \neq 0$. If $K'$ is the compact attractor with respect to $\mathcal{S}'$ and $K"$ is the set of centers of the degenerated similitudes, we have $K = K' \cup K"$. Because $K"$ is a finite set, the upper fractal dimension will be the same as that of $K'$. By the first case, it is overestimated by $e$.

Since the Hausdorff dimension of a set is always overestimated by its upper fractal dimension, we obtain the following corollary.

**Corollary 1 (Edgar [5], (6.3.8)).** *Let $K$ be a self-similar attractor in the plane with respect to a finite set of contractive similitudes $\mathcal{S} = \{S_1, S_2, \ldots, S_N\}$. Then the Hausdorff dimension of $K$ is overestimated by $e(\mathcal{S})$.*

## 4 Examples of equality of two dimensions

In this section, we will describe some situations where there is equality between the fractal dimension of an attractor and its similarity exponent from a finite set of similitudes which may be the origin of this attractor. We give here two results on this subject.

**Theorem 3 (Barnsley, pp. 184-185 [1]).** *Let $K$ be a self-similar attractor in the plane with respect to a finite set of contractive similitudes $\mathcal{S} = \{S_1, S_2, \ldots, S_N\}$. If the sets $S_i(K)$ are totally disconnected or just-touching, then $\Delta(K) = e(\mathcal{S})$.*

**Theorem 4 (Tricot, pp. 190 [12]).** *Let $\Gamma$ be a self-similar simple curve defined by a finite set $\mathcal{S}$ of $N$ contractive similitudes of ratio $\rho_1, \ldots, \rho_N$. Then $\Delta(\Gamma) = \delta(\Gamma) = e(\mathcal{S})$.*

Despite our interest in the last theorem, excessive importance should not be atributed to the hypothesis that the attractor is a *simple* curve. To emphasize this, we remind you of a result due to Lévy.

**Theorem 5 (Lévy, p. 271 [9]).** *Let $C$ be a self-similar curve whose generator is a polygonal line with three vertices: $(0,0), (1/2, 1/2), (1,0)$. The plane Lebesgue measure of $C$ is $1/4$.*

The curve $C$ that we just mentioned is called the Lévy curve. Its graph is reproduced in Figure 2. Because the plane measure of the region occupied by $C$ is positive, its Hausdorff dimension is the same as the fractal dimension of $C$ and is equal to 2. However a very simple computation shows that the similarity dimension of $C$ is also 2. Thus these two dimensions are equal but the curve is far from being simple.

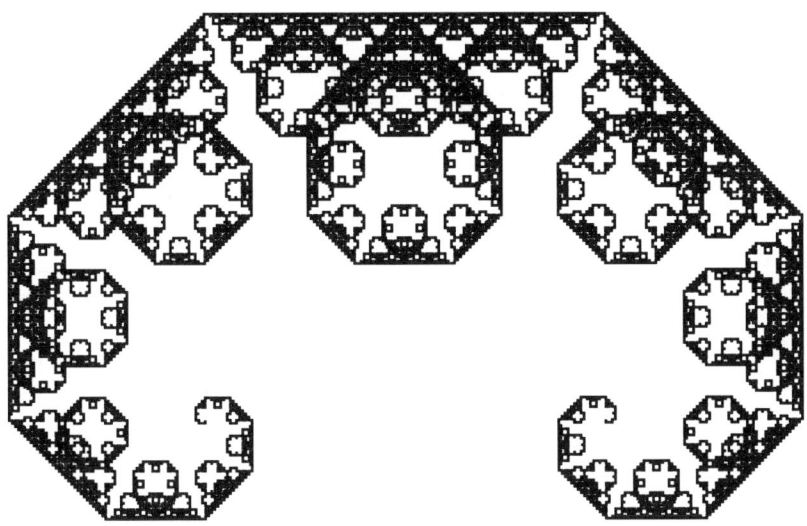

**Fig. 2.** Lévy curve

## 5 Inequality between two dimensions

We give some examples of attractors whose fractal dimension is less than their similarity dimension. We begin by two kinds of examples.

If the similarity exponent of a finite set $\mathcal{S}$ of contractive similitudes in the plane is larger than two, the fractal dimension of the attractor generated by $\mathcal{S}$, which can not exceed two, is evidently different from its similarity dimension.

Let $\mathcal{S}$ and $\mathcal{S}'$ be two distinct finite sets of contractive similitudes with the characterictic that any similitude of one system also occurs in the other system and that the number of times that a similitude appears in $\mathcal{S}$ is always less than or equal to the number of times it appears in $\mathcal{S}'$. The attractors associated with these two systems coincide and we may write $K$ for the common attractor. Moreover we have $e(\mathcal{S}) < e(\mathcal{S}')$. It is known that the fractal dimension of $K$ will never be larger than its similarity dimension given by $\mathcal{S}$. Thus the fractal dimension of $K$ will be different from the similarity dimension given by $\mathcal{S}'$. The example that we give here generalizes a remark of Edgar [5], p.159.

The next sections are concerned with another example of an attractor whose fractal dimension is less than its similarity dimension.

## 6 The forest curve and its associated graph

We consider the self-similar curve whose initiator is $\mathcal{P}^0 = \{(0,0),(1,0)\}$ and whose generator is the polygonal line $\mathcal{P}^1 = \{(0,0),(1/2,0),(1,0),(1,1/2),(1,0)\}$. The forest curve can be obtained as the limit of a sequence of polygonal lines

$\mathcal{P}^n, n = 0, 1, 2, \ldots$ where $\mathcal{P}^n$ is transformed to $\mathcal{P}^{n+1}$ as $\mathcal{P}^0$ is transformed to $\mathcal{P}^1$. This curve is the attractor of a finite set of four similitudes: $S_1(x,y) = (x/2, y/2), S_2(x,y) = (x/2 + 1/2, y/2), S_3(x,y) = (-y/2 + 1, x/2), S_4(x,y) = (y/2 + 1, -x/2 + 1/2)$.

It is obvious that the similarity dimension of $\{S_1, S_2, S_3, S_4\}$ is 2. Looking at Fig. 1, we might be confident that the fractal dimension of the forest curve is less than 2. However a visual estimate is not a proof. For such a proof, some mathematical machinery is needed; it will be developed in the next 3 sections. We invite the reader's interest in this curve because its fractal dimension will be evaluated using graph theory, which is an uncommon tool in fractal geometry.

It is convenient to give special attention to the sequence of polygonal lines $\mathcal{Q}^n$ with the initiator $\mathcal{Q}^0 = \{(0,0), (1,0), (0,0)\}$ and no change of generator. Figure 3 gives the first 4 polygonal lines. This sequence of polygonal lines converges to a curve $\Gamma'$ that can be easily deduced from $\Gamma$: $\Gamma' = \Gamma \cup S(\Gamma)$ where $S(x,y) = (1-x, -y)$. It is clear that the fractal dimension of $\Gamma$ is the same as that of $\Gamma'$. All following computations will be based on the sequence $\mathcal{Q}^n$.

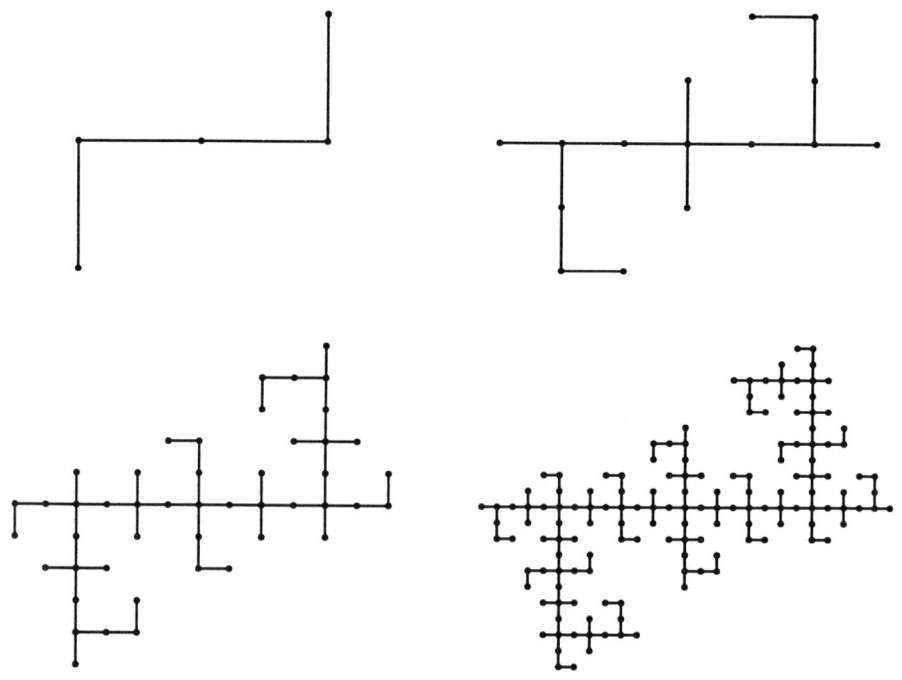

Fig. 3. The first 4 polygonal lines approaching $\Gamma'$.

For the evaluation of the fractal dimension of $\Gamma'$, we need the help of a sequence of graphs $G_n$. The graph $G_n$ can be obtained by the intermediary of the polygonal line $\mathcal{Q}^n$. For terminology related to the theory of graphs, we will use Berge [2] and Harary [7]. The vertices of $G_n$ are the same vertices as of $Q^n$; the edges of $G_n$ are the pairs of consecutive vertices of the polygonal line $Q^n$. However we will not consider repetitions or orientation. Every edge corresponds to a horizontal or vertical segment of length $1/2^n$ that joins its two vertices.

We first note that every graph $G_n$ is connected. We verify some other properties of these graphs.

**Lemma 1.** *Let P and P' be two vertices of $G_n$ with distance $1/2^n$. Then $\{P,P'\}$ is an edge of $G_n$.*

*Proof.* We proceed by induction on $n$. The lemma is trivial for $n = 0$. We suppose that the lemma has been already verified for a given integer $n$. Consider two vertices $P$ and $P'$ of $G_{n+1}$ such that $||PP'|| = 1/2^{n+1}$. The distance between two distinct points of $G_n$ is greater than or equal to $1/2^n$, so $P$ or $P'$ is not in $G_n$. Without loss of generality, we can suppose that $P \notin G_n$. We distinguish two cases.

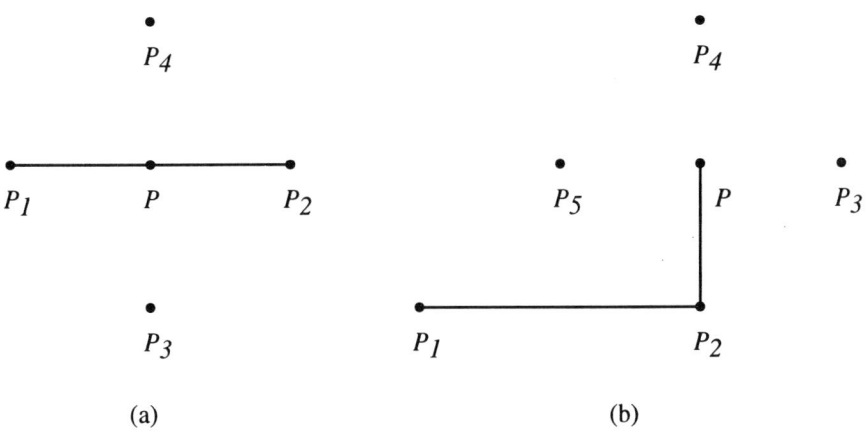

**Fig. 4.** Two cases for the point $P$

Case 1: $P$ is the midpoint of an edge of $G_n$. If $\{P_1, P_2\}$ is an edge of $G_n$ with $P$ as its midpoint, denote by $P_3$ and $P_4$ the two points such that $P_1P_3P_2P_4$ is a square with diagonal $P_1P_2$ (see Fig. 4a). $P'$ is necessarily one of the points $P_1, P_2, P_3, P_4$. By checking the construction, it is impossible that $P_3$ or $P_4$ belongs to $G_{n+1}$. (In fact, if $(x, y) \in G_{n+1}$, then one of the two numbers $x2^{n+1}, y2^{n+1}$ is an even number.) Therefore $\{P, P'\}$ must be an edge of $G_{n+1}$.

Case 2: $P$ is not the midpoint of an edge of $G_n$. There exists an edge $\{P_1, P_2\}$ of $G_n$ such that $P_1P_2P$ is a right triangle (right angle at $P_2$) and $||PP_2|| = 1/2^{n+1}$. $P_2$ and three other points $P_3, P_4, P_5$ give four horizontal or vertical segments of length $1/2^{n+1}$ when they are connected to $P$ (see Fig. 4b). In fact, $P'$ is one of these four points $P_2, P_3, P_4, P_5$. From what we have seen, it is impossible that $P_3$ or $P_5$ belongs to $G_{n+1}$. We may also exclude the case that $P' = P_4$ because this would make $P_2$ and $P_4$ two vertices of $G_n$ at a distance of $1/2^n$. By induction, we would have that $\{P_2, P_4\}$ is an edge which contradicts the hypothesis that $P$ is not the midpoint of any edge of $G_n$. So $P' = P_2$ and $\{P, P'\}$ is an edge of $G_{n+1}$.

**Theorem 6.** *For every integer $n$, the graph $G_n$ has no cycle.*

*Proof.* We proceed by induction on $n$. The theorem is trivial for $n = 0$; $G_0$ with only one edge is necessarily without a cycle. Suppose that for an integer $n$, the graph $G_n$ has no cycle. We reason by the absurd by assuming that the graph $G_{n+1}$ has at least one elementary cycle, $[P_0, P_1, \ldots, P_q]$. We notice that two consecutive vertices $P_i, P_{i+1}$ of the cycle cannot belong to $G_n$ at the same time. (This comes from the fact that if $P, P'$ are two distinct vertices of $G_n$, then $||PP'|| \geq 1/2^n$.) Moreover if $P_i, P_{i+1}$ are two consecutive vertices of the cycle, only one can belong to $G_n$. Without loss of generality, we suppose that $P_0$ is a vertex of the graph $G_n$; otherwise one makes a circular shift within the cycle to produce this property. $q$ is necessarily even and the points of the cycle of even rank $P_0, P_2, \ldots, P_q$ are all in $G_n$. Since $||P_i - P_{i+2}|| \leq 1/2^n$ and $P_i \neq P_{i+2}$ (the cycle is elementary), $||P_i - P_{i+2}|| = 1/2^n$. According to Lemma 1, $P_i P_{i+2}$ is an edge of graph $G_n$. $[P_0, P_2, \ldots, P_q]$ would then be a cycle in $G_n$ which contradicts the induction hypothesis. So the theorem is proved.

We distinguish 5 types of vertices:
1. a vertex which belongs to only one edge;
2. a vertex which belongs to exactly two edges where one of these is a prolongation of the other;
3. a vertex which belongs to exactly two edges where these two edges are perpendicular;
4. a vertex which belongs to exactly three edges;
5. a vertex which belongs to exactly four edges.

We introduce 5 sequences of integers: $a_n, b_n, c_n, d_n, e_n$; they are the numbers of vertices from $G_n$ which are respectively of type 1, 2, 3, 4, 5. Obviously, $a_0 = 2, b_0 = c_0 = d_0 = e_0 = 0$. The recurrence formulas for these sequences are as follows.

**Proposition 1.** *The numbers of vertices of various types in the graph $G_{n+1}$ can be obtained from those of $G_n$ by the following formulas*

$$a_{n+1} = a_n + 2b_n + c_n + d_n \quad (1)$$
$$b_{n+1} = a_n + b_n + c_n + d_n + e_n - 1 \quad (2)$$
$$c_{n+1} = a_n \quad (3)$$
$$d_{n+1} = c_n \quad (4)$$
$$e_{n+1} = b_n + d_n + e_n \quad (5)$$

*Proof.* For the proof, it is convenient to consult Figure 5 which illustrates how the edges around a vertex of a certain type change from one generation to the next. Consider a vertex $P$ in the graph $G_n$. In the figure, we indicate this point by a small circle. There are 5 possible types enumerated in the upper part of the figure. We must discuss how the type of $P$ will be transformed in the graph $G_{n+1}$ of the next generation. Moreover, we have to determine the type of the new vertices of $G_{n+1}$ in a neighbourhood of $P$. These are the vertices $P'$ of $G_{n+1}$ such that $\{P, P'\}$ is an edge of $G_{n+1}$; the new vertices are indicated by a black dot. The transformations of the various types of old vertices and the types of the new vertices in the graph $G_{n+1}$ are shown in the lower part of the figure. From 1 and Figure 5, it follows that:

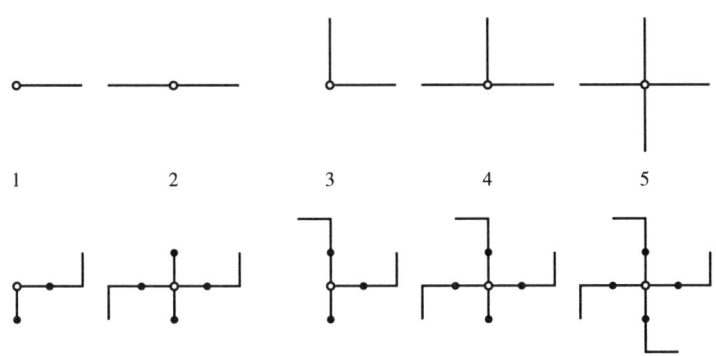

Fig. 5. The various types of vertices with transformations of the edges that contain them

If $P$ is of type 1 in the graph $G_n$, $P$ is of type 3 in the graph $G_{n+1}$ and produces a new vertex of type 1 and a new vertex of type 2. If $P$ is of type 2 in the graph $G_n$, $P$ is of type 5 in the graph $G_{n+1}$ and produces two new vertices of type 1 and two other new vertices of type 2. If $P$ is of type 3 in the graph $G_n$, $P$ is of type 4 in the graph $G_{n+1}$ and produces one new vertex of type 1 and two new vertices of type 2. If $P$ is of type 4 or 5 in the graph $G_n$, $P$ is of type 5 in the graph $G_{n+1}$ and produces four new vertices of type 2.

A vertex of type 2 in the $n$-th generation will generate two vertices of type 1 in the next generation; a vertex of type 1, 3 or 5 will each produce a vertex of type 1, but a vertex of type 4 makes no contribution to the vertices of type 1. Thus we have Formula (1).

$b_{n+1}$ is also equal to the number of edges of the graph $G_n$. Theorem 6 states that the graph $G_n$ has no cycle, and it is well known that in a connected nonempty graph without cycle, the number of edges is equal to the number of vertices decreased by 1 (cf Berge, p. 22, [2] or Harary, p. 32-33, [7]). Since the number of vertices of $G_n$ is $a_n + b_n + c_n + d_n + e_n$, Formula (2) follows.

The vertices of $G_n$ which become type 3 vertices in the next generation, are only those of type 1. This gives Formula (3).

The vertices of $G_n$ which become type 4 vertices in the next generation, are only those of type 3. This gives Formula (4).

The vertices of $G_n$ which become vertices of type 5 in the next generation, are those of type 2, 4 or 5. This gives Formula (5).

## 7 Asymptotic behaviour of the order of the graphs

Let us denote by $t_n$ the number of vertices of $G_n$. In the terminology of graph theory, the number of vertices of a graph $G$ is *the order of $G$*. To find the fractal dimension of $\Gamma'$, an analysis of the asymptotic behaviour of $t_n$ is essential. That is what we examine now.

**Theorem 7.** *Let $A$ be a square matrix of order $m$ and $B$ and $V_0$ two vectors in $\mathbb{R}^m$. We consider the sequence of vectors $V_n$ obtained by the recurrence relation*

$$V_n = A \cdot V_{n-1} + B.$$

*We suppose that $A$ admits $m$ distinct eigenvalues $\{\lambda_i\}$ such that $|\lambda_1| \leq |\lambda_2| \leq \cdots \leq |\lambda_{m-1}| \leq 1 < |\lambda_m|$.*

*Suppose that $X_m$ is a right eigenvector such that $AX_m = \lambda_m X_m$ and $U_m$ a right eigenvector such that $U_m A = \lambda_m U_m$, normalized such that $U_m X_m = 1$. Then as $n$ goes to infinity, $V_n = c\lambda_m^n X_m + O(n)$ where $c = U_m(V_0 + B/(\lambda_m - 1))$.*

*Proof.* Since the eigenvectors of $A$ are distinct, $A$ is similar to a diagonal matrix, i.e. there exists a matrix $P$ with order $m$ such that

$$A = P \cdot \begin{pmatrix} \lambda_1 & 0 & \cdots & 0 & 0 \\ 0 & \lambda_2 & \cdots & 0 & 0 \\ \cdots & \cdots & \cdots & \cdots & \cdots \\ 0 & 0 & \cdots & \lambda_{m-1} & 0 \\ 0 & 0 & \cdots & 0 & \lambda_m \end{pmatrix} \cdot P^{-1}.$$

where $P = (X_1, X_2, \cdots, X_m)$ and $X_i$ is the eigenvector corresponding to $\lambda_i$.

By the recurrence, we have
$V_n = A \cdot V_{n-1} + B = A^n \cdot V_0 + \sum_{i=0}^{n-1} A^i \cdot B$

$$= P \cdot \begin{pmatrix} \lambda_1^n & 0 & \cdots & 0 & 0 \\ 0 & \lambda_2^n & \cdots & 0 & 0 \\ \cdots & \cdots & \cdots & \cdots & \cdots \\ 0 & 0 & \cdots & \lambda_{m-1}^n & 0 \\ 0 & 0 & \cdots & 0 & \lambda_m^n \end{pmatrix} \cdot P^{-1} \cdot V_0 + P \cdot \begin{pmatrix} \frac{1-\lambda_1^n}{1-\lambda_1} & 0 & \cdots & 0 & 0 \\ 0 & \frac{1-\lambda_2^n}{1-\lambda_2} & \cdots & 0 & 0 \\ \cdots & \cdots & \cdots & \cdots & \cdots \\ 0 & 0 & \cdots & \frac{1-\lambda_{m-1}^n}{1-\lambda_{m-1}} & 0 \\ 0 & 0 & \cdots & 0 & \frac{1-\lambda_m^n}{1-\lambda_m} \end{pmatrix} \cdot$$

$P^{-1} \cdot B$.

(If $\lambda_i = 1$, we have to replace $\frac{1-\lambda_i^n}{1-\lambda_i}$ by $n$).

$$= \lambda_m^n \cdot P \cdot \begin{pmatrix} 0 & \cdots & 0 & 0 \\ \cdots & \cdots & \cdots & \cdots \\ 0 & \cdots & 0 & 0 \\ 0 & \cdots & 0 & 1 \end{pmatrix} \cdot P^{-1} \cdot (V_0 + \tfrac{1}{\lambda_m - 1} \cdot B) + E_n V_0 + E'_n B$$

where $E_n$ and $E'_n$ are two sequences of matrices whose coefficients are uniformly bounded in $n$ or are at most $O(n)$.

We have seen that the matrix $P$ may be obtained by placing the left eigenvectors $X_1 \ldots, X_m$ one by one. In addition, we can verify that the $i$-th row of the matrix $P^{-1}$ is a left eigenvector of $A$ corresponding to the eigenvalue $\lambda_i$, and that this vector $U_i$ is normalized such that $U_i X_i = 1$. A simple calculation shows that the matrix $P \cdot \begin{pmatrix} 0 & \cdots & 0 & 0 \\ \cdots & \cdots & \cdots & \cdots \\ 0 & \cdots & 0 & 0 \\ 0 & \cdots & 0 & 1 \end{pmatrix} \cdot P^{-1}$ is exactly the matrix $X_m U_m$.

This yields the result of the theorem.

**Proposition 2.** $t_n$ has the following asymptotic behavior: There exists a $c \neq 0$ such that

$$t_n \sim c \cdot 3.152757602^n \tag{6}$$

*Proof.* If we put $V_n = (a_n, b_n, c_n, d_n, e_n)^T$, we can write the formulas of the Proposition 1 in a matrix form

$$V_n = A \cdot V_{n-1} + B$$

where

$$A = \begin{pmatrix} 1 & 2 & 1 & 1 & 0 \\ 1 & 1 & 1 & 1 & 1 \\ 1 & 0 & 0 & 0 & 0 \\ 0 & 0 & 1 & 0 & 0 \\ 0 & 1 & 0 & 1 & 1 \end{pmatrix}, B = \begin{pmatrix} 0 \\ -1 \\ 0 \\ 0 \\ 0 \end{pmatrix}.$$

By using Maple, we find that $A$ has five distinct eigenvalues: $\lambda_1 = 0$; $\lambda_2 = -0.5763788013 + 0.549684246551i$; $\lambda_3 = -0.5763788013 - 0.549684246551i$; $\lambda_4 = 1$; $\lambda_5 = 3.152757602$. Thus the condition on eigenvalues of Theorem 7 is satisfied.

On the other hand, if we denote the eigenvector corresponding to the eigenvalue $\lambda_5$ by $X = (x_1, x_2, x_3, x_4, x_5)^T$, we can always normalize $X$ such as $\sum_{i=1}^{5} x_i = 1$. (By Frobenius' Theorem, all entries of $X$ are positive.) By Theorem 7, there exists a constant $c$ such that every entry of $V_n$ is asymptotic to $c\lambda_5^n$ times the corresponding entry of $X$ when $n$ goes to infinity. Thus we have $t_n \sim c\lambda_5^n$. To verify that $t_n$ grows at least exponentially, it is enough to observe that $t_n > 2^n$, because all points $(i/2^n, 0), i = 0, 1, \ldots, 2^n$ are vertices of $G_n$.

# 8 Computation of the forest curve's fractal dimension

Everything is in place for the computation of the fractal dimension of $\Gamma$.

**Theorem 8.** *The fractal dimension of $\Gamma'$ is* $1.656614253$.

*Proof.* Denote by $\omega_n$ the number of elementary squares of length $1/2^n$ determined by the grid of points $(i/2^n, j/2^n), i \in \mathbb{Z}, j \in \mathbb{Z}$, and which intersect $\Gamma'$. The union of all squares centered on a vertex of $G_n$ and of edge $2^{1-n}$ will cover $\Gamma'$. So we have $\omega_n \leq 4t_n$. On the other hand, it is clear that $t_n \leq \omega_n$. Thus, $t_n$ and $\omega_n$ have the same order of magnitude and either can be used for the computation of the fractal dimension of $\Gamma'$ (the asymptotic behaviour of $\omega_n$ determines the dimension of $\Gamma'$):

$$\Delta(\Gamma') = \lim_{n \to \infty} \frac{\log \omega_n}{n \log 2}$$

(cf. Tricot, p.125 [12]).

Finally, by using Proposition 2, we have

$$\Delta(\Gamma') = \lim_{n \to \infty} \frac{\log t_n}{n \log 2} = \frac{\log 3.152757602}{\log 2} = 1.656614253.$$

**Remark:** We have already indicated that the fractal dimension of $\Gamma$ is that of $\Gamma'$. Since the similarity dimension is 2, we have that $\Delta(\Gamma) < e(\Gamma)$.

## 9 Conclusion

We have shown that the fractal dimension of an attractor of a finite set $S$ of similitudes is always overestimated by the similarity exponent of $S$. We hoped to characterize the situations where the fractal dimension coincides with the similarity dimension, or where these two dimensions differ. For the time being, it does not seem possible to give a general explanation of the difference between these two dimensions for fractal curves.

The situation is a bit different for self-similar sets of the line, as can be seen in a recent paper of Simon and Solomyak [11]. In this work they compare the box-counting dimension and the similarity dimension for Cantor sets of overlapping construction on $\mathbb{R}$. They are able to explain why the two dimensions differ in two specific families of such Cantor sets.

The following interesting remark was made by one of the referees. If $S$ is a contractive similitude whose center is the origin, then the system of similitudes $\{S, -S\}$ has the origin as an attractor; the similarity dimension is positive but the fractal dimension of the attractor is 0. The only drawback of this example is that the attractor is not a curve.

We finish with a question. If a self-similar curve is obtained with the help of a polygonal line with the property that at least two of its segments meet in more than one point, is it true that the curve's fractal dimension is always less than its similarity dimension?

## References

1. M. F. Barnsley (1988): *Fractals Everywhere*. Academic Press, San Diego.

2. C. Berge (1985): *Graphs.* North Holland, Amsterdam, 2nd ed..
3. G. Bouligand (1928): Ensembles impropres et nombre dimensionnel. *Bull. Sc. math.* **52**, 320–334.
4. S. Dubuc (1982): *Une foire de courbes sans tangentes.* In Actualités Mathématiques. Actes du VI ième Congrès du Groupement des Mathématiciens d'Expression Latine, pp. 99–123. Gauthier-Villars, Paris.
5. G. A. Edgar (1990): *Measure, Topology, and Fractal Geometry.* Springer-Verlag, New York.
6. K. Falconer (1990): *Fractal Geometry - Mathematical Foundations and Applications.* John Wiley, New York.
7. F. Harary (1969): *Graph Theory.* Addison-Wesley, Reading.
8. J. E. Hutchinson (1981): Fractals and self-similarity. *Indiana Univ. Math. J.* **30**, 713–747.
9. P. Lévy (1938): Les courbes planes ou gauches et les surfaces composées de parties semblables au tout. *J. Ecole Poly., Série III* **7-8**, 227–292. Translated in G. A. Edgar, *Classics on Fractals.* Selection 12. Addison-Wesley, Menlo Park, 1993.
10. B. B. Mandelbrot (1983): *The Fractal Geometry of Nature.* W.H. Freeman, San Francisco.
11. K. Simon and B. Solomyak (1998): Correlation dimension for self-similar Cantor sets with overlaps. *Fund. Math.* **155** 293-300.
12. C. Tricot (1995): *Curves and Fractal Dimension.* Springer-Verlag , Paris.

# FRACTIONAL CALCULUS

# Vector Analysis on Fractal Curves

Massimiliano Giona

Dipartimento di Ingegneria Chimica, Universitá di Roma "La Sapienza",
via Eudossiana 18, 00184 Roma - ITALY
max@giona.ing.uniroma1.it

**Abstract.** This article introduces a constructive definition of contour integrals over fractal curves in the plane by making use of the notion of oriented Iterated Function Systems and directional pseudo-measures. An expression for the contour integral of continuous functions over fractal interfaces is obtained through renormalization. As a result, a vector calculus on fractal interfaces which are boundaries of regular two-dimensional domains is developed by extending Green's theorem in the plane also to fractal curves.

## 1 Introduction

There are many important physical implications of the non-rectifiability of fractal interfaces [1]. Dynamic phenomena evolving on and across fractal interfaces possess different qualitative (scaling) properties from the corresponding properties evolving in the presence of regular Euclidean interfaces, as demostrated by the electric (impedance scaling with frequency) and transport properties of fractal electrodes [2, 3], by the anomalous behaviour in adsorption kinetics [4, 5], and by the eigenvalue properties of the Laplace equation on a Euclidean domain bounded by closed fractal curves (fractal-drum problem) [6–8]. Many important results have been derived on the spectral properties of the Laplace equation in the presence of fractal boundaries. A general mathematical physics on fractal curves, encompassing the generalization of standard vector calculus (e.g. Stokes, Gauss, Green theorems), has not yet been developed, however, mostly because of the impossibility of establishing a local (i.e. pointwise) definition of the notion of tangent and normal vectors. It is in fact only recently that some preliminary but interesting results have been obtained on the existence and unicity of the solution of the Dirichlet problem in the presence of fractal boundaries [9–11].

This article shows that the iterative construction of fractal curves can be used to obtain a definition of the notion of curvilinear integrals, generalizing for fractal curves the notion of the curvilinear integral of differential forms (1-forms),

$$\int_{+\Gamma} M(x,y)dx + N(x,y)dy , \qquad (1)$$

of the standard vector calculus on rectifiable curves.

The starting points of this analysis are 1) the concepts of *Oriented* Iterated Function Systems (IFSO) generating fractal curves, and 2) the definition of the resulting *directional pseudo-measures* associated with an IFSO, which are introduced in order to obtain a definition of the notion of contour integrals of differential forms over fractal curves.

A renormalization approach is applied to derive recursive relations for contour integrals involving algebraic functions (moment hierarchies), which make it possible to obtain analytic expressions for the contour integrals of any continuous function that can be expressed in the form of a power series.

## 2 IFS theory: a brief summary

IFS theory provides a simple and self-consistent mathematical framework for the generation of fractal sets and singular (multifractal) measures, and for the quantitative study of their properties [12, 13].

An IFS with probability (referred for short to as IFSP) is a system of $N$ contractive transformations $\{\mathbf{w}_h\}_{h=1}^N$ of a complete metric space $(X,d)$ onto itself ($X$ is the base space, $d$ the distance function), each of which is equipped with a probability weight $p_h > 0$, $\sum_{h=1}^N p_h = 1$. The limit set $\mathcal{C}$ of the IFS satisfies the so-called Hutchinson equation, $\mathcal{C} = \cup_{h=1}^N \mathbf{w}_h(\mathcal{C})$.

In the space of probability measures on the $\sigma$-algebra of the Borel subset of $X$, the Markov operator associated with the IFSP can be defined as $M[\mu] = \sum_{h=1}^N p_h \mu \circ \mathbf{w}_h^{-1}$, where $\circ$ indicates composition and $\mu$ is a probability measure. A basic result of IFSP theory states that there exists a unique ergodic invariant measure $\mu^*$ which is a fixed point of the Markov operator i.e. $M[\mu^*] = \mu^*$. Moreover, for any continuous function $f(\mathbf{x})$, the following identity holds:

$$\int_{\mathcal{C}} f(\mathbf{x}) d\mu^*(\mathbf{x}) = \sum_{h=1}^N p_h \int_{\mathcal{C}} f(\mathbf{w}_h(\mathbf{x})) d\mu^*(\mathbf{x}) \ . \qquad (2)$$

Throughout this article we consider exclusively IFS generating self-similar fractal curves on the plane, i.e. such that the maps $\mathbf{w}_h(\mathbf{x})$ are linear and can be expressed as

$$\mathbf{w}_h(\mathbf{x}) = s\mathbf{T}_{\theta_h}\mathbf{x} + \mathbf{b}_h \ , \qquad (3)$$

where $0 < s < 1$ is a scaling factor, which is unique for all the maps of the IFS, $\mathbf{T}_{\theta_h}$ is the rotation matrix of angle $\theta_h$,

$$\mathbf{T}_{\theta_h} = \begin{pmatrix} \cos\theta_h & -\sin\theta_h \\ \sin\theta_h & \cos\theta_h \end{pmatrix} \ , \qquad (4)$$

$\mathbf{x} = (x,y)^t$ and $\mathbf{b}_h = (b_{x,h}, b_{y,h})^t$ (the superscript $t$ indicates transpose), which is a particular case of a linear IFS.

As a model structure, the Koch curve depicted in Fig. 1 is analyzed throughout this article. This self-similar fractal curve is generated by means of an 8-map IFS of the form Eq. (3), with $s = 1/4$, $\theta_h = 0$, $h = 1, .., 4$, $\theta_5 = \theta_8 = \pi/2$, $\theta_6 = \theta_7 = -\pi/2$, and $\mathbf{b}_1 = (0,0)^t$, $\mathbf{b}_2 = (1/4, 1/4)^t$, $\mathbf{b}_3 = (1/2, -1/4)^t$, $\mathbf{b}_4 = (3/4, 0)^t$,

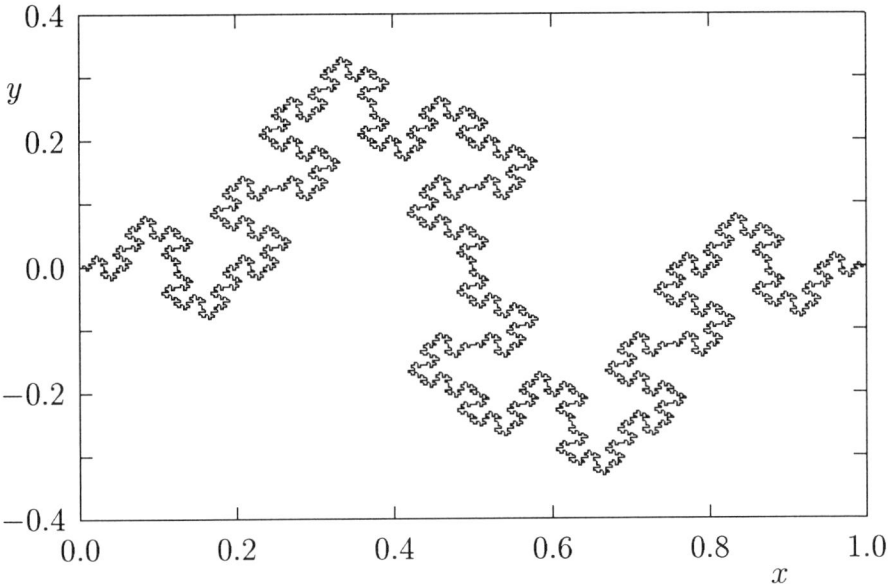

**Fig. 1.** Koch fractal curve obtained by means of an 8-map IFS.

$\mathbf{b}_5 = (1/4, 0)^t$, $\mathbf{b}_6 = (1/2, 1/4)^t$, $\mathbf{b}_7 = (1/2, 0)^t$, $\mathbf{b}_8 = (3/4, -1/4)^t$. To simplify the notation, this IFS is indicated as

$$\mathbf{w}_h(\mathbf{x}) = s\mathbf{x} + \mathbf{b}_h \,, \ h = 1,..,4 \,, \quad \mathbf{w}_h(\mathbf{x}) = s \begin{pmatrix} 0 & c_h \\ -c_h & 0 \end{pmatrix} \mathbf{x} + \mathbf{b}_h \,, \ h = 5,..,8 \,,$$

where $c_5 = c_8 = -1$, $c_6 = c_7 = 1$. The fractal (Hausdorff) dimension of the resulting curve $\mathcal{C}$ is $d_H = \log 8 / \log 4 = 3/2$.

By changing the translation vectors, and by considering four different IFS, it is possible to generate a closed fractal curve, the union of the limit sets of the four IFS, which is the boundary of a two-dimensional compact set $\mathcal{D}$, Fig. 2. The translation vectors for the first IFS have been given above, while those for the other three IFS are as follows: (IFS 2) $\mathbf{b}_1 = (3/4, 0)^t$, $\mathbf{b}_2 = (1/2, 1/4)^t$, $\mathbf{b}_3 = (1, 1/2)^t$, $\mathbf{b}_4 = (3/4, 3/4)^t$, $\mathbf{b}_5 = (1, 0)^t$, $\mathbf{b}_6 = (3/4, 3/4)^t$, $\mathbf{b}_7 = (1, 3/4)^t$, $\mathbf{b}_8 = (5/4, 1/2)^t$; (IFS 3) $\mathbf{b}_1 = (0, 3/4)^t$, $\mathbf{b}_2 = (1/4, 1)^t$, $\mathbf{b}_3 = (1/2, 1/2)^t$, $\mathbf{b}_4 = (3/4, 3/4)^t$, $\mathbf{b}_5 = (1/2, 1)^t$, $\mathbf{b}_6 = (1/4, 5/4)^t$, $\mathbf{b}_7 = (1/4, 1)^t$, $\mathbf{b}_8 = (1, 3/4)^t$; (IFS 4) $\mathbf{b}_1 = (0, 0)^t$, $\mathbf{b}_2 = (-1/4, 1/4)^t$, $\mathbf{b}_3 = (1/4, 1/2)^t$, $\mathbf{b}_4 = (0, 3/4)^t$, $\mathbf{b}_5 = (0, 1/4)^t$, $\mathbf{b}_6 = (-1/4, 1/2)^t$, $\mathbf{b}_7 = (0, 1/2)^t$, $\mathbf{b}_8 = (1/4, 3/4)^t$. The values of the parameters characterizing these IFS are given in full as this structure is extensively used in the numeric exemplification of the theory.

The two-dimensional compact set $\mathcal{D}$, the boundary of which is given by the four Koch curves described above (see Fig. 2), can be obtained by means of a 16-map IFS of the form $\mathbf{w}_h = s\mathbf{x} + \mathbf{b}_h$, with $\mathbf{b}_h$ given by: $(1/2, -1/4)^t$, $(0, 0)^t$, $(1/2, 0)^t$, $(3/4, 0)^t$, $(-1/4, 1/4)^t$, $(0, 1/4)^t$, $(1/4, 1/4)^t$, $(1/2, 1/4)^t$, $(1/4, 1/2)^t$,

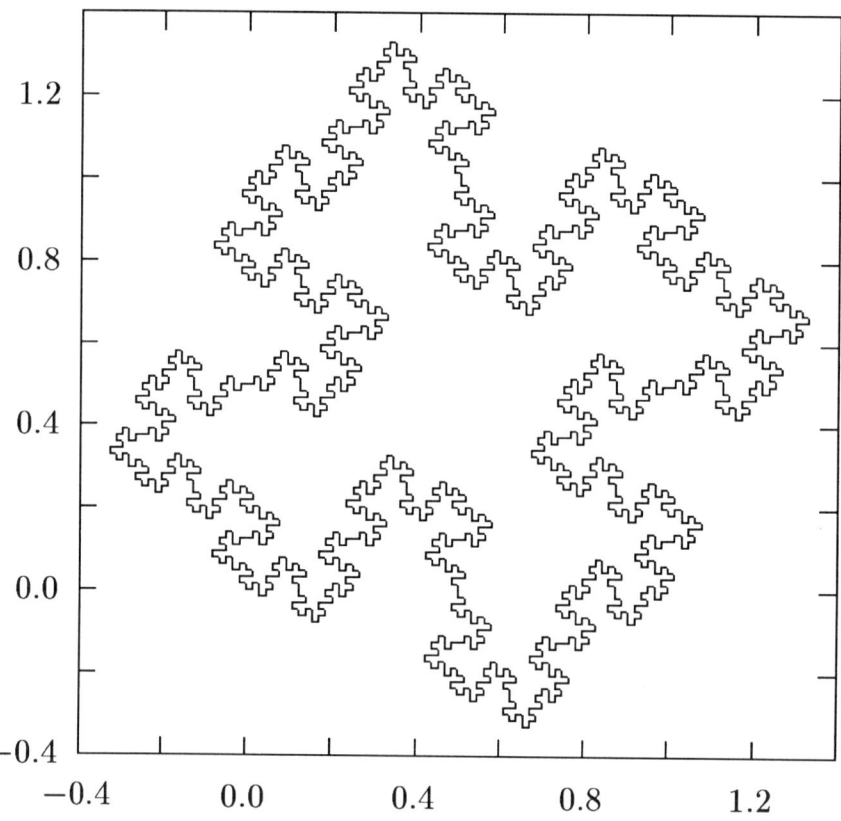

**Fig. 2.** Compact set $\mathcal{D}$, the boundary of which is given by 4 translated copies of the curve depicted in Fig. 1.

$(1/2, 1/2)^t$, $(3/4, 1/2)^t$, $(1, 1/2)^t$, $(0, 3/4)^t$, $(1/4, 3/4)^t$, $(3/4, 3/4)^t$, $(1/4, 1)^t$. For $p_h = 1/N = 1/16$, the resulting invariant measure $\mu^*$ on $\mathcal{D}$ is uniform, i.e. $d\mu^*(\mathbf{x}) = d\mathbf{x} = dx\,dy$.

## 3 Oriented IFS and directional pseudo-measures

This section provides an introduction of the basic notions that will be used for the definition of contour integrals over fractal curves, and in particular the concept of *oriented IFS* and *directional pseudo-measures*. As discussed in section 2, we focus exclusively on those fractal curves $\mathcal{C}$ which can be generated by means of similar IFS, for which each map takes the form $\mathbf{w}_h(\mathbf{x}) = s\mathbf{T}_{\theta_h}\mathbf{x} + \mathbf{b}_h$.

Let $\{\mathbf{w}_h(\mathbf{x})\}_{h=1}^N$ be an affine IFS of the above-mentioned form and $\{\mathbf{d}_h\}_{h=1}^N$ a set of vectors. The system $\{\mathbf{w}_h(\mathbf{x}), \mathbf{d}_h\}_{h=1}^N$ is called an oriented IFS or IFSO (i.e. an IFS with Orientation). The choice of the vectors $\mathbf{d}_h$ specifying an IFSO follows from the iterative generation process of the fractal interface, which is equivalent to its IFS construction. To clarify this point, let us consider the model fractal curve introduced in section 2. Let $\mathbf{d}^{(0)}$ by the vector representing the zero-th order approximation of the fractal curve, $\mathbf{d}^{(0)} = (1,0)^t$. Given $\mathbf{d}^{(0)}$, the orientation vectors $\{\mathbf{d}_h\}_{h=1}^N$ can be defined as

$$\mathbf{d}_h = \mathbf{T}_{\theta_h}[\mathbf{d}^{(0)}]\,,\quad h = 1, .., N\,,\qquad(5)$$

where $\mathbf{T}_{\theta_h}$ are the rotation matrices characterizing each map $\mathbf{w}_h$. In the case of the 8-map Koch curve (Fig. 1), it follows that $\mathbf{d}_h = (1,0)^t$ for $h = 1, .., 4$, $\mathbf{d}_5 = \mathbf{d}_8 = (0,1)^t$, $\mathbf{d}_6 = \mathbf{d}_7 = (0,-1)^t$.

At iteration $n = 0$, the $n$-th approximant of the curve is given by a single segment, the orientation of which is specified by the vector $\mathbf{d}^{(0)}$. At iteration $n = 1$, the approximant $\mathcal{C}^{(1)}$ is given by $N$ segments of size $s\|\mathbf{d}^{(0)}\|$, and the distribution of orientations is given by $\mathbf{d}_h^{(1)} = \mathbf{T}_{\theta_h}[\mathbf{d}^{(0)}]$, $h = 1, .., N$.

In order to follow the natural ordering along the curve, a table of correspondance should be defined for the $N$ map forming the IFS. This can be achieved by defining an ordering vector $\mathbf{q}$, the $i$-th entry of which, $q(i)$, is the label of the IFS map attaining the $i$-th position in the ordering along the curve. If the map is already ordered, it obviously follows that $q(i) = i$. To give an example, in the case of the fractal curve of Fig. 1, it follows that $q(1) = 1$, $q(2) = 5$, $q(3) = 2$, $q(4) = 6$, $q(5) = 7$, $q(6) = 3$, $q(7) = 8$, $q(8) = 4$.

In this way, the ordered orientations $\overline{\mathbf{d}}_h^{(n)}$ can be defined. For $n = 0$ $\overline{\mathbf{d}}^{(0)} = \mathbf{d}^{(0)}$, and for $n = 1$ $\overline{\mathbf{d}}_h^{(1)} = \mathbf{d}_{q(h)}^{(1)}$. In general, given the ordered set of orientations $\{\overline{\mathbf{d}}_h^{(n)}\}_{h=1}^{N^n}$ at iteration $n$, the ordered orientations at iteration $n+1$ are given by

$$\overline{\mathbf{d}}_h^{(n+1)} = \begin{cases} \mathbf{T}_{\theta_{q(1)}}[\overline{\mathbf{d}}_j^{(n)}] & h = j = 1, .., N^n\,, \\ \mathbf{T}_{\theta_{q(2)}}[\overline{\mathbf{d}}_j^{(n)}] & h = N^n + j, \quad j = 1, .., N^n\,, \\ \cdots\cdots\cdots & \cdots\cdots\cdots \quad \cdots\cdots \\ \mathbf{T}_{\theta_{q(N)}}[\overline{\mathbf{d}}_j^{(n)}] & h = (N-1)N^n + j, j = 1, .., N^n\,. \end{cases} \qquad(6)$$

Geometrically, the vectors $\overline{\mathbf{d}}_h^{(n)}$ are tangent to the $h$-th segment forming the $n$-approximant $\mathcal{C}^{(n)}$. The spatial distribution of $\{\overline{\mathbf{d}}_h^{(n)}\}$ attains a singular structure. In order to treat the distribution of the orientations (tangent vectors) along a fractal curve analytically, it is convenient to introduce a global formalism, which in the manuscript shall be referred to as directional pseudo-measures.

The directional pseudo-measures $\mu_{d,x}^{(n)}(\xi)$, $\mu_{d,y}^{(n)}(\xi)$ ($\xi \in [0,1]$ is the curvilinear abscissa normalized to unity) at iteration $n$ along the $x$- and $y$-axes can be defined as

$$\mu_{d,x}^{(n)}(h/N^n) = s^n \sum_{j=1}^{h} \overline{d}_{x,j}^{(n)}, \quad \mu_{d,y}^{(n)}(h/N^n) = s^n \sum_{j=1}^{h} \overline{d}_{y,j}^{(n)}, \quad h = 1,..,N^n, \tag{7}$$

where $\overline{d}_{x,h}^{(n)}, \overline{d}_{y,h}^{(n)}$ are the two entries of $\overline{\mathbf{d}}_h^{(n)}$. The physical (geometrical) meaning of the directional pseudo-measures is fairly evident. Let $\xi \in [0,1]$ be the normalized curvilinear abscissa for the $n$-order approximant $\mathcal{C}^{(n)}$, and $\mathcal{C}^{(n)}(\xi) \subseteq \mathcal{C}^{(n)}$ be the portion of the curve $\mathcal{C}^{(n)}$ up to a value $\xi$ of the normalized abscissa. Given $\overline{\mathbf{d}}_h^{(n)}$, we can define an equivalent function $\overline{\mathbf{d}}^{(n)}(\xi)$ of the continuous variable $\xi$ defined as $\overline{\mathbf{d}}^{(n)}(\xi) = (\overline{d}_x^{(n)}(\xi), \overline{d}_y^{(n)}(\xi))^t = \overline{\mathbf{d}}_h^{(n)}$ for $(h-1)/N^n \leq \xi < h/N^n$. Therefore

$$\mu_{d,x}^{(n)}(\xi) = \int_{\mathcal{C}^{(n)}} \overline{d}_x^{(n)}(\xi)\, d\xi, \quad \mu_{d,y}^{(n)}(\xi) = \int_{\mathcal{C}^{(n)}} \overline{d}_y^{(n)}(\xi)\, d\xi. \tag{8}$$

From Eq. (8) it follows that $\mu_{d,x}^{(n)}(\xi)$, $\mu_{d,y}^{(n)}(\xi)$, are by no means measures in any mathematical sense as they may be non-monotonic functions of $\xi$ and may attain negative values. Indeed the pseudo-measures so introduced correspond to sign-measure as defined by Halmos [14]. The diction "pseudo-measures" has nevertheless been adopted in order to point out that these quantities are related to the tangent length element along $\mathcal{C}^{(n)}$. Indeed, if $\mathbf{t}$ indicates the unit tangent vector and $d\sigma$ the length element, then

$$\mathbf{t}\, d\sigma = d\mu_d^{(n)}(\xi), \quad \xi = \sigma/L_n \quad (h-1)/N^n \leq \xi < h/N^n, \tag{9}$$

where $L_n$ is the length of $\mathcal{C}^{(n)}$. Fig. 3 shows the behaviour of the two directional pseudo-measures for the curve considered. Numerical evidence indicates that for $n$ tending to infinity both $\mu_{d,x}^{(n)}(\xi)$ and $\mu_{d,y}^{(n)}(\xi)$ tend towards two limit-invariant pseudo-measures $\mu_{d,x}^*(\xi)$ and $\mu_{d,y}^*(\xi)$, which are continuous - albeit almost everywhere not differentiable - functions of the argument $\xi \in [0,1)$. This observation is important in connection with the definition of integrals of differentiable forms over fractal curves, as discussed in the next section.

## 4 Contour integrals over fractal curves

Consider the directional pseudo measures as a vector, $\mu_d^{(n)} = (\mu_{d,x}^{(n)}, \mu_{d,y}^{(n)})^t$, and let $\mathbf{f}(\mathbf{x}) = (f_1(\mathbf{x}), f_2(\mathbf{x}))^t$ be a generic continuous function of position. In order to

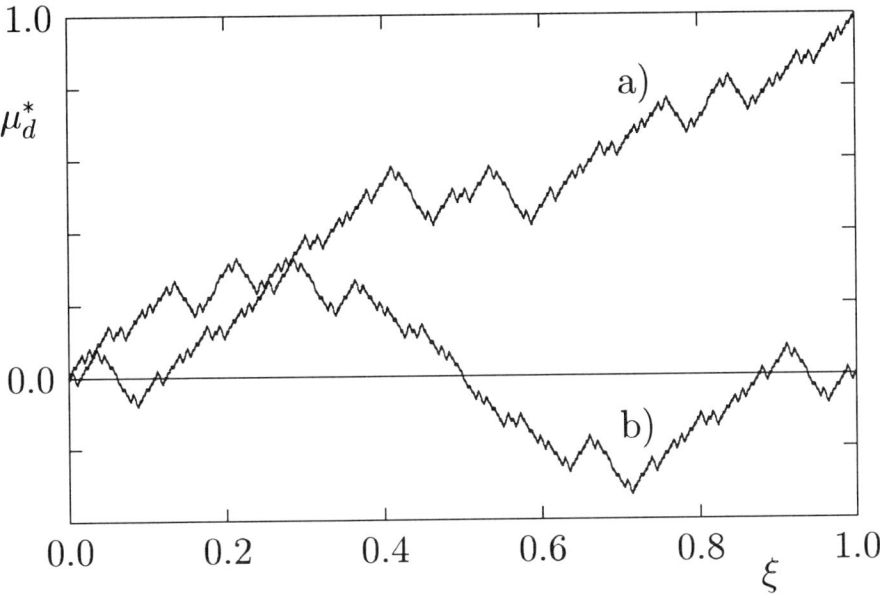

**Fig. 3.** Directional pseudo-measures vs $\xi$ for the Koch curve depicted in Fig. 1. a) refers to $\mu^*_{d,x}(\xi)$, and b) to $\mu^*_{d,y}(\xi)$.

define contour integrals over fractal curves, it is convenient to follow the iterative scheme for the generation process of the structure, as discussed in section 3. With no loss of generality, we assume that the IFSO is ordered in such a way that the corresponding ordering vector $\mathbf{q}$, defined in section 3, returns to $q(i) = i$, $i = 1, .., N$. This assumption is a purely formal point serving to reduce and simplify the notational complexity. Indeed, all the results derived in this section are independent of the labelling.

For finite $n$, the $n$-th approximation $\mathcal{C}^{(n)}$ of $\mathcal{C}$ is a piecewise differentiable curve. The contour integral of $\mathbf{f}$ over $\mathcal{C}$ can be expressed as

$$\int_{\mathcal{C}^{(n)}} \mathbf{f}(\mathbf{x}) \cdot \mathbf{t}\, d\sigma = \int_{\mathcal{C}^{(n)}} f_1(\mathbf{x}) dx + f_2(\mathbf{x}) dy , \qquad (10)$$

where $\cdot$ indicates scalar product. Let $\{\mathbf{z}_j\}_{j=1}^{n_o}$ be an ensemble of points, e.g. $n_o = 2$, and $\mathbf{z}_j$, $j = 1, 2$ the two endpoints of the curve. The integral in Eq. (10) can be approximated up to terms of the order of $O(N_n^{-1})$ by the following expression:

$$I^{(n)}[\mathbf{f}(\mathbf{x})] = \int_{\mathcal{C}^{(n)}} \mathbf{f}(\mathbf{x}) \cdot d\mu_d^{(n)} = s^n \sum_{j=1}^{n_o} \sum_{h_n=1}^{N} \cdots \sum_{h_1=1}^{N} \mathbf{f}(\mathbf{w}_{h_n} \circ \cdots \circ \mathbf{w}_{h_1}(\mathbf{z}_j))$$
$$\cdot \mathbf{T}_{\theta_{h_n}} \circ \cdots \circ \mathbf{T}_{\theta_{h_1}}[\mathbf{d}^{(0)}] . \qquad (11)$$

Although $\mathbf{T}_{\theta_k}[\mathbf{d}^{(0)}]$ indicates just matrix multiplication, i.e. the image of the vector $\mathbf{d}^{(0)}$ through the rotation matrix $\mathbf{T}_{\theta_k}$, the notation with square brackets has been used for the sake of clarity. At iteration $n+1$, $I^{(n+1)}[\mathbf{f}(\mathbf{x})]$ becomes

$$I^{(n+1)}[\mathbf{f}(\mathbf{x})] = s^{n+1} \sum_{j=1}^{n_o} \sum_{h_{n+1}=1}^{N} \sum_{h_n=1}^{N} \cdots \sum_{h_1=1}^{N} \mathbf{f}(\mathbf{w}_{h_{n+1}} \circ \mathbf{w}_{h_n} \circ \cdots \circ \mathbf{w}_{h_1}(\mathbf{z}_j))$$
$$\cdot \mathbf{T}_{\theta_{h_{n+1}}} \circ \mathbf{T}_{\theta_{h_n}} \circ \cdots \circ \mathbf{T}_{\theta_{h_1}}[\mathbf{d}^{(0)}]$$
$$= s \sum_{k=1}^{N} s^n \sum_{j=1}^{n_o} \sum_{h_n=1}^{N} \cdots \sum_{h_1=1}^{N} \mathbf{T}_{\theta_k}^t [\mathbf{f}(\mathbf{w}_k \circ \mathbf{w}_{h_n} \circ \cdots \circ \mathbf{w}_{h_1}(\mathbf{z}_j))]$$
$$\cdot \mathbf{T}_{\theta_{h_n}} \circ \cdots \circ \mathbf{T}_{\theta_{h_1}}[\mathbf{d}^{(0)}] , \qquad (12)$$

where we have put $k = h_{n+1}$. By comparing Eq. (12) with Eq. (11), the following recursion scheme for $I^{(n)}[\mathbf{f}(\mathbf{x})]$ is obtained:

$$I^{(n+1)}[\mathbf{f}(\mathbf{x})] = s \sum_{k=1}^{N} I^{(n)}[\mathbf{T}_{\theta_k}^t[\mathbf{f}(\mathbf{w}_k(\mathbf{x}))]] . \qquad (13)$$

Eq. (13) represents a recursive relation for $I^{(n)}[\mathbf{f}]$ converging towards a limit for $n \to \infty$ independently of the initial choice of $\{\mathbf{z}_j\}_{j=1}^{n_o}$. The limit of $I^{(n)}[\mathbf{f}]$ for $n \to \infty$ will be indicated as $I^*[\mathbf{f}]$ (strong numerical evidence indicates the existence of this limit) or equivalently as

$$I^*[\mathbf{f}] = \int_\mathcal{C} \mathbf{f} \cdot d\mu_d^* , \qquad (14)$$

and represents the curvilinear integral of $\mathbf{f}$ over the fractal curve $\mathcal{C}$. From Eq. (13), the following identity holds for the limit value $I^*[\mathbf{f}]$:

$$I^*[\mathbf{f}] = s \sum_{k=1}^{N} I^*[\mathbf{T}_{\theta_k}^t[\mathbf{f}(\mathbf{w}_k)]] , \qquad (15)$$

or equivalently

$$\int_\mathcal{C} \mathbf{f} \cdot d\mu_d^* = s \sum_{k=1}^{N} \int_\mathcal{C} \mathbf{T}_{\theta_k}^t[\mathbf{f}(\mathbf{w}_k)] \cdot d\mu_d^* = s \sum_{k=1}^{N} \int_\mathcal{C} \mathbf{f}(\mathbf{w}_k) \cdot \mathbf{T}_{\theta_k}[d\mu_d^*] . \qquad (16)$$

The latter expression displays close formal analogies with the definition of the Markov operator and with the expression for the average of a continuous function with respect to the invariant measure $\mu^*$, Eq. (2). There are, however, some fundamental differences between Eq. (16) and Eq. (2), as discussed in the next section. In any case, Eqs. (15)-(16) are the fundamental expressions for representing curvilinear integrals over fractal curves (under the limitations discussed in section 2 regarding the functional form of the IFS generating $\mathcal{C}$), and enable us to compute these integrals in closed form (see next section).

To complete the analysis of curvilinear integrals, let us consider another important case, namely the integral of the normal component of a vector field over a curve. In many physical problems associated with transport phenomena and more generally with field theory, it is necessary to evaluate, e.g. as a consequence of the divergence theorem, the integral

$$\int_\Gamma \mathbf{f} \cdot \nu \, d\sigma \,,$$

where $\nu$ is the normal unit vector (oriented in some way, $\nu$ usually being the outer normal), and $\Gamma$ a curve (a close curve in most applications).

In the case of fractal curves, the analysis developed for $I^{(n)}[\mathbf{f}]$ can be extended in a straightforward way to this kind of contour integral by simply replacing the directional pseudo-measures with the normal pseudo-measures $\mu_\nu^{(n)}$. The quantities $\mu_\nu^{(n)}$ can be defined in exactly the same way as $\mu_d^{(n)}$ by replacing $\mathbf{d}^{(0)}$ with $\nu^{(0)} = \mathbf{d}^{(0)} \times \mathbf{e}_3$, (here $\times$ indicates vector product, and $\mathbf{e}_3$ is the unit vector orthogonal to the $xy$-plane in a Cartesian reference system), i.e. with the normal vector to the zero-th order approximation: $\nu^{(0)} = (0, -1)^t$ for the 8-map Koch curve. Indeed, the vector element $\nu d\sigma$ corresponds in the recursive generation of fractal curves to $d\mu_\nu^{(n)}$ up to the order $O(N_n^{-1})$.

The same recursive relation obtained for $I^{(n)}[\mathbf{f}]$ holds for $I_\nu^{(n)}[\mathbf{f}]$, i.e.

$$I_\nu^{(n+1)}[\mathbf{f}(\mathbf{x})] = s \sum_{k=1}^N I_\nu^{(n)}[\mathbf{T}_{\theta_k}^t[\mathbf{f}(\mathbf{w}_k(\mathbf{x}))]] \,, \tag{17}$$

and in the limit for $n \to \infty$, $I_\nu^{(n)}[\mathbf{f}] \to I_\nu^*[\mathbf{f}] = \int_C \mathbf{f} \cdot d\mu_\nu^*$ with

$$\int_C \mathbf{f} \cdot d\mu_\nu^* = s \sum_{k=1}^N \int_C \mathbf{T}_{\theta_k}^t[\mathbf{f}(\mathbf{w}_k)] \cdot d\mu_\nu^* = s \sum_{k=1}^N \int_C \mathbf{f}(\mathbf{w}_k) \cdot \mathbf{T}_{\theta_k}[d\mu_\nu^*] \,. \tag{18}$$

In the next section we shall use Eqs. (16), (18) to obtain closed-form expressions for $I^*[\mathbf{f}]$ and $I_\nu^*[\mathbf{f}]$.

## 5 Closed-form results

Before further development of Eqs. (15) and (16), let us briefly analyze the structure and the properties of these expressions for curvilinear integrals over fractal curves. At first sight, Eqs. (15), (16) look very similar to Eq. (2), and there is indeed a close formal analogy between these expressions. In particular, the same mathematical approach applied to derive the moment hierarchy associated with the invariant measure $\mu^*$ of affine IFSP can be applied to derive closed-form expressions for the curvilinear integral of continuous functions, as developed below.

Nevertheless, there are some basic differences between the theory underlying Eq. (2) and Eqs. (15), (16). The most important difference is that, while the ensemble average performed with respect to the invariant measure of the IFSP coincides with the temporal average along a trajectory $\{\mathbf{x}_i\}$ of the IFSP (Elton's ergodic theorem [12]),

$$\int_C f(\mathbf{x})d\mu^*(\mathbf{x}) = \lim_{M \to \infty} \frac{1}{M} \sum_{i=1}^M f(\mathbf{x}_i), \qquad (19)$$

the same property does not hold for curvilinear integrals defined for an IFSO since the directional pseudo-measures are not measures in a formal sense and ergodicity does not hold. This implies that the curvilinear integrals defined by means of Eqs. (15) and (16) cannot be evaluated from temporal averages as in Eq. (19). This result does not limit the applicability of Eqs. (15) and (16) since they can be still used to obtain closed-form expressions for the curvilinear integrals, or at least to evaluate these quantities by means of numerical procedures.

As regards the analysis of ensemble averages of continuous functions over the invariant measure $\mu^*$, it is convenient to consider for $f_1$, $f_2$ functional forms such as $x^n y^m$, with $n$, $m$ integers, thus obtaining a closed-form expression for the moment hierarchy.

Let us consider $\mu_d^*$, since the analysis for $\mu_\nu^*$ is identical. Due to the vectorial nature of $\mu_d^*$, it is useful to regard $I^*[\mathbf{f}]$ as the superposition of two terms: $I^*[\mathbf{f}] = I_1^*[f_1] + I_2^*[f_2]$, where

$$I_1^*[f_1] = \int_C f_1 d\mu_{d,x}^*, \qquad I_2^*[f_2] = \int_C f_2 d\mu_{d,y}^*. \qquad (20)$$

To exemplify the analysis on a fractal curve, the 8-map Koch curve in Fig. 1, is taken as a model structure. For this model fractal curve, Eq. (16) becomes

$$\int_C f_1(\mathbf{x})d\mu_{d,x}^* = s \sum_{h=1}^4 \int_C f_1(\mathbf{w}_h(\mathbf{x}))d\mu_{d,x}^* + s \sum_{h=5}^8 c_h \int_C f_1(\mathbf{w}_h(\mathbf{x}))d\mu_{d,y}^*,$$

$$\int_C f_2(\mathbf{x})d\mu_{d,y}^* = s \sum_{h=1}^4 \int_C f_2(\mathbf{w}_h(\mathbf{x}))d\mu_{d,y}^* - s \sum_{h=5}^8 c_h \int_C f_2(\mathbf{w}_h(\mathbf{x}))d\mu_{d,x}^*,$$

or equivalently

$$I_1^*[f_1] = s \sum_{h=1}^4 I_1^*[f_1(\mathbf{w}_h)] + s \sum_{h=5}^8 c_h I_2^*[f_1(\mathbf{w}_h)],$$

$$I_2^*[f_2] = s \sum_{h=1}^4 I_2^*[f_2(\mathbf{w}_h)] - s \sum_{h=5}^8 c_h I_1^*[f_2(\mathbf{w}_h)]. \qquad (21)$$

Due to the presence of a rotation of angle $\theta_h \neq 0$ (for $h = 5,..,8$), the expressions for $I_1^*$ and $I_2^*$ become coupled together. Let us focus on the moment hierarchies, by defining $m_{i,x}(n) = I_i^*[x^n]$, $m_{i,y}(n) = I_i^*[y^n]$ and $m_i(n,m) = I_i^*[x^n y^m]$, $i =$

1, 2. The values of the zero-th order moments are given in this case by

$$m_{1,x}(0) = m_{1,y}(0) = m_1(0,0) = \int_C d\mu^*_{d,x} = 1$$
$$m_{2,x}(0) = m_{2,y}(0) = m_2(0,0) = \int_C d\mu^*_{d,y} = 0 \, . \qquad (22)$$

By applying Eq. (21) to the hierarchy $m_i(n,m)$ $i = 1, 2$, it is possible to obtain a system of linear equations that can be solved in closed form recursively. The algebra for deriving closed-form expression of $m_i(n,m)$ is formally equivalent to the technique adopted to obtain the moment hierarchies of the invariant measure of IFSP [15–17], and is developed in detail elsewhere. To give a numerical example Fig. 4 shows several moment hierarchies for the 8-map Koch curve obtained from the closed-form expression derived from Eq. (21), compared with the direct numerical definition Eq. (11) for finite $n$.

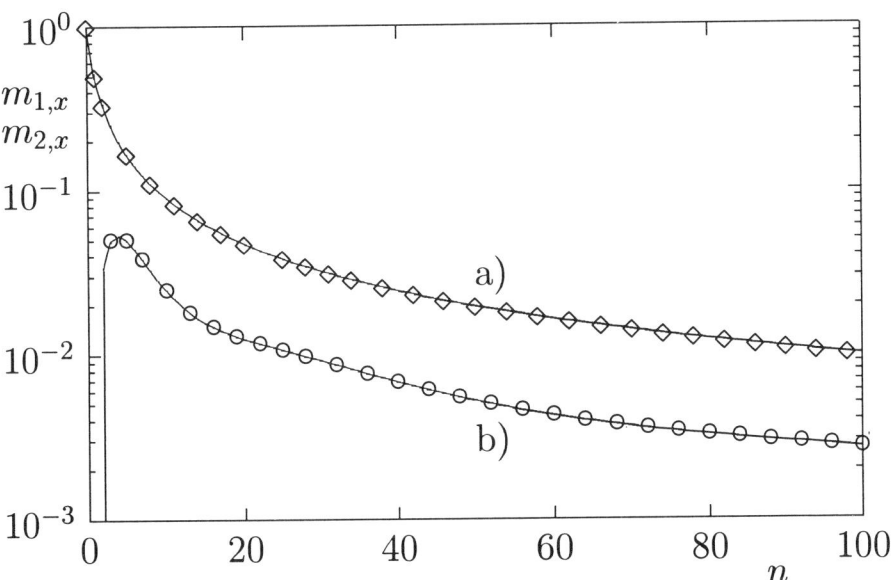

**Fig. 4.** Moment hierarchies associated with contour integrals over the fractal curve in Fig. 1. $m_{1,x}(n)$ (curve a) and $m_{2,x}(n)$ (curve b). The solid line a) is the curve $m_{1,x}(n) = (1+n)^{-1}$, the dots are the results of the closed-form expression derived from Eq. (21). The solid line b) refers to the numerical evaluation of $m_{2,x}(n)$, obtained by applying Eq. (11) up to $n = 5$ starting from two intial points ($n_o = 2$) located at the endpoints of the curve.

## 6 Vector calculus on fractal curves

One interesting result associated with the formal development of contour integrals over fractal curves is the possibility of defining a vector calculus. In particular, it is useful to develop integral theorems in the presence of fractal curves connecting, for two-dimensional structures, surface integrals over a closed domain and contour integrals over its boundary.

In the classical vector calculus in two dimensions, the most significant result is Green's theorem [18]: if $\mathcal{D}$ is a regular domain of the plane, $\Gamma$ its boundary, composed by union of regular curve arcs, and $\mathbf{f} = (f_1, f_2)^t$ a continuous vector function of the position $\mathbf{x}$ with continuous partial derivatives, then

$$\int_{+\Gamma} (f_1 dx + f_2 dy) = \int_{+\Gamma} \mathbf{f} \cdot \mathbf{t} d\sigma = \int_{\mathcal{D}} \left( \frac{\partial f_2}{\partial x} - \frac{\partial f_1}{\partial y} \right) dx dy , \qquad (23)$$

where $+\Gamma$ indicates the positive (counterclockwise) orientation, $\mathbf{t}$ is the tangent vector according to the positive orientation of $+\Gamma$, and $d\sigma$ the line element along the curve. In two-dimensions, Green's theorem Eq. (23) can be viewed either as a consequence of the Stokes theorem in the plane or as a corollary of the Gauss divergence theorem.

In this section we analyze how Green's theorem can be extended to the case of fractal boundaries. Let $\mathcal{D}$ be a two-dimensional finite domain, the boundary of which is a closed fractal curve $\mathcal{C}$. We further assume that $\mathcal{D}$ is the limit set of an IFSP and let $\mathcal{D}^{(n)}$ be the $n$-th order approximation of $\mathcal{D}$ associated with the generation process, the boundary of which is $\mathcal{C}^{(n)}$, see Fig. 2. We assume that the probability weights of the IFSP generating $\mathcal{D}$ are such that the resulting invariant measure is uniform, i.e. $d\mu^* = dx dy / V(\mathcal{D})$, where $V(\mathcal{D})$ is the Lebesgue measure (in this case the area) of $\mathcal{D}$. One further assumption concerns the boundary of $\mathcal{C}$, namely that it can be decomposed into the union of $N_\mathcal{C}$ disjoint (just-touching) curve arcs $\mathcal{C}_i$, each of which can be generated by means of an IFS, as in Fig. 2. By assuming the counterclockwise orientation, the orientation vector $\mathbf{d}_i^{(0)}$ for each IFSO generating $\mathcal{C}_i$ is given by $\mathbf{d}_i^{(0)} = \mathbf{x}_i^{(2)} - \mathbf{x}_i^{(1)}$, $i = 1, .., 4$, where $\mathbf{x}_i^{(1)}$, $\mathbf{x}_i^{(2)}$ are the two end-points of $\mathcal{C}_i$ in the ordering induced by the counterclockwise orientation. As a consequence, the normal direction $\nu_i^{(0)}$ are given by $\nu_i^{(0)} = (\mathbf{x}_i^{(2)} - \mathbf{x}_i^{(1)}) \times \mathbf{e}_3$.

Green's theorem Eq. (23) applies to each $\mathcal{D}^{(n)}$, the boundary of which is $\mathcal{C}^{(n)}$, to yield

$$\int_{+\mathcal{C}^{(n)}} \mathbf{f} \cdot d\mu_d^{(n)} + O(N_n^{-1}) = V(\mathcal{D}^{(n)}) \int_{\mathcal{D}^{(n)}} \left( \frac{\partial f_2}{\partial x} - \frac{\partial f_1}{\partial y} \right) d\mu^{(n)} , \qquad (24)$$

where $\mu^{(n)}$ is the uniform measure normalized to unity within $\mathcal{D}^{(n)}$. In the limit of $n \to \infty$, Eq. (24) therefore becomes

$$\int_{+\mathcal{C}} \mathbf{f} \cdot d\mu_d^* = V(\mathcal{D}) \int_{\mathcal{D}} \left( \frac{\partial f_2}{\partial x} - \frac{\partial f_1}{\partial y} \right) d\mu^* , \qquad (25)$$

which is Green's theorem for a closed fractal curve $\mathcal{C}$ expressed in terms of the invariant measure $\mu^*$ associated with the IFSP generating $\mathcal{D}$ and of the directional pseudo-measures associated with its boundary $\mathcal{C}$. Eq. (25) can be generalized to the case where the domain $\mathcal{D}$ cannot be generated by means of an IFSP as

$$\int_{+\mathcal{C}} \mathbf{f} \cdot d\mu_d^* = \int_{\mathcal{D}} \left( \frac{\partial f_2}{\partial x} - \frac{\partial f_1}{\partial y} \right) dx dy . \tag{26}$$

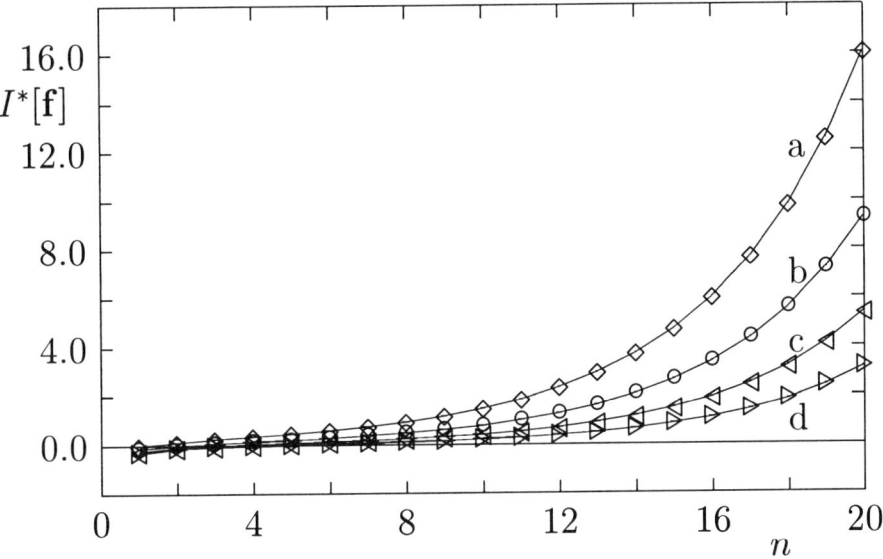

**Fig. 5.** Numerical validation of Green's theorem, Eq. (25). The function **f** considered is given by $\mathbf{f} = (x^n y^m, x^n y^m)^t$ vs $n$ for different values of the integer $m$. a) $m = 1$, b) $m = 2$, c) $m = 3$, d) $m = 4$.

In order to give some numerical examples of the validity of Green's theorem for closed fractal curves, it is convenient to consider the case in which both the inner domain and its closed boundary can be generated by means of a family of IFSP, and Eq. (25) can therefore be applied. A typical case is the bounded domain $\mathcal{D}$, as shown in Fig. 2, generated by means of a 16-map self-similar IFSP, the boundary of which is given by the union of four Koch curves, as discussed in section 2. For the four IFSO generating the boundary $\mathcal{C}$, the directional vectors are given by: $\mathbf{d}_1^{(0)} = (1,0)^t$, $\mathbf{d}_2^{(0)} = (0,1)^t$, $\mathbf{d}_3^{(0)} = (-1,0)^t$, $\mathbf{d}_4^{(0)} = (0,-1)^t$.

By inserting $\mathbf{f} = (-y, x)^t$ into Eq. (25), the measure of $\mathcal{D}$ can be expressed as

$$V(\mathcal{D}) = \frac{1}{2} \int_{+\mathcal{C}} x d\mu^*_{d,y} - y d\mu^*_{d,x} = \frac{1}{2} \sum_{i=1}^{N_C} m^{(i)}_{2,x}(1) - m^{(i)}_{1,y}(1) , \qquad (27)$$

where $N_C = 4$, and $\mu_d^{(i)*}$ are the directional pseudo-measures of each of the four IFSO generating the boundary.

It is important to stress that the recursion schemes for the moment hierachies for each of the four IFSO generating the boundary are exactly the same. The only feature that characterizes each IFSO (and therefore the behaviour of the moment hierarchies) is the value of the zero-th order moments, which are given by

$$m^{(i)}_{1,x}(0) = m^{(i)}_{1,y}(0) = m^{(i)}_1(0,0) = d^{(0)}_{x,i}, \qquad i = 1,..,4$$
$$m^{(i)}_{2,x}(0) = m^{(i)}_{2,y}(0) = m^{(i)}_2(0,0) = d^{(0)}_{y,i}, \qquad i = 1,..,4 . \qquad (28)$$

In the case of the structure shown in Fig. 2, since $\sum_{i=1}^{N_C} m^{(i)}_{2,x}(1) = 1$ and $\sum_{i=1}^{N_C} m^{(i)}_{1,y}(1) = -1$, the application of Eq. (27) yields $V(\mathcal{D}) = 1$. This result can be readily obtained also by means of more elementary arguments. The integral of continuous functions over $\mathcal{D}$ with respect to the invariant measure $\mu^*$ can be evaluated analytically by deriving a set of recursions for the moment hierarchy, i.e. for the quantities $v(n,m) = \int_\mathcal{D} x^n y^m d\mu^*$.

Fig. 5 shows the comparison and the perfect agreement of contour integrals over $\mathcal{C}$ and surface integrals over $\mathcal{D}$ for several families of moments, both obtained by applying the iterative recursions. The dots are the results of the contour integrals $I^*[\mathbf{f}]$ obtained by applying Eq. (25), and the lines are the volume integrals of $(\partial f_2/\partial x - \partial f_1/\partial y)$ averaged with respect to the uniform invariant measure associated with the IFSP generating the inner domain (since $V(\mathcal{D}) = 1$). The vector functions $\mathbf{f}$ considered are given by $\mathbf{f}(\mathbf{x}) = (x^n y^m, x^n y^m)^t$. The result expressed by Fig. 5 represents a further numerical exemplification of the applicability of Eq. (25) to a two-dimensional domain possessing fractal self-similar boundaries.

## 7 Concluding remarks

This article has developed a method to approach contour integrals over fractal (self-similar) boundaries in two-dimensions. The analysis has developed for structures generated by self-similar IFS characterized by a single scaling factor, although the method could be extended to more general cases in the presence of IFS with different scaling factors. The style of the presentation has been kept as simple as possible in order to highlight the connections with renormalization and with the physics of fractals. More technical details regarding convergence and other strictly mathematical properties are left to further, more mathematically oriented publications by following the paths outlined here. An important consequence of the analysis developed is that it is possible to define, and in certain cases to evaluate in closed form, contour integrals over fractal curves and to extend to fractal boundaries a vector calculus analogous to the vector calculus on differentiable structures. This result is by no means surprising, since many results of classical vector calculus are indeed more general topological properties, as stressed by modern differential geometry on manifolds (integration on manifolds). At a first sight, the most impressive result of the present theory is that it is possible to define by renormalization (i.e. following the construction principle of fractal self-similar structures) curvilinear integrals for curves for which tangent and normal vectors, defined in the classical way, do not exist almost everywhere. Indeed, the proper definition of such quantites lies in the recursive generation process characterizing fractal structures, leading to recursive relations for contour integrals of continuous functions, converging for $n$ tending to infinity towards a limit. The starting point of the analysis is the definition of the directional pseudo-measures. Retrospectively, this definition is motivated by the fact that the properties of tangent and normal vectors to fractal curves can be defined only in tems of integral quantities, so that the corresponding contour integrals can be obtained as Stieltjes integrals over these pseudo-measures. Several observations are also important. First of all, the theory developed in two-dimensions can be extended with no special difficulty to three-dimensional spaces. To this end, it is sufficient to consider three-dimensional similar IFS generating fractal surfaces, and to patch together different IFS in order to generate a closed fractal surface, the boundary of a three-dimensional Lebesgue-measurable finite domain. This extension is simply a matter of patience and can be regarded as the natural consequence of the analysis developed in this article. A further generalization of the theory consists in its extension to more general fractal but not self-similar curves, and in particular to self-affine structures also generated by means of IFS. It is also important to stress once again the usefulness of IFS theory which can rightly be considered as the natural framework to approach many mathematical physical problems involving fractal structures.

## References

1. C. Tricot (1995): Curves and fractal dimensions. Springer Verlag, Berlin.
2. T. Pajkossy (1991). J. Electroanal. Chem. **300** 1.
3. S.H. Liu (1985). Phys. Rev. Lett. **55** 529.
4. A. Seri-Levy and A. Avnir (1993). J. Phys. Chem. **97** 10380.
5. M. Giona and M. Giustiniani (1996). J. Phys. Chem. **100** 16690.
6. M. V. Berry (1979). In Structural Stability in Physics. Guttinger W. and Elkeimer H. (Eds). Springer Verlag, Berlin.
7. J. Kigami and M. L. Lapidus (1993). Comm. Math. Phys. **158** 93.
8. S. Russ, B. Sapoval and Haeberlé (1997). Phys. Rev. E **55** 1413.
9. H. Wallin (1991). Manuscripta Math. **73** 117.
10. A. Jonsson and H. Wallin (1995). Studia Mathematica **112** 285.
11. A. Jonsson and H. Wallin (1997): Chaos Solitons & Fractals. **8** 191.
12. M. F. Barnsley (1988): Fractals Everywhere. Academic Press, Boston.

13. K. Falconer (1993): Fractal Geometry Mathematical Foundations and Applications. J. Wiley & Sons, New York.
14. P. R. Halmos (1950): Measure Theory. Van Nostrand, New York.
15. D. Bessis and S. Demko (1991). Physica **D47** 427.
16. S. Abenda, S. Demko and G. Turchetti (1992): Inverse Problems. **8** 737.
17. B. Forte and E. R. Vrscay (1994): Fractals. **2** 325.
18. R. Courant and D. Hilbert (1962): Methods of Mathematical Physics. J. Wiley & Sons, New York.

# Local Fractional Calculus: a Calculus for Fractal Space-Time

Kiran M. Kolwankar[1] and Anil D. Gangal[2]

[1] Department of Mathematics, Indian Institute of Science,
Bangalore 560 012 - INDIA
kirkol@math.iisc.ernet.in
[2] Department of Physics, University of Pune,
Pune 411 007 - INDIA
adg@physics.unipune.ernet.in

**Abstract.** Recently, new notions such as local fractional derivatives and local fractional differential equations were introduced. Here we argue that these developments provide a possible calculus to deal with phenomena in fractal space-time. We show how the usual calculus is generalized to deal with non Lipschitz functions. We also indicate how a definition of a fractal measure arises from these developments much the same way as the Lebesgue measure from ordinary calculus

**Key Words:** Fractals, Fractal measures, Fractional Calculus, Diffusion Equation, Subdiffusion.

## 1 Introduction

After Mandelbrot [1] wrote his seminal book, fractals have been found useful in Science [2] as well as Engineering [3, 4]. Many applications of fractals in the fields ranging from growth phenomena [5, 6], turbulence, to chaotic systems [7], etc. have been found. Recently (see [8] and references therein) even space-time has been considered to be fractal (i.e., nondifferentiable manifold rather than smooth manifold) at small scales. The characteristic of fractals is their irregularity on all length scales. This irregularity of fractals make them very difficult to handle analytically. It is well-known that the usual calculus is inadequate to treat such structures and processes. Fractals are too irregular to have any smooth differentiable structure defined on them. Fractal functions do not possess first order derivative at any point [9, 10]. Such functions have been encountered at number of places [11–15, 36]. This suggests that a new calculus should be developed which incorporates fractal structure and processes. In this paper we investigate the possibility of using fractional calculus, a field which deals with derivatives and integrals of arbitrary order, for this purpose.

The relation between ordinary calculus and measures on $I\!R^n$ is well-known. For example, an $n$-fold integration gives an $n$-dimensional volume. Or the length of a curve $y(x)$ is given by an integral $\int \sqrt{(1+y'^2)}dx$ which involves first order derivative of the curve. Also, the solution of the differential equation of the type $df/dx = 1_{[0,x]}$, where $1_{[0,x]}$ is an indicator function of $[0,x]$, gives length of the interval $[0,x]$.

The possible connection between fractals and fractional calculus raises various formal questions [16] of the type mentioned in the above paragraph. In particular, could an appropriately defined "fractional integral" yield a fractal measure? Or is there some differential equation of fractional order whose solution, analogous to an example above, gives a measure of the fractal set involved? We try to answer such questions in this paper.

The plan of the paper is as follows. In the next section we begin by reviewing the conventional fractional calculus and then we devote section 3 to local fractional calculus. A definition of fractal measures arising from local fractional calculus is introduced in section 4. In section 5 we demonstrate, with the help of local fractional diffusion equation, that the concepts discussed in this work can be used to describe phenomena involving fractals.

## 2 Conventional fractional calculus

There are various equivalent ways in which an operator $D^n$, where $D = d/dx$, can be constructed which gives derivative of order $n$ for positive integer $n$ and/or an integral of order $n$ for negative integer $n$. Once one has such an operator it raises an interesting question if this operator can be generalized to all real values (or even complex) of the order. The branch of mathematics which deals with such a generalization is called fractional calculus.

In this section we recall one of the definitions of fractional derivatives and integrals and study some of its properties. The details and many other definitions can be found in Miller and Ross [17], Oldham and Spanier [18] and Samko et al. [19]. For possible applications the reader is referred to [20, 21].

### 2.1 Definitions

There are various definitions of derivatives and integrals of fractional order not necessarily equivalent to each other. All these definitions have different origin. The definition introduced in this section will be used in this work.

The most frequently used definition of a fractional integral is the Riemann–Liouville definition. According to this definition, a fractional integral of order $q$ of a function $f$ is given by

$$\frac{d^q f(x)}{[d(x-a)]^q} = \frac{1}{\Gamma(-q)} \int_a^x \frac{f(y)}{(x-y)^{q+1}} dy \quad \text{for} \quad q < 0, \tag{1}$$

where the lower limit $a$ is some real number. The fractional derivative of order $q$ is, where $n - 1 < q < n$,

$$\frac{d^q f(x)}{[d(x-a)]^q} = \frac{1}{\Gamma(n-q)} \frac{d^n}{dx^n} \int_a^x \frac{f(y)}{(x-y)^{q-n+1}} dy \qquad (2)$$

$$= \sum_{k=0}^{n-1} \frac{(x-a)^{-q+k} f^{(k)}(a)}{\Gamma(-q+k+1)} + \frac{d^{q-n} f^{(n)}}{[d(x-a)]^{q-n}}. \qquad (3)$$

As is clear, this definition uses the concepts of ordinary derivatives and integration. Since it amounts to evaluating an integral, it is more convenient to use.

## 2.2 Some properties and examples

Many properties of the fractional integrals and derivatives, like chain rule, Leibniz rule, composition law, etc. have been studied (see Oldham and Spanier [18]). Here we note one property which is useful in the context of scaling functions. When the argument of the function is scaled by a factor $\beta$, the differintegral (differentiation or integration of arbitrary order) satisfies

$$\frac{d^q f(\beta x)}{[dx]^q} = \beta^q \frac{d^q f(\beta x)}{[d(\beta x)]^q}. \qquad (4)$$

For a more general formula, with nonzero lower limit $a$, see [18].

Except in a few cases, it is difficult to evaluate a derivative or integral of fractional order. We consider one example below which will be used later. If we choose $f(x) = x^p$ then using the Riemann-Liouville definition it can be shown that

$$\frac{d^q x^p}{dx^q} = \frac{\Gamma(p+1)}{\Gamma(p-q+1)} x^{p-q}, \quad p > -1. \qquad (5)$$

## 2.3 Fractional Taylor series

In [22] Osler generalized the Taylor series to include fractional derivative. He proved a general result whose special case gives a series:

$$f(z) = \sum_{-\infty}^{\infty} \frac{(z-z_0)^{n+\gamma}}{\Gamma(n+\gamma+1)} \frac{d^{n+\gamma} f(z)}{[d(z-b)]^{n+\gamma}} \bigg|_{z=z_0}, \qquad (6)$$

where $b \neq z_0$ and $f(z)$ is analytic. Osler used this series to study special functions of mathematical physics. Since this series contains terms with negative power of $(z - z_0)$ it is not suitable for local approximations.

## 2.4 Fractional differential equations

Once we have a definition of fractional derivatives and integrals it is natural, in analogy with ordinary calculus, to write equations in terms of such quantities and find possible applications of such equations. In the last decade or so, many fractional differential equations (FDEs) have been proposed. Many of them are generalizations of the differential equations of mathematical physics [23–28]. These include relaxation equations, wave equations, diffusion equations, Fokker-Planck equations, etc. In these generalizations one replaces the usual integer order derivative by a fractional one. This replacement is either ad-hoc or involves some plausible arguments. Although these equations are nothing but integral equations writing them as FDEs has its own advantages.

Fractional differential equations have been used for the purpose of studying phenomena involving fractals. Giona and Roman [29, 30] proposed fractional diffusion equation to study diffusion on fractals. Zaslavsky [31, 32] argued that fractional analog of Fokker-Planck equation can be used to study transport in the phase space of chaotic Hamiltonian systems. Nonnenmacher [33] showed that a class of Lévy type processes satisfy a integral equation of fractional order. The fractional order is the same as the Lévy index which also happens to be the fractal dimension of the set visited by the random walker performing random walk with jump size distribution following the given Lévy distribution.

## 3 Local fractional calculus

Recently, a new notion called local fractional derivative (LFD) was introduced with the motivation of studying the local properties of fractal structures and processes. An interesting feature of the LFD is that it naturally appears in fractional Taylor expansion suitable for local approximations of scaling functions.

### 3.1 Definition

The definition of Riemann-Liouville fractional derivative was discussed in the last section. This derivative, along with others, differs in some aspects from integer order derivatives. In order to see this, one may note, from eqn (2), that except when $q$ is a positive integer, the $q^{th}$ derivative is nonlocal as it depends on the lower limit '$a$'. The same feature is also shown by other definitions. However, if one wants to study local scaling properties then those definitions are not suitable and one has to modify them accordingly. Secondly from equation (5) it is clear that the fractional derivative of a constant function is not zero. Therefore adding a constant to a function alters the value of the fractional derivative. Such a dependence on origin is again undesirable. While constructing an LFD operator, we have to correct for these two features. This forces one to choose the lower limit as well as the additive constant before hand. The most natural choices are as follows. i) We subtract, from the function, the value of the function at the point where we want to study the local scaling property. This makes the value of the function zero at that point, canceling the effect of any constant term. ii) Natural choice of a lower limit will again be that point itself, where we intend to examine the local scaling.

**Definition 1.** *If, for a function* $f : [0,1] \to \mathbb{R}$*, the limit*

$$I\!\!D^q f(y) = \lim_{x \to y} \frac{d^q(f(x) - f(y))}{d(x-y)^q}, \quad 0 < q \leq 1 \tag{7}$$

*exists and is finite, then we say that the LFD of order q (denoted by* $I\!\!D^q f(y)$*), at y, exists.*

This defines the LFD for $0 < q \leq 1$. It was first introduced in [39], and later generalized [44] to include all positive values of $q$ as follows.

**Definition 2.** *If, for a function* $f : [0,1] \to \mathbb{R}$*, the limit*

$$I\!\!D^q f(y) = \lim_{x \to y} \frac{d^q \left( f(x) - \sum_{n=0}^{N} \frac{f^{(n)}(y)}{\Gamma(n+1)} (x-y)^n \right)}{[d(x-y)]^q} \tag{8}$$

*exists and is finite, where N is the largest integer for which* $N^{th}$ *derivative of* $f(x)$ *at y exists and is finite, then we say that the local fractional derivative (LFD) of order q* $(N < q \leq N+1)$*, at* $x = y$*, exists.*

We subtract the Taylor series term in the above definition for the same reason as one subtracts $f(y)$ in the definition 1. We do this to suppress any regular behavior that may mask the local singularity. This definition was also generalized for functions of several variables [45].

This definition was subsequently used in Ref. [39] to study fractional differentiability properties of nowhere differentiable functions. It was shown that a continuous everywhere but differentiable nowhere function is locally fractionally differentiable up to a "critical order" between 0 and 1. It was further shown that this critical order is equivalent to local Hölder exponent of the function. It was also demonstrated that LFD can be used to study local scaling properties of multifractal functions.

### 3.2 Local fractional Taylor expansion

Following the usual procedure to derive Taylor expansion with a remainder [46] one arrives at the fractional Taylor expansion for $N < q \leq N+1$ (provided $I\!\!D^q$ exists), given by,

$$f(x) = \sum_{n=0}^{N} \frac{f^{(n)}(y)}{\Gamma(n+1)} (x-y)^n + \frac{I\!\!D^q f(y)}{\Gamma(q+1)} (x-y)^q + R_q(x,y) \tag{9}$$

where

$$R_q(x,y) = \frac{1}{\Gamma(q+1)} \int_0^{x-y} \frac{dF(y,t;q,N)}{dt} (x-y-t)^q dt \tag{10}$$

where

$$F(y, x-y; q, N) = \frac{d^q(f(x) - \sum_{n=0}^{N} \frac{f^{(n)}(y)}{\Gamma(n+1)}(x-y)^n)}{[d(x-y)]^q} \qquad (11)$$

We note that the local fractional derivative (not the usual nonlocal fractional derivative) as defined above provides the coefficient $A$ in the approximation of $f(x)$ by the function $f(y) + A(x-y)^q/\Gamma(q+1)$, for $0 < q < 1$, in the vicinity of $y$. We further note that the terms on the RHS of eqn(9) are nontrivial and finite only in the case $q = \alpha$. An interesting point to be noted here is that the LFD appears in Taylor expansion naturally and independently of reasoning given in section 3.1. This leads us to believe that there must be something more to the LFD than just a quantity introduced to study fractional differentiability property.

The existence of such Taylor expansion assigns a geometrical interpretation to the LFD. In order to see this note that when $q$ is set equal to unity in the equation (9) one gets the equation of the tangent. It may be recalled that all the curves passing through a point $y$ and having the same tangent form an equivalence class (which is modeled by a linear behavior). Analogously, all the functions (curves) with the same critical order $\alpha$ and the same $I\!\!D^\alpha$ will form an equivalence class modeled by $x^\alpha$. This is how one may generalize the geometric interpretation of derivatives in terms of 'tangents'.

## 3.3 Local fractional differential equations

The most important equations of physical sciences are differential or integral equations, two themes emerging from the seventeenth-century calculus. The derivatives of integer order are coefficients of integer powers in the Taylor expansion. In particular the first order derivative models a local linear behaviour and the existence of the first order derivative implies Lipschitz continuity of the function. Therefore it is not possible to have Hölder continuous functions as solutions to ordinary differential equations. Such functions arise while dealing with phenomena involving fractals. Hence it becomes necessary to develop calculus which would go beyond ordinary calculus and incorporate Hölder continuous functions as well.

The LFDs that we have studied in section 3.1 preserve local nature of the derivative operator. Moreover they characterize the Hölder exponent of an irregular function. This suggests these LFDs may provide a much needed tool for doing calculus for fractal space-time. Hence it becomes pertinent to ask the questions such as: what is the inverse of the LFD operator (if it exists), or can one write and solve equations involving LFDs? Also what meaning and applications these local fractional differential equations will potentially have? In this section we attempt to answer some of these questions. It should be emphasized that these are new kind of equations and unlike FDEs considered in the last section are not just integrodifferential equations.

In order to understand the meaning of these equations we consider a simple equation

$$I\!D_x^q f(x) = g(x). \tag{12}$$

We note that [42] the equation $I\!D_x^q f(x) = \text{const}$, does not have a finite solution when $0 < q < 1$. Interestingly, the solutions to (12) can exist, when $g(x)$ has a fractal support. For instance, when $g(x) = \chi_C(x)$, the membership function of a cantor set $C$ (i.e. $g(x) = 1$ if $x$ is in $C$ and $g(x) = 0$ otherwise), the solution with initial condition $f(0) = 0$ exists if $q = \alpha \equiv \dim_H C$. Explicitly, generalizing the Riemann integration procedure,

$$f(x) \equiv \frac{P_C(x)}{\Gamma(\alpha+1)} = \lim_{N \to \infty} \sum_{i=0}^{N-1} \frac{(x_{i+1} - x_i)^\alpha}{\Gamma(\alpha+1)} F_C^i \tag{13}$$

where $x_i$ are subdivision points of the interval $[x_0 = 0, x_N = x]$ and $F_C^i$ is a flag function which takes value 1 if the interval $[x_i, x_{i+1}]$ contains a point of the set $C$ and 0 otherwise. Note that $P_C(x)$ is a Lebesgue-Cantor (staircase) function and satisfies the bounds $ax^\alpha \leq P_C(x) \leq bx^\alpha$ where $a$ and $b$ are suitable positive constants. The above procedure of integration works only when the box dimension of $C$ is the same as the Hausdorff dimension.

## 4 A fractal measure

As mentioned in introduction, with proper choice of the known function a solution of a simple differential equation gives us the measure of the set involved. The solution of the differential equation in that case is nothing but the integral of the known function. Here we try to extend this idea to fractal measures and corresponding fractional integrals. It can be seen immediately that the Riemann-Liouville fractional integral can not give us a desired measure owing to its non-trivial kernel. It will fail to be additive.

But the inverse of the LFD given by equation (13), henceforth called fractal integral, is a right candidate for "fractional" integral we are looking for. We now define the fractional measure of a subset $A \cap [0, x]$ (assuming it to be measurable) as

$$\mathcal{F}^\alpha(A \cap [0, x]) = {}_0 I\!D_x^{-\alpha} 1_A(x). \tag{14}$$

It can be seen that this definition satisfies the additivity property, i.e., $\mathcal{F}^\alpha(A \cup B) = \mathcal{F}^\alpha(A) + \mathcal{F}^\alpha(B)$ if A and B are disjoint. In the case that $A \subset [0, x]$ and $B \subset [x, y]$ it is clear that the sum in equation (13) breaks into two parts implying the additivity. Of course, this remains to be proved for a general case. Consider an example of a one-third Cantor set $C$. For this set $\mathcal{F}$ can be written as

$$\mathcal{F}^\alpha(C) = {}_0 I\!D_1^{-\alpha} 1_C(x) \tag{15}$$

$$= \lim_{N \to \infty} \sum_{i=0}^{N-1} F_C^i \frac{(x_{i+1} - x_i)^\alpha}{\Gamma(\alpha+1)}, \tag{16}$$

where $F_C^i$ is a flag function which is 1 if a point of set $C$ belongs to the interval $[x_i, x_{i+1}]$ and zero otherwise. Clearly this measure will be infinite if $\alpha < \ln(2)/\ln(3)$ and zero if $\alpha > \ln(2)/\ln(3)$. At $\alpha = \ln(2)/\ln(3)$ we have $\mathcal{F}^{\ln(2)/\ln(3)}(C) = 1/\Gamma(1 + \ln(2)/\ln(3))$. This shows that at least for simple sets like Cantor sets this definition gives the same value of dimension as the Hausdorff dimension. It differs from the Hausdorff measures in the value of the normalization constant. The normalization constant used usually [1] in the definition of the Hausdorff measure is $[\Gamma(1/2)]^\alpha/\Gamma(1 + \alpha/2)$, which is an $\alpha$-dimensional volume of a ball of unit radius, whereas it is $1/\Gamma(\alpha + 1)$ in the definition above. The normalization constant in the Hausdorff measure has its origin in the geometry of the covering elements. On the other hand the above new normalization constant has its origin in the calculus. Rigorous results establishing that such a construction defines measure on a Borel $\sigma$-algebra will be given elsewhere [43].

## 5 Local fractional diffusion equation

The purpose of this section is to demonstrate that the ideas studied in this paper find applications in modeling phenomena involving fractals. To this end we consider a local fractional diffusion equation which involves LFD with respect to time only. We solve this equation by discretizing the LFD operator with the help of fractional Taylor series. In other words we invert the LFD operator using the algorithm used in section 3.3. Interestingly the fractal measure as introduced in section 4 appears naturally in the solution.

In this section we consider a local fractional diffusion equation given by (compare Ref. [42])

$$\mathbb{D}_t^\alpha W(x,t) = \frac{1_C(t)}{2} \frac{\partial^2}{\partial x^2} W(x,t), \qquad (17)$$

where $W(x,t)$ is a probability density for finding a particle in neighbourhood of $x$ at time $t$. This equation contains LFD with respect to time and the usual second order derivative with respect to space. The diffusion constant is replaced by an indicator function of a Cantor like set with dimension $\alpha$. This means that the transition is very rare and takes place only when the time instance lies in $C$. This can be used to model many phenomena in physical and biological sciences where the diffusion is slow and there is a power law in the time instances on which events occur. We feel that the examples may include diffusion in presence of traps, transport in the phase space of chaotic Hamiltonian systems, avalanches in the sandpile models, subdiffusive transport across biomembranes etc.

The equation given above is a special case of a more general local fractional Fokker-Planck equation derived recently [42]. The solution [42] of this equation can be written by discretizing it as follows:

$$W(x,t) = P_{t-t_0} W(x,t_0) \qquad (18)$$

where

$$P_{t-t_0} = \lim_{N\to\infty} \prod_{i=1}^{N} [1 + \frac{1}{2}(t_i - t_{i-1})^\alpha \frac{F_C^i}{\Gamma(\alpha+1)} \frac{\partial^2}{\partial x^2}]. \quad (19)$$

After taking the limit and using equation (16) this gives us

$$W(x,t) = \exp(\frac{\mathcal{F}(C \cap [0,t])}{2} \frac{\partial^2}{\partial x^2}) W(x,0). \quad (20)$$

If we choose $W(x,0) = \delta(x)$ we get

$$W(x,t) = \frac{1}{\sqrt{2\pi \mathcal{F}(C \cap [0,t])}} \exp(\frac{-x^2}{2\mathcal{F}(C \cap [0,t])}) \quad (21)$$

We can immediately find its mean square displacement as

$$<x^2> = 2\mathcal{F}(C \cap [0,t]). \quad (22)$$

It is clear that $<x^2>$ is proportional to the fractal measure, in the sense defined in section 4, of the 'effective' evolution time. When the dimension of C becomes one we get back the usual result for diffusion, i.e., mean square displacement is proportional to $t$. Since $\mathcal{F}(C \cap [0,t])$ is proportional to $t^\alpha$ the equation (21) gives a subdiffusive solution.

# 6 Conclusions

Since the ordinary calculus can not deal with fractal structures and processes there is a need to develop a new calculus. Recent surge in applications of fractional calculus to fractals suggests the use of the fractional calculus for this purpose. In this paper we demonstrated that recent developments leading to introduction of LFD and LFDEs give rise to a new calculus which we call local fractional calculus. We indicated which are all structures of ordinary calculus get generalized in local fractional calculus. Though there are no analogues of Leibnitz rule, chain rule, etc. the local fractional derivative satisfies a fractional Taylor expansion which is useful in approximation schemes. Considering the uses of Taylor series in ordinary calculus we feel this is an important development. A detailed study comparing ordinary calculus and local fractional calculus should be carried out.

We have indicated how the inverse of LFD gives rise to a fractal measure. Local fractional differential equations appear to be suitable to study phenomena in fractal space-time. We have demonstrated this with the help of a local fractional diffusion equation in which diffusion is very rare and takes place in fractal time. This example shows that with the help of local fractional calculus we can incorporate not only the phenomena in fractal space-time but also the fractal measure into an equation and its solution.

## Acknowledgements

One of the authors (KMK) would like to thank DST (India) (DST: PAM: GR: 381) for financial assistance.

## References

1. B.B. Mandelbrot (1977): *The Fractal Geometry of Nature.* Freeman, New York.
2. A. Bunde and S. Havlin, Eds (1995): *Fractals in science.* Springer.
3. S. Baldo, F. Normant and C. Tricot, Eds. (1994): *Fractals in Engineering.* World Scientific, Singapore.
4. J. Lévy-vehel, E. Lutton and C. Tricot, Eds. (1997): *Fractals in Engineering.* Springer.
5. A.-L. Barabási and H. E.Stanley (1995): *Fractal concepts in surface growth.* Cambridge University Press.
6. T. Vicsek (1989): *Fractal growth phenomenon.* World Scientific.
7. H. Peitgen, H. Jurgens and D. Saupe (1992): *Chaos and fractals: New frontiers of science.* Springer, New York.
8. L. Nottale (1996): *Chaos, Solitons & Fractals.* **7**, 877.
9. K. Falconer (1990): *Fractal Geometry.* John Wiley, New York.
10. C. Tricot (1993): *Curves and fractal dimension.* Springer, New York.
11. J. L. Kaplan, J. Malet-Peret and J. A. Yorke (1994). Ergodic Th. and Dyn. Syst. **4**, 261.
12. R. P. Feynmann and A. R. Hibbs (1965): *Quantum Mechanics and Path Integrals.* McGraw-Hill, New York.
13. P. Constantin, I. Procaccia and K. R. Sreenivasan (1991). Phys. Rev. Lett. **67**, 1739.
14. P. Constantin and I. Procaccia (1994). Nonlinearity **7**, 1045.
15. L. F. Abott and M. B. Wise (1981): Am. J. Phys. **49**, 37.
16. K. M. Kolwankar (1997). Ph. D. thesis, University of Pune.
17. K. S. Miller and B. Ross (1993): *An Introduction to the Fractional. Calculus and Fractional Differential Equations.* John Wiley, New York.
18. K. B. Oldham and J. Spanier (1974): *The Fractional Calculus.* Academic Press, New York.
19. S. G. Samko, A. A. Kilbas and O. I. Marichev (1993): *Fractional integrals and derivatives, theory and applications.* Gordon & Breach.
20. A. Carpinteri and F. Mainardi, eds. (1997): *Fractals and Fractional Calculus in Continuum Mechanics.* Springer, New York.
21. R. Hilfer, ed. (1998): *Applications of Fractional Calculus in Physics.* World Scientific, Singapore.
22. T. J. Osler (1971). *SIAM J. Math. Anal.* **2**, 37–48.
23. M. Caputo and F. Mainardi (1971). *Pure Appl. Geophys.,* **91**, 134–147.
24. Nigmatullin R. (1986). *Phys. Stat. Sol.* **B133**, 425–430.
25. T. F. Nonnenmacher and G. W. Glöckle (1991). *Phil. Mag. Lett.* **B64**, 89–93.
26. H. Schiessel and A. Blumen (1993). *J. Phys. A: Math. Gen.* **26**, 5057–5069.
27. F. Mainardi (1996). *Chaos, Solitons and Fractals* **7**, 1461–1477.
28. A. Compte and M. O. Cáceres (1998). *Phys. Rev. Lett,* **81**, 3140–43.
29. M. Giona and H. E. Roman (1992). *J. Phys. A: Math Gen.* **25**, 2093.
30. H. E. Roman and M. Giona (1992). *J. Phys. A: Math. Gen.* **25**, 2107.

31. G. M. Zaslavsky (1994). *Physica D* **76**, 110.
32. G. M. Zaslavsky (1994). *Chaos*, **4**, 25.
33. T. F. Nonnenmacher (1990). *J. Phys. A: Math. gen.* **23**, L697–L700.
34. J. P. Bouchaud and A. Georges (1990). *Phys. Rep.* **195**, 127.
35. B. Souillard (1993). In *Chance and Matter* edited by J. Souletie, J. Vannimenus ans R. Stora, North Holland, Amsterdam.
36. K. Sarkar and C. Meneveau (1993). *Phys. Rev. E* **47** 957.
37. B. B. Mandelbrot and J. W. Van Ness (1968). *SIAM Rev.* **10**, 422.
38. K. L. Sebastian (1995). *J. Phys.* **A28**, 4305.
39. K. M. Kolwankar and A. D. Gangal (1996). *Chaos* **6**, 505.
40. H. Risken (1984): *The Fokker-Planck Equation*. Springer-Verlag, Berlin.
41. W. Feller (1968): *An Introduction to Probability Theory and its Applications*. Wiley, New York, Vol 2.
42. K. M. Kolwankar and A. D. Gangal (1998). *Phys. Rev. Lett.* **80** 214.
43. K. M. Kolwankar and A. D. Gangal. Unpublished.
44. K. M. Kolwankar and A. D. Gangal (1997). *Pramana-J. Phys.* **48**, 49.
45. K. M. Kolwankar and A. D. Gangal (1997). In proceedings of 'Fractals in Engineering', Arcachon, France.
46. R. Courant and F. John (1965): *Introduction to calculus and analysis*. John Wiley, Vol 1.
47. H. L. Royden, *Real Analysis 3e* (Macmillan, New York, 1988).
48. K. J. Falconer, *The Geometry of Fractal Sets* (Cambridge University Press, Cambridge, 1986).
49. P. R. Halmos, *Measure Theory* (Springer, New York, 1986).
50. Feder J., *Fractals*, Pergamon, 1988.

# PHYSICAL SCIENCES

# Conformal Multifractality of Random Walks, Polymers, and Percolation in Two Dimensions

Bertrand Duplantier

[1] Service de Physique Théorique de Saclay,
F-91191 Gif-sur-Yvette Cedex - FRANCE
[2] Institut Henri Poincaré,
11, rue Pierre et Marie Curie, F-75231 Paris Cedex 05 - FRANCE
[3] Isaac Newton Institute for Mathematical Sciences,
20 Clarkson Road, CB3 OEH - U. K.
bertrand@spht.saclay.cea.fr

**Abstract.** Our aim is to derive from conformal invariance the multifractal spectrum of the harmonic measure near a random fractal, such as the frontier of a random walk, i.e., a Brownian motion, a self-avoiding walk, or a percolation cluster. First we consider the related problem of $L$ planar random walks (or Brownian motions) of large time $t$, starting at neighboring points, and the probability $P_L(t) \approx t^{-\zeta_L}$ that their paths do not intersect. By a $2D$ *quantum gravity* method, i.e., a non linear map onto a *random Riemann surface*, the former conjecture that $\zeta_L = \frac{1}{24}\left(4L^2 - 1\right)$ is established. This also applies to the half-plane where $\tilde{\zeta}_L = \frac{L}{3}(1 + 2L)$. The non-intersection exponents of *unions* of independent paths are obtained from generalization of the above formulae to non integer or non rational values of $L$. In particular, Mandelbrot's conjecture for the Hausdorff dimension $D_H = 4/3$ of the frontier of a Brownian path follows from $L = \frac{3}{2}$, as $D_H = 2 - 2\zeta_{3/2}$. The same techniques apply to the harmonic measures (or electrostatic potential, or diffusion field) near a RW or a SAW, or near a critical *percolation* cluster, whose moments exhibit a multifractal spectrum. The generalized dimensions $D(n)$ as well as the multifractal functions $f(\alpha)$ are derived, and are shown to be all identical for a Brownian motion, a polymer, or a percolation cluster. These are examples of exact conformal multifractality. They are generalized to Potts clusters.

## 1 Introduction

Quantum mechanics and, more importantly, interacting quantum field theory can be described very generally in terms of the statistics of Brownian paths and of their intersections [1]. This equivalence is used in polymer theory [1] and in rigorous studies of second-order phase transitions and field theories [2]. In probability theory, non trivial properties of Brownian paths have led to intriguing conjectures. Mandelbrot [3] suggested for instance that in two dimensions, the external frontier of a planar Brownian path has a Hausdorff dimension $D = 4/3$, identical to that of a planar sef-avoiding walk, i.e., a polymer. Families of universal critical exponents are associated with *intersection* properties of sets of random walks[4-9].

On another hand, the concepts of generalized dimensions and associated multifractal (Mf) measures have been developed more than a decade ago [10-13]. Universal geometrical fractals, e.g., random walks, polymers, Ising or percolation models are essentially related to standard critical phenomena and field theory, for which conformal invariance in two dimensions (2D) has brought a wealth of exact results [14-18]. By contrast, few connections between multifractals and field theory have been found, although the algebras of their respective correlation functions reveal intriguing similarities [19]. The moments of the harmonic measure, i.e., the Laplacian diffusion field near an absorber, the latter taken as a simple random walk (RW, i.e., Brownian motion), or self-avoiding walk (SAW, i.e., polymer), exhibit multifractal scaling [20].

Consider a two-dimensional very large "absorber" $\mathcal{S}$, which can be a random walk, or a self-avoiding walk. Define $H(w)$ as the probability that another random walker (RW) launched from infinity, *first* hits the outer "hull" or (accessible) frontier $\mathcal{H}(\mathcal{S})$ at point $w \in \mathcal{H}(\mathcal{S})$. One then considers a covering of $\mathcal{H}$ by balls $\mathcal{B}(w,a)$ of radius $a$, and centered at points $w \in \mathcal{H}/\{a\}$ forming a discrete subset of $\mathcal{H}$. Let $H(\mathcal{H} \cap \mathcal{B}(w,a))$ be the harmonic measure of the points of $\mathcal{H}$ in the ball $\mathcal{B}(w,a)$. We are especially interested in the moments of $H$, averaged over all realizations of RW's and $\mathcal{S}$ (*annealed* average)

$$\mathcal{Z}_n = \left\langle \sum_{w \in \mathcal{H}/\{a\}} H^n \left( \mathcal{H} \cap \mathcal{B}(w,a) \right) \right\rangle, \quad (1)$$

where $n$ can be, *a priori*, a real number. For very large absorbers $\mathcal{S}$ and hulls $\mathcal{H}(\mathcal{S})$ of average size $R$, one expects these moments to scale as

$$\mathcal{Z}_n \approx (a/R)^{\tau(n)}, \quad (2)$$

where the radius $a$ serves, in physics, as a microscopic cut-off, reminiscent of the lattice structure, and where the multifractal scaling exponents $\tau(n)$ encode *generalized dimensions*

$$D(n) = \frac{\tau(n)}{n-1}, \quad (3)$$

which vary in a non-linear way with $n$[10-13]. Several *a priori* results are known. $D(0)$ is the Hausdorff dimension of the accessible frontier of the fractal. By construction, $H$ is a normalized probability measure, so that $\tau(1) = 0$. Makarov's theorem [21], here applied to the Hölder regular curve describing the frontier [22], gives the *non trivial* information dimension $\tau'(1) = D(1) = 1$. The multifractal formalism [10-13] further involves characterizing subsets $\mathcal{H}_\alpha$ of sites of the hull $\mathcal{H}$ by a Hölder exponent $\alpha$, such that the H-measure of the frontier points in the ball $\mathcal{B}(w,a)$ of radius $a$ centered at $w$ scales as

$$H(\mathcal{H} \cap \mathcal{B}(w,a), w \in \mathcal{H}_\alpha) \approx (a/R)^\alpha. \quad (4)$$

The "fractal dimension" $f(\alpha)$ of the set $\mathcal{H}_\alpha$, such that

$$\text{Card}\mathcal{H}_\alpha \approx R^{f(\alpha)}, \tag{5}$$

is given by the symmetric Legendre transform of $\tau(n)$ :

$$\alpha = \frac{d\tau}{dn}(n), \quad \tau(n) + f(\alpha) = \alpha n, \quad n = \frac{df}{d\alpha}(\alpha). \tag{6}$$

Because of the statistical ensemble average (1), values of $f(\alpha)$ can become negative for some domains of $\alpha$ [19].

As we shall see, the associated exponents $\tau(n)$ above can be recast as those of star copolymers made of independent RW's in a bunch, diffusing away from a generic point of the absorber.

Percolation theory is another archetypal model for critical phenomena. Its scaling (continuum) limit is assumed to enjoy conformal invariance, which present a mathematical challenge[23-25]. The harmonic measure near a percolation cluster also possesses multifractal exponents.

I report here recent works which show that in two dimensions *all* these universal exponents and multifractal functions can be derived from conformal invariance methods involving *quantum gravity*, i.e., conformal invariance on a random Riemann surface [26-28]. This establishes former conjectures concerning the intersections of random walks [7], including the conjecture above for the Hausdorff dimension of the Brownian frontier. Actually, these Brownian intersection exponents are linked with the exponents governing the multifractal moments of the harmonic measure. The conformal approach thus also provides the exact multifractal harmonic spectrum $f(\alpha)$ of a RW or a SAW, or a *percolation cluster*, an example of exact *conformal multifractality*.

Several caveats are in order here, especially for an audience in applied mathematics or engineering sciences. The results given below are obtained by theoretical physics methods. This means that some points are not fully rigorous, and involve some assumptions. However, the body of knowledge in the statistical mechanics of critical phenomena, or in conformal invariance in two-dimensions, accumulated over the last twenty years, allows multiple cross-checks which leave no doubt concerning the validity of the methods implied. Thus, the results given here, if not *stricto sensu* rigorous, are expected to be fully exact. When this is possible, I will make contact with related rigorous results in probability theory of Brownian motion, or conformal invariants [29].

Let me add that the latest *quantum gravity* techniques used here are not yet widely known, even in statistical mechanics, but rather belong to string field theory. This hints at deep connections between these two apparently remote fields, probability theory and string field theory. In particular, the correspondence extensively used here, which exists between scaling laws in the plane, and on a random Riemann surface is fundamental, and I have no doubt that it will finally make his way into the theory of probability.

## 2 Intersections of Random Walks

I shall first define intersection exponents for random walks or Brownian motions, which are simpler than the multifractal exponents considered above, but which, in fact, generate the latter. Consider a number $L$ of independent random walks (or Brownian paths) $B^{(l)}, l = 1, .., L$ in $\mathbf{Z}^d$ (or $\mathbf{R}^d$), starting at fixed neighboring points, and the probability

$$P_L(t) = P\left\{\cup_{l,l'=1}^{L}(B^{(l)}[0,t] \cap B^{(l')}[0,t]) = \emptyset\right\}, \quad (7)$$

that the intersection of their paths up to time $t$ is empty[4, 6]. At large times and for $d < 4$, one expects this probability to decay as

$$P_L(t) \approx t^{-\zeta_L}, \quad (8)$$

where $\zeta_L(d)$ is a *universal* exponent depending only on $L$ and $d$. Above the upper critical dimension $d = 4$, RWs almost surely do not intersect. The existence of exponents $\zeta_L$ in $d = 2, 3$ and their universality have been proven[9], and they can be calculated near $d = 4$ by renormalization theory [6]. A generalization was introduced [7] for $L$ walks constrained to stay in a half-plane, and starting at neighboring points on the boundary, with a non-intersection probability $\tilde{P}_L(t)$ of their paths governed by a "surface" critical exponent $\tilde{\zeta}_L$ such that

$$\tilde{P}_L(t) \approx t^{-\tilde{\zeta}_L}. \quad (9)$$

We have conjectured from conformal invariance arguments and numerical simulations that in 2D [7]

$$\zeta_L = h_{0,L}^{(c=0)} = \frac{1}{24}\left(4L^2 - 1\right), \quad (10)$$

and for the half-plane

$$2\tilde{\zeta}_L = h_{1,2L+2}^{(c=0)} = \frac{1}{3}L(1+2L), \quad (11)$$

where $h_{p,q}^{(c)}$ denotes the Kač conformal weight

$$h_{p,q}^{(c)} = \frac{[(m+1)p - mq]^2 - 1}{4m(m+1)}, \quad (12)$$

of a minimal conformal field theory of central charge $c = 1 - 6/m(m+1)$, $m \in \mathbf{N}^*$ [15]. For Brownian motions $c = 0$, and $m = 2$. For $L = 1$, the intriguing $\zeta_1 = 1/8$ is actually the disconnection exponent governing the probability that the origin of a single walk remains accessible from infinity without crossing the walk.

To derive the conjectured intersection exponents above, the idea [26] is to map the original random walk problem in the plane onto a random lattice with planar geometry, or, in other words, in presence of two-dimensional *quantum gravity* [30]. The key point is that the random walk intersection exponents on the random lattice are related to those in the plane. Furthermore, the RW intersection problem can be solved in quantum gravity. Thus, the exponents $\zeta_L$ (Eq. (10)) and $\tilde{\zeta}_L$ (Eq. (11)) in the standard Euclidean plane are derived from this mapping to a random lattice or Riemann surface.

Random surfaces, in relation to string theory [31], have been the subject and source of important developments in statistical mechanics in two-dimensions. In particular, the discretization of string models led to the consideration of abstract random lattices $G$, the connectivity fluctuations of which represent those of the metric, i.e. pure 2D quantum gravity [32]. One can then put any 2D statistical model (like Ising model [33], self-avoiding walks [34]) on the random planar graph $G$, thereby obtaining a new critical behavior, corresponding to the confluence of the criticality of the random surface $G$ with the critical point of the original model. The critical system "dressed by gravity" has a larger conformal symmetry which allowed Knizhnik, Polyakov, and Zamolodchikov (KPZ) [30] to establish the existence of a *relation* between the conformal dimensions $\Delta^{(0)}$ of scaling operators in the plane and those in presence of gravity, $\Delta$ :

$$\Delta^{(0)} = \Delta\left[1 - (1-\Delta)/\kappa\right], \tag{13}$$

where $\kappa$ is a parameter related to the central charge of the statistical model in the plane:

$$c = 1 - 6(1-\kappa)^2/\kappa; \tag{14}$$

for a minimal model of the series (12), with $\kappa = 1 + 1/m$, and $\Delta_{p,q}^{(0)} \equiv h_{p,q}^{(c)}$.

Let us now consider as a statistical model *random walks* on a *random graph*. We know [7] that their central charge $c = 0$, whence $m = 2$, $\kappa = 3/2$. Thus the KPZ relation becomes

$$\Delta^{(0)} = U(\Delta) \equiv \frac{1}{3}\Delta(1+2\Delta), \tag{15}$$

which has exactly the same analytical form as the conjecture (11)! Thus, from the KPZ equation one infers that the planar Brownian intersection exponents Eqs. (10,11) are equivalent to Brownian intersection exponents in quantum gravity:

$$\Delta_L = \frac{1}{2}(L - \frac{1}{2}), \tag{16}$$

$$\tilde{\Delta}_L = L. \tag{17}$$

Let us now sketch the derivation of the latter quantum gravity exponents.

Consider the set of planar random graphs $G$, built up with, e.g., trivalent vertices tied together in a *random way*. The topology is fixed here to be that of a sphere ($\mathcal{S}$) or a disc ($\mathcal{D}$). The partition function is defined as

$$Z_\chi(\beta) = \sum_G \frac{1}{S(G)} e^{-\beta|G|}, \tag{18}$$

where $\chi$ denotes the Euler characteristic $\chi = 2\,(\mathcal{S})\,,1\,(\mathcal{D})\,; |G|$ is the number of vertices of $G$, $S\,(G)$ its symmetry factor. The partition sum converges for all values of the parameter $\beta$ larger than some critical $\beta_c$. At $\beta \to \beta_c^+$, a singularity appears due to the presence of infinite graphs in (18)

$$Z_\chi(\beta) \sim (\beta - \beta_c)^{2-\gamma_{\text{str}}(\chi)}, \qquad (19)$$

where $\gamma_{\text{str}}(\chi)$ is the string susceptibility exponent. For pure gravity as described in (18), the embedding dimension $d = 0$ coincides with the central charge $c = 0$, and $\gamma_{\text{str}}(\chi) = 2 - \frac{5}{4}\chi$ [36].

Now, put a set of $L$ random walks $\mathcal{B} = \{B_{ij}^{(l)}, l = 1, ..., L\}$ on the *random graph* $G$ with the special constraint that they start at the same vertex $i \in G$, end at the same vertex $j \in G$, and have no intersections in between. We introduce the $L$−walk partition function on the random lattice [26]:

$$Z_L(\beta,z) = \sum_{\text{planar } G} \frac{1}{S(G)} e^{-\beta|G|} \sum_{i,j \in G} \sum_{\substack{B_{ij}^{(l)} \\ l=1,...,L}} z^{|\mathcal{B}|}, \qquad (20)$$

where a fugacity $z$ is associated with the total number $|\mathcal{B}| = \left|\cup_{l=1}^L B^{(l)}\right|$ of vertices visited by the walks.

We generalize this to the *boundary* case where $G$ now has the topology of a disc and where the random walks connect two sites $i$ and $j$ now on the boundary $\partial G$:

$$\tilde{Z}_L(\beta,\beta',z) = \sum_{\text{disc } G} e^{-\beta|G|} e^{-\beta'|\partial G|} \sum_{i,j \in G} \sum_{\substack{B_{ij}^{(l)} \\ l=1,...,L}} z^{|\mathcal{B}|}, \qquad (21)$$

where $e^{-\beta'}$ is the fugacity associated with the boundary's length. The double grand canonical partition function (20) associated with non-intersecting RW's on a random lattice can be calculated exactly [26]. The critical behavior of $Z_L(\beta, z)$ is then obtained by taking the double scaling limit $\beta \to \beta_c$ (infinite random surface) and $z \to z_c$ (infinite RW's). The analysis of this singular behavior in terms of conformal dimensions is performed by using *finite size scaling* (FSS) [34], where one must have $|\mathcal{B}| \sim |G|^{\frac{1}{2}}$. One obtains [26]:

$$Z_L(\beta,z) \sim (\beta - \beta_c)^L \sim |G|^{-L}. \qquad (22)$$

$Z_L$ (20) represents a random surface with two *punctures* where two conformal operators of dimension $\Delta_L$ are located (here two vertices of $L$ non-intersecting RW's), and in a graphical way scales as

$$Z_L \sim Z[\,\boxed{\bullet\ \bullet}\,] \times |G|^{-2\Delta_L} \qquad (23)$$

where the partition function of the two-puncture surface is the second derivative of $Z_{\chi=2}(\beta)$ (19). The latter two equations yield

$$2\Delta_L - \gamma_{\text{str}}(\chi = 2) = L, \tag{24}$$

where $\gamma_{\text{str}}(\chi = 2) = -1/2$. We thus get the announced result

$$\Delta_L = \frac{1}{2}(L - \frac{1}{2}). \tag{25}$$

For the boundary partition function $\tilde{Z}_L$ (21) a similar analysis can be performed near the triple critical point where the boundary length also diverges. The boundary partition function $\tilde{Z}_L$ corresponds to two boundary operators of dimensions $\tilde{\Delta}_L$, integrated over $\partial G$, on a random surface with the topology of a disc, or in graphical terms:

$$\tilde{Z}_L \sim Z(\bigcirc) \times |\partial G|^{-2\tilde{\Delta}_L}. \tag{26}$$

From the exact calculation of the *boundary* partition function (21), one gets the further equivalence to the *bulk* one:

$$\tilde{Z}_L / Z(\bigcirc) \sim Z_L, \tag{27}$$

where the equivalences hold true in terms of scaling behavior. Comparing eqs. (26), (27), and (22), and using the FSS $|\partial G| \sim |G|^{1/2}$ gives

$$\tilde{\Delta}_L = L. \tag{28}$$

Applying the quadratic KPZ relation (15) to $\Delta_L$ and $\tilde{\Delta}_L$ above yields at once the values in the plane $\mathbf{R}^2, \Delta_L^{(0)} \equiv \zeta_L$ (Eq. (10)), and $\tilde{\Delta}_L^{(0)} \equiv 2\tilde{\zeta}_L$ (Eq. (11)).

Consider now the exponents $\zeta(n_1, .., n_L) = \Delta^{(0)}\{n_l\}$, as well as $2\tilde{\zeta}(n_1, .., n_L) = \tilde{\Delta}^{(0)}\{n_l\}$, describing $L$ mutually-avoiding bunches $l = 1, .., L$, each made of $n_l$ *independent* walks, i.e., mutually "transparent"[37], with possible mutual intersections in a bunch. In presence of gravity each bunch will contribute its own *normalized boundary partition function* as a factor, and yield a natural generalization of (27)

$$Z\{n_l\} \sim \frac{\tilde{Z}\{n_l\}}{Z(\bigcirc)} \sim \prod_{l=1}^{L} \frac{\tilde{Z}(n_l)}{Z(\bigcirc)}, \tag{29}$$

to be identified with $|\partial G|^{-2\tilde{\Delta}\{n_l\}}$. The *factorization* property (29) immediately implies the *additivity of boundary conformal dimensions in presence of gravity*

$$\tilde{\Delta}\{n_1, .., n_L\} = \sum_{l=1}^{L} \tilde{\Delta}(n_l), \tag{30}$$

where $\tilde{\Delta}(n)$ is now the boundary dimension of a *single* bunch of $n$ transparent walks on the random surface. We know $\tilde{\Delta}(n)$ exactly since it corresponds in the standard plane to a trivial surface conformal dimension $\tilde{\Delta}^{(0)}(n) = n$. It thus suffices to *invert* (15) to get

$$\tilde{\Delta}(n) = U^{-1}(n) = \frac{1}{4}(\sqrt{24n+1} - 1). \tag{31}$$

One notes the identification (29), on a random surface, of the bulk partition function with the ratio of boundary ones. In the plane, using once again the KPZ relation (15) for $\tilde{\Delta}\{n_l\}$ gives the general results [26]

$$\zeta(n_1,..,n_L) = V(x) \equiv \frac{1}{24}(4x^2 - 1), \tag{32}$$

$$2\tilde{\zeta}(n_1,..,n_L) = U(x) = \frac{1}{3}x(1+2x), \tag{33}$$

$$x = \sum_{l=1}^{L} U^{-1}(n_l) = \sum_{l=1}^{L} \frac{1}{4}(\sqrt{24n_l+1} - 1). \tag{34}$$

Lawler and Werner [29] proved by probabilistic means, using the geometrical conformal invariance of Brownian motions, the *existence* of two (unspecified) functions $U$ and $V$ satisfying the structure (32-34). The quantum gravity approach here explains this structure in terms of linear equation (30), and yields the explicit functions $U(x)$ and $V(x) \equiv U(\frac{1}{2}(x - \frac{1}{2}))$ of (32)(33).

Let us remark that the above equations yield for $\zeta(2, 1^{(L)})$ describing a two-sided walk and $L$ one-sided walks, all mutually non-intersecting,

$$\zeta(2, 1^{(L)}) = \zeta_{L+\frac{3}{2}} = V(L + \frac{3}{2}) = \frac{1}{6}(L+1)(L+2). \tag{35}$$

For $L = 1, \zeta(2,1) = \zeta_{5/2} = 1$ gives correctly the escape probability of a RW from another RW. For $L = 0, \zeta(2, 1^{(0)}) = \zeta_{3/2} = 1/3$ is related to the Hausdorff dimension of the frontier by $D = 2 - 2\zeta[38]$. Thus we obtain

$$D = 2 - 2\zeta_{\frac{3}{2}} = \frac{4}{3}, \tag{36}$$

i.e., *Mandelbrot's conjecture*. The quantum geometric structure explicited here allows several generalizations, which we now describe [27].

## 3 Random walks and Copolymers

Consider a general star copolymer $\mathcal{S}$ in the plane $\mathbf{R}^2$ (or in $\mathbf{Z}^2$), made of an arbitrary mixture of Brownian paths or RW's (set $\mathcal{B}$), and polymers or

SAW's (set $\mathcal{P}$), all starting at neighboring points. Any pair $(A, B)$ of such paths, $A, B \in \mathcal{B}$ or $\mathcal{P}$, can be constrained in a specific way: either they avoid each other $(A \cap B = \emptyset$, noted $A \wedge B)$, or they are independent, i.e., "transparent" and can cross each other (noted $A \vee B$)[27, 40, 39]. This notation allows any *nested* interaction structure [27]; one can decide for instance that the branches $\{P_\ell \in \mathcal{P}\}_{\ell=1,...,L}$ of an $L$-star polymer, all mutually avoiding, further avoid a bunch of Brownian paths $\{B_k \in \mathcal{B}\}_{k=1,...,n}$, all transparent to each other:

$$\mathcal{S} = \left(\bigwedge_{\ell=1}^{L} P_\ell\right) \wedge \left(\bigvee_{k=1}^{n} B_k\right). \tag{37}$$

In 2D the order of the branches of the star copolymer *does* matter and is intrinsic to our $(\wedge, \vee)$ notation.

To each *specific* star copolymer center $\mathcal{S}$ is attached a conformal scaling operator with a scaling dimension $x(\mathcal{S})$. To obtain proper scaling we consider the partition functions of Brownian paths and polymers having the same mean size $R$. When the star is constrained to stay in a *half-plane* with its core placed near the *boundary*, its partition function will scale with new boundary scaling dimension $\tilde{x}(\mathcal{S})$ [7, 17, 18].

Any scaling dimension $x$ in the bulk is twice the *conformal dimension* (c.d.) $\Delta^{(0)}$ of the corresponding operator, while near a boundary (b.c.d.) they are identical:

$$x = 2\Delta^{(0)}, \quad \tilde{x} = \tilde{\Delta}^{(0)}. \tag{38}$$

As above, the idea is to use the representation where the RW's or SAW's are on a 2D random lattice, or a random Riemann surface, i.e., in the presence of 2D *quantum gravity* [30]. The general relation (15) depends only on the central charge, and is valid for polymers, for which $c = 0$. Let us summarize the results [27], expressed here in terms of the scaling dimensions in the standard plane. For a critical system with central charge $c = 0$, the two universal functions:

$$U(x) = \frac{1}{3}x(1 + 2x), \quad V(x) = \frac{1}{24}(4x^2 - 1), \tag{39}$$

with $V(x) \equiv U\left(\frac{1}{2}\left(x - \frac{1}{2}\right)\right)$, generate all the scaling exponents. The scaling exponents $x(A \wedge B)$, and $\tilde{x}(A \wedge B)$, of two *mutually avoiding* stars $A, B$, with proper scaling exponents $x(A), x(B)$, or boundary exponents $\tilde{x}(A), \tilde{x}(B)$, obey the *star algebra* [26, 27]

$$x(A \wedge B) = 2V\left[U^{-1}(\tilde{x}(A)) + U^{-1}(\tilde{x}(B))\right]$$
$$\tilde{x}(A \wedge B) = U\left[U^{-1}(\tilde{x}(A)) + U^{-1}(\tilde{x}(B))\right], \tag{40}$$

where $U^{-1}(x)$ is the inverse function of $U$

$$U^{-1}(x) = \frac{1}{4}\left(\sqrt{24x + 1} - 1\right). \tag{41}$$

On a random surface, $U^{-1}(\tilde{x})$ is the boundary dimension corresponding to the value $\tilde{x}$ in $\mathbf{R} \times \mathbf{R}^+$, and the sum of $U^{-1}$ functions in Eq. (40) represents linearly the juxtaposition $A \wedge B$ of two sets of random paths near their random frontier, i.e., the product of two "boundary operators" on the random surface. The latter sum is mapped by the functions $U$, $V$, into the scaling dimensions in $\mathbf{R}^2$ [27].

The rules (40), which mix bulk and boundary exponents, come from simple factorization properties on a random Riemann surface, i.e., in quantum gravity [26, 27], (and are also recurrence relations in $\mathbf{R}^2$ between conformal Riemann maps of the successive mutually avoiding paths onto the line $\mathbf{R}$[29]).

If, on the contrary, $A$ and $B$ are *independent* and can overlap, then by trivial factorization of probabilities their dimensions are additive[27]

$$x(A \vee B) = x(A) + x(B),$$
$$\tilde{x}(A \vee B) = \tilde{x}(A) + \tilde{x}(B). \qquad (42)$$

It is clear at this stage that the set of equations above is *complete*. It allows for the calculation of any conformal dimensions associated with a star structure $\mathcal{S}$ of the most general type, as in (37), involving $(\wedge, \vee)$ operations separated by nested parentheses [27].

*Brownian-polymer exponents:* The single extremity scaling dimensions are for a RW or a SAW near a Dirichlet boundary in $\mathbf{R}^2$ [18, 41]

$$\tilde{x}_B(1) = \tilde{\Delta}_B^{(0)}(1) = 1, \ \tilde{x}_P(1) = \tilde{\Delta}_P^{(0)}(1) = \tfrac{5}{8} \qquad (43)$$

or on $G$, $\tilde{\Delta}_B(1) = U^{-1}(1) = 1$, $\tilde{\Delta}_P(1) = U^{-1}\left(\tfrac{5}{8}\right) = \tfrac{3}{4}$. Because of the star algebra described above these are the only numerical seeds, i.e., generators, we need.

Stars can include bunches of $n$ copies of transparent RW's or $m$ transparent SAW's. Their b.c.d.'s in $\mathbf{R}^2$ are respectively, by using (42) and (43), $\tilde{\Delta}_B^{(0)}(n) = n$ and $\tilde{\Delta}_P^{(0)}(m) = \tfrac{5}{8}m$, from which the inverse mapping to the random surface yields $\tilde{\Delta}_B(n) = U^{-1}(n)$ and $\tilde{\Delta}_P(m) = U^{-1}\left(\tfrac{5}{8}m\right)$. The star made of $L$ bunches $\ell \in \{1, ..., L\}$, each of them made of $n_\ell$ transparent RW's and of $m_\ell$ transparent SAW's, and the $L$ bunches being mutually avoiding, has planar scaling dimensions

$$\tilde{\Delta}^{(0)}\{n_\ell, m_\ell\} = U\left(\tilde{\Delta}\right), \ \Delta^{(0)}\{n_\ell, m_\ell\} = V\left(\tilde{\Delta}\right),$$
$$\tilde{\Delta}\{n_\ell, m_\ell\} = \sum_{\ell=1}^{L} U^{-1}\left(n_\ell + \tfrac{5}{8}m_\ell\right).$$

This encompasses all previously known exponents for RW's and SAW's [7, 17, 18].

## 4  Conformal Multifractality and the Harmonic Measure

The *harmonic measure*, i.e., the diffusion or electrostatic field near an equipotential fractal boundary[42], or, equivalently, the electric charge appearing on the frontier of a perfectly conducing fractal, possesses a self-similarity property, which is reflected in a *multifractal* (Mf) behavior. Cates and Witten [20] considered the case of the Laplacian diffusion field near a simple random walk, or near a self-avoiding walk. The associated exponents can be recast as those of star copolymers made of a bunch of independent RW's diffusing away from a generic point of the absorber. We recently gave the exact solution to this problem in two dimensions, as follows [27].

The two-dimensional "absorber" $\mathcal{S}$ can be a random walk, or a self-avoiding walk. Define the harmonic measure $H(w)$ as the probability that another random walker (RW) launched from infinity, *first* hits the outer "hull" or (accessible) frontier $\mathcal{H}(\mathcal{S})$ at point $w \in \mathcal{H}(\mathcal{S})$. A covering of $\mathcal{H}$ by balls $\mathcal{B}(w,a)$ of radius $a$ is centered at points $w \in \mathcal{H}/\{a\}$ forming a discrete subset $\mathcal{H}/\{a\}$ of $\mathcal{H}$. Let $H(\mathcal{H} \cap \mathcal{B}(w,a))$ be the harmonic measure of the intersection of $\mathcal{H}$ and the ball $\mathcal{B}(w,a)$. The moments of $H$, averaged over all realizations of RW's and $\mathcal{S}$ are defined as

$$\mathcal{Z}_n = \left\langle \sum_{w \in \mathcal{H}/\{a\}} H^n \left(\mathcal{H} \cap \mathcal{B}(w,a)\right) \right\rangle, \qquad (44)$$

where $n$ can be, *a priori*, a real number. In the limit of large absorbers $\mathcal{S}$ and hulls $\mathcal{H}(\mathcal{S})$ of average size $R$, or small covering radius $a$, i.e, $a/R \to 0$, one expects these moments to scale as

$$\mathcal{Z}_n \approx (a/R)^{\tau(n)}, \qquad (45)$$

where the multifractal scaling exponents $\tau(n)$ encode generalized dimensions $D(n)$, $\tau(n) = (n-1) D(n)$, which vary in a non-linear way with $n$ [10–13].

*A priori* results are known: $D(0)$ is the Hausdorff dimension of the set accessible to the Brownian motions representing the harmonic measure; by construction, the measure $H$ is normalized, so that $\tau(1) = 0$. Makarov's theorem [21], here applied to the connected frontier, gives the information dimension $\tau'(1) = D(1) = 1$. The multifractal formalism [10–13] further involves characterizing subsets $\mathcal{H}_\alpha$ of sites of the hull $\mathcal{H}$ by a Hölder exponent $\alpha$, such that the H-measure, i.e., the local charge appearing on the frontier points contained in the ball $\mathcal{B}(w,a)$ of radius $a$ and centered at $w$, scales as

$$H(\mathcal{H} \cap \mathcal{B}(w,a), w \in \mathcal{H}_\alpha) \approx (a/R)^\alpha. \qquad (46)$$

The "fractal dimension" $f(\alpha)$ of the set $\mathcal{H}_\alpha$, such that

$$\mathrm{Card}\mathcal{H}_\alpha \approx R^{f(\alpha)}, \qquad (47)$$

is given by the symmetric Legendre transform of $\tau(n)$:

$$\alpha = \frac{d\tau}{dn}(n), \quad \tau(n) + f(\alpha) = \alpha n, \quad n = \frac{df}{d\alpha}(\alpha). \qquad (48)$$

As said above, because of the statistical ensemble average (44), values of $f(\alpha)$ can become negative for some domains of $\alpha$ [20]. The existence of the harmonic multifractal spectrum $f(\alpha)$ for a Brownian path has been recently established [43].

By the very definition of the H-measure, $n$ independent RW's diffusing away from the absorber give a geometric representation of the $n^{th}$ moment $H^n$, for $n$ integer, and convexity arguments give the whole continuation to real values. When the absorber is a RW or a SAW of size $R$, the site average of its moments $H^n$ is represented by a copolymer star partition function $\mathcal{Z}_R(\mathcal{S}_\wedge n)$, where we have introduced the short-hand notation $\mathcal{S}_\wedge n \equiv \mathcal{S} \wedge (\vee B)^n$ describing the copolymer star made by the absorber $\mathcal{S}$ hit by the bunch $(\vee B)^n$ at the apex only [20, 27]. More precisely one has

$$\mathcal{Z}_n \approx R^2 \mathcal{Z}_R (\mathcal{S}_\wedge n) \tag{49}$$

where the absorber $\mathcal{S}$ is either the two-RW star $B \vee B$ or the two-SAW star $P \wedge P$, made of two non-intersecting SAW's. Owing to Eq.(45), we get the scaling relation

$$\tau(n) = x(\mathcal{S}_\wedge n) - 2. \tag{50}$$

Our formalism (40) immediately gives the scaling dimensions

$$x(\mathcal{S}_\wedge n) = 2V\left(\tilde{\Delta}(\mathcal{S}) + U^{-1}(n)\right), \tag{51}$$

where $\tilde{\Delta}(\mathcal{S})$ is as above the quantum gravity boundary dimension of the absorber $\mathcal{S}$ alone. For a RW absorber, we have $\tilde{\Delta}(B \vee B) = U^{-1}(2) = \frac{3}{2}$, while for a SAW $\tilde{\Delta}(P \wedge P) = 2\tilde{\Delta}_{P,1} = 2U^{-1}\left(\frac{5}{8}\right) = \frac{3}{2}$. The coincidence of these two values tells us that *in 2D the harmonic multifractal spectra $f(\alpha)$ of a random walk or a self-avoiding walk are identical.* The calculation gives [27]

$$\tau(n) = \frac{1}{2}(n-1) + \frac{5}{24}\left(\sqrt{24n+1} - 5\right), \tag{52}$$

$$\alpha = \frac{d\tau}{dn}(n) = \frac{1}{2} + \frac{5}{2}\frac{1}{\sqrt{24n+1}}, \tag{53}$$

$$D(n) = \frac{1}{2} + \frac{5}{\sqrt{24n+1}+5}, \quad n \in \left[-\tfrac{1}{24}, +\infty\right), \tag{54}$$

$$f(\alpha) = \frac{25}{48}\left(3 - \frac{1}{2\alpha - 1}\right) - \frac{\alpha}{24}, \quad \alpha \in \left(\tfrac{1}{2}, +\infty\right). \tag{55}$$

The corresponding universal curves are shown in Fig. 1: $\tau(n)$ is half a parabola, and $f(\alpha)$ a hyperbola. $D(1) = \tau'(1) = 1$ is Makarov's theorem. The singularity at $\alpha = \frac{1}{2}$ in the multifractal functions $f(\alpha)$ corresponds to points on the fractal boundary $\mathcal{F}$ where the latter has the local geometry of a needle. The mathematical version of this statement is given by Beurling's theorem [44], which states

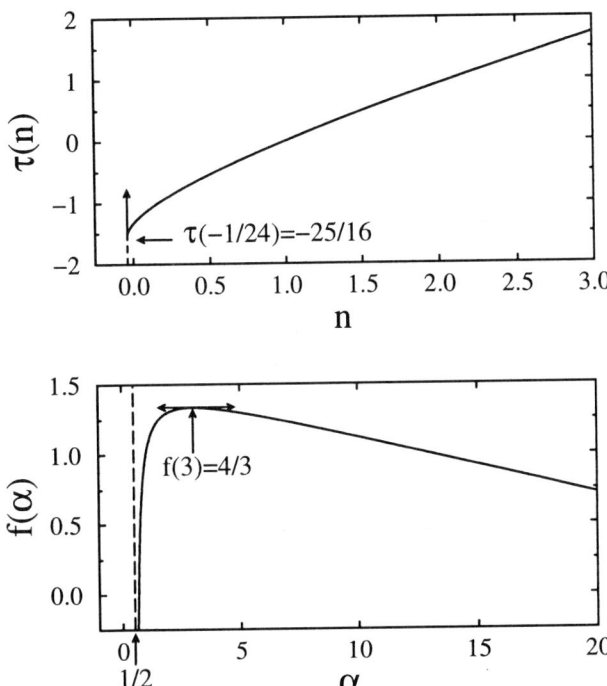

**Fig. 1.** Harmonic multifractal dimensions $\tau(n)$ and spectrum $f(\alpha)$ of a two-dimensional RW or SAW.

that at distance $\epsilon$ from the boundary, the harmonic measure is bounded above by

$$H\left(z : \inf_{w \in \mathcal{F}} |z - w| \leq \epsilon\right) \leq C\epsilon^{1/2}, \tag{56}$$

where $C$ is a constant. This insures that the spectrum of multifractal Hölder exponents $\alpha$ is bounded below by $\frac{1}{2}$. The right branch of the $f(\alpha)$ curve has a linear asymptote

$$\lim_{\alpha \to +\infty} \frac{1}{\alpha} f(\alpha) = -\frac{1}{24}. \tag{57}$$

Its linear shape is quite reminiscent of that of the multifractal function of the growth probability as in the case of a 2D DLA cluster [45]. The domain of large values of $\alpha$ corresponds to the lowest part $n \to n^* = -\frac{1}{24}$ of the spectrum of dimensions, which is dominated by almost inaccessible sites, and the existence of a linear asymptote to the multifractal function $f$ implies a peculiar behaviour for the number of those sites in a lattice setting. Indeed define $\mathcal{N}(H)$ as the number of sites having a probability $H$ to be hit:

$$\mathcal{N}(H) = \text{Card}\{w \in \mathcal{H} : H(w) = H\}. \tag{58}$$

Using the Mf formalism to change from variable $H$ to $\alpha$ (at fixed value of $a/R$), shows that $\mathcal{N}(H)$ obeys, for $H \to 0$, a power law behavior

$$\mathcal{N}(H)|_{H \to 0} \approx H^{-\tau^*} \tag{59}$$

with an exponent

$$\tau^* = 1 + \lim_{\alpha \to +\infty} \frac{1}{\alpha} f(\alpha) = 1 + n^*. \tag{60}$$

Thus we predict

$$\tau^* = \frac{23}{24}. \tag{61}$$

One remarks that $-\tau(0) = \max_\alpha f(\alpha) = f(3) = \frac{4}{3}$ is the Hausdorff dimension of the *Brownian frontier* or of a SAW. Thus Mandelbrot's classical conjecture identifying the latter two is generalized and proven for the *whole* $f(\alpha)$ harmonic spectrum.

**An Invariance Property of $f(\alpha)$**

The expression of $f(\alpha)$ simplifies if one considers the combination:

$$f(\alpha) - \alpha = \frac{25}{24}\left[1 - \frac{1}{2}\left(2\alpha - 1 + \frac{1}{2\alpha - 1}\right)\right]. \tag{62}$$

Thus the multifractal function possesses the invariance symmetry [46]

$$f(\alpha) - \alpha = f(\alpha') - \alpha', \tag{63}$$

for $\alpha$ and $\alpha'$ satisfying the duality relation:

$$(2\alpha - 1)(2\alpha' - 1) = 1, \tag{64}$$

or, equivalently

$$\alpha^{-1} + \alpha'^{-1} = 2. \tag{65}$$

When associating a wedge angle $\theta = \pi/\alpha$ to each local singularity exponent $\alpha$, one recovers the complementary rule for angles in the plane [46]

$$\theta + \theta' = \frac{\pi}{\alpha} + \frac{\pi}{\alpha'} = 2\pi. \tag{66}$$

It is interesting to note that, owing to the explicit forms (53) of $\alpha$ and (54) of $D(n)$, the condition (65) reads also after a little algebra

$$D(n) + D(n') = 2. \tag{67}$$

This basic symmetry (63) reflects that of the cluster boundary itself under the *exchange of interior and exterior domains*, as studied in ref.[46].

**Polymultifractality**

It is interesting to note that one can also define *polymultifractal* spectra as those depending on several $\alpha$ variables [47]. An exemple is given by the double moments of the harmonic measure on *both* sides of a random fractal, taken here as a Brownian motion or a self-avoiding walk. Let us define:

$$\mathcal{Z}_{n,n'} = \left\langle \sum_{w \in \mathcal{H}/\{a\}} [H_+(w)]^n [H_-(w)]^{n'} \right\rangle, \tag{68}$$

where $H_+(w) \equiv H_+ \left( \mathcal{H} \cap \mathcal{B}(w,a) \right)$ and $H_-(w) \equiv H_- \left( \mathcal{H} \cap \mathcal{B}(w,a) \right)$ are respectively the harmonic measures on "left"or "right" sides of the random fractal. These moments have a multifractal scaling behavior

$$\mathcal{Z}_n \approx (a/R)^{\tau(n,n')}, \tag{69}$$

where the exponents $\tau(n,n')$ now depend on two moment orders. The generalization of the Legendre transform Eq. (48) reads

$$\alpha = \frac{\partial \tau}{\partial n}(n,n'), \quad \alpha' = \frac{\partial \tau}{\partial n'}(n,n'),$$
$$f(\alpha,\alpha') = \alpha n + \alpha' n' - \tau(n,n'),$$
$$n = \frac{\partial f}{\partial \alpha}(\alpha,\alpha'), \quad n' = \frac{\partial f}{\partial \alpha'}(\alpha,\alpha'). \tag{70}$$

We find the $\tau$ exponents from the star algebra (40):

$$\tau(n,n') = 2V \left( a' + U^{-1}(n) + U^{-1}(n') \right) - 2, \tag{71}$$

or, after performing the double Legendre transform

$$f(\alpha, \alpha') = 2 + \frac{1}{12} - \frac{1}{3}\alpha'^2 \left[1 - \frac{1}{2}\left(\frac{1}{\alpha} + \frac{1}{\alpha'}\right)\right]^{-1}$$
$$- \frac{1}{24}(\alpha + \alpha'), \tag{72}$$

$$\alpha = 2\frac{1}{\sqrt{24n+1}}\left[a' + \frac{1}{4}\left(\sqrt{24n+1} + \sqrt{24n'+1}\right)\right], \tag{73}$$

and a similar equation for $\alpha'$. Here $a' = 1$ for a random walk, and $a' = \frac{3}{2}$ for a self-avoiding walk. This doubly multifractal spectrum possesses the requested properties, like $\max_{\alpha'} f(\alpha, \alpha') = f(\alpha)$, where $f(\alpha)$ is (55) above. This can be generalized to a star made of $m$ random walks or $m$ self-avoiding walks, with the *polymultifractal* results:

$$f\{\alpha_{i=1,\ldots,m}\} = 2 + \frac{1}{12} - \frac{1}{3}\alpha'^2 \left[1 - \frac{1}{2}\left(\sum_{i=1}^{m}\alpha_i^{-1}\right)\right]^{-1}$$
$$- \frac{1}{24}\sum_{i=1}^{m}\alpha_i, \tag{74}$$

with

$$\alpha_i = 2\frac{1}{\sqrt{24n_i+1}}\left(a' + \frac{1}{4}\sum_{j=1}^{m}\sqrt{24n_j+1}\right), \tag{75}$$

and where $a' = \frac{1}{2}m$ for $m$ random walks in a star configuration, and $a' = \frac{3}{4}m$ for $m$ self-avoiding walks. The two-sided case above is recovered for $m = 2$. The domain of definition of the polymultifractal function $f$ is given by

$$1 - \frac{1}{2}\left(\sum_{i=1}^{m}\alpha_i^{-1}\right) \geq 0, \tag{76}$$

as verified by Eq. (75).

## 5 Percolation clusters

Consider now a two-dimensional very large incipient cluster $\mathcal{C}$, at the percolation threshold $p_c$. Define $H(w)$ as the probability that a random walker (RW) launched from infinity, *first* hits the outer (accessible) percolation hull $\mathcal{H}(\mathcal{C})$ at point $w \in \mathcal{H}(\mathcal{C})$. The moments of $H$ are averaged over all realizations of RW's and $\mathcal{C}$, as in Eq.(44) above. For very large clusters $\mathcal{C}$ and hulls $\mathcal{H}(\mathcal{C})$ of average size $R$, one expects again these moments to scale as in Eq. (45): $\mathcal{Z}_n \approx (a/R)^{\tau(n)}$. These exponents $\tau(n)$ have been obtained recently [28].

We consider site percolation on the 2D triangular lattice. Fig. 2 depicts $n$ independent random walks, in a bunch, *first* hitting the external hull of a percolation cluster at a site $w = (\bullet)$. This site, to belong to the *accessible* hull, must remain, in the *continuous scaling limit*, the source of at least *three nonintersecting crossing paths*, noted $\mathcal{S}_3$, reaching to a (large) distance $R$ (Fig. 2) [48]. (Notice that the definition of the *standard* hull requires only a pair of dual lines). The $n$ independent RW's, or Brownian paths $B$ in the scaling limit, in a bunch noted $(\vee B)^n$, *avoid* the set $\mathcal{S}_3 \equiv (\wedge \mathcal{P})^3$ of three *non-intersecting* connected paths in the percolation system, and this system is governed by a new critical exponent $x(\mathcal{S}_3 \wedge n)$ depending on $n$. In terms of these definitions, the

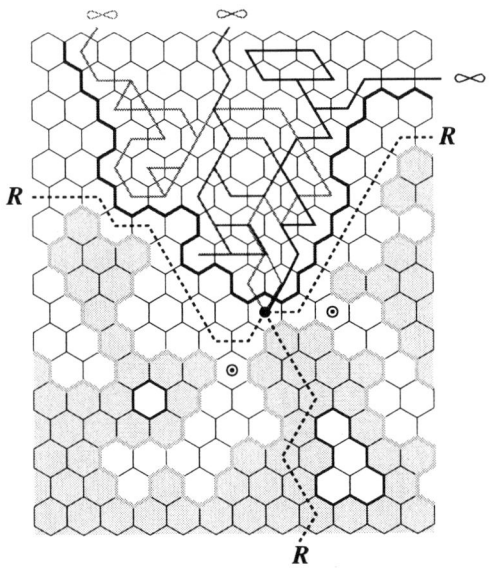

**Fig. 2.** An accessible site ($\bullet$) on the external perimeter for site percolation on the triangular lattice. It is defined by the existence, in the *scaling limit*, of three nonintersecting, and connected paths $\mathcal{S}_3$ (dotted lines), one on the incipient cluster, the other two on the dual empty sites. The entrances of fjords $\odot$ close in the scaling limit. Point ($\bullet$) is first reached by three independent RW's (red, green, blue), contributing to $H^3(\bullet)$. The hull of the incipient cluster (golden line) avoids the outer frontier of the RW's (thick blue line).

harmonic measure moments simply scale with an exponent [27]

$$\tau(n) = x(\mathcal{S}_3 \wedge n) - 2. \qquad (77)$$

For percolation, two values of half-plane crossing exponents $\tilde{x}_\ell$ are known by *elementary* means: $\tilde{x}_2 = 1, \tilde{x}_3 = 2$ [24]. We fuse the two objects $\mathcal{S}_3$ and $(\vee B)^n$

into a new star $\mathcal{S}_3 \wedge (\vee B)^n$, and use (40) to obtain

$$x(\mathcal{S}_3 \wedge n) = 2V\left(U^{-1}(\tilde{x}_3) + U^{-1}(n)\right). \tag{78}$$

Specifying $U^{-1}(\tilde{x}_3) = \frac{3}{2}$ finally gives from (39)(41)

$$x(\mathcal{S}_3 \wedge n) = 2 + \frac{1}{2}(n-1) + \frac{5}{24}\left(\sqrt{24n+1} - 5\right).$$

From this $\tau(n)$ (77) is found to be *identical* to (52) for RW's and SAW's; $D(n)$ follows as:

$$D(n) = \frac{1}{2} + \frac{5}{\sqrt{24n+1}+5}, \quad n \in \left[-\frac{1}{24}, +\infty\right), \tag{79}$$

valid for all values of moment order $n, n \geq -\frac{1}{24}$. The Legendre transform reads again exactly as in Eq. (55):

$$f(\alpha) = \frac{25}{48}\left(3 - \frac{1}{2\alpha-1}\right) - \frac{\alpha}{24}, \quad \alpha \in \left(\frac{1}{2}, +\infty\right). \tag{80}$$

Only in the case of percolation has the harmonic measure been systematically studied numerically, by Meakin et al. [49]. We give in Fig. 3 the exact curve $D(n)$ (79) [28] together with the numerical results for $n \in \{2, ..., 9\}$ [49], showing fairly good agreement.

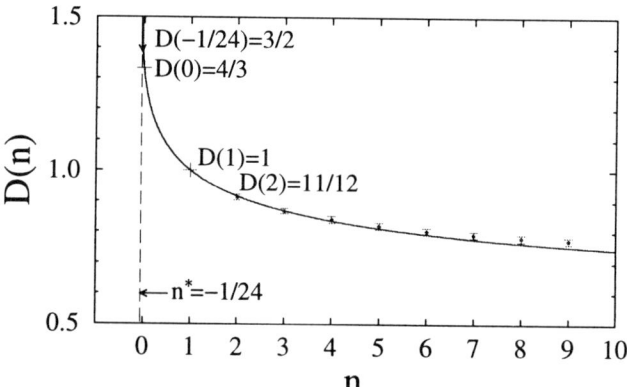

**Fig. 3.** Universal generalized dimensions $D(n)$ as a function of $n$, corresponding to the harmonic measure near a percolation cluster, or to self-avoiding or random walks, and comparison with the numerical data obtained by Meakin et al. (1988) for percolation.

The average number $\mathcal{N}(H)$ (59) has been also determined numerically for percolation clusters in [49], and for $c = 0$, our prediction (61) $\tau^* = \frac{23}{24} =$

0.95833... compares very well with the result $\tau^* = 0.951 \pm 0.030$, obtained for $10^{-5} \leq H \leq 10^{-4}$.

The dimension of the support of the measure $D(0) = \frac{4}{3} \neq D_H$, where $D_H = \frac{7}{4}$ is the Hausdorff dimension of the standard hull, i.e., the outer boundary of critical percolating clusters [51]. The value $D(0) = \frac{4}{3}$ corresponds to dimension of the *accessible external perimeter*. A direct derivation of its exact value is given in [48]. The complement of the accessible perimeter in the hull is made of deep fjords, which do close in the scaling limit and are not probed by the harmonic measure. This is in agreement with the instability phenomenon observed on a lattice for the hull dimension [52]. A striking fact is the complete identity of the multifractal spectrum for percolation to the corresponding results, Eqs.(52-55), *both* for random walks and self-avoiding walks. Seen from outside, these three fractal curves are not distinguished by the harmonic measure. As we have seen, this fact is linked to the presence of a universal conformal field theory (with a vanishing so-called central charge $c = 0$) and to the underlying presence of quantum gravity, which structures the associated conformal dimensions.

## 6   Multifractal spectrum of the frontier of a Potts cluster

The formalism above can be generalized to universal multifractal properties of other well-known random fractals [47], like, e.g., the harmonic measure of single *Potts clusters*. We simply give here the results:

$$f(\alpha) = \frac{25-c}{48}\left(3 - \frac{1}{2\alpha - 1}\right) - \frac{1-c}{24}\alpha,$$
$$\alpha \in \left(\tfrac{1}{2}, +\infty\right), \quad (81)$$

where the parameter $c$ is the so-called *central charge* of the conformal field theory describing the Potts model. It reads explicitely

$$c = 1 - 6\frac{(1-g)^2}{g}, \quad (82)$$

where $g$ is a continuous function of the number $Q$ of states of the Potts model

$$\sqrt{Q} = -2\cos\pi g, \quad g \in [\tfrac{1}{2}, 1]. \quad (83)$$

The corresponding functions are illustrated in Fig. 4. It is interesting to note that the general multifractal function (81) possesses the invariance property (63), since it also reads

$$f(\alpha) - \alpha = \frac{25-c}{24}\left[1 - \frac{1}{2}\left(2\alpha - 1 + \frac{1}{2\alpha - 1}\right)\right]. \quad (84)$$

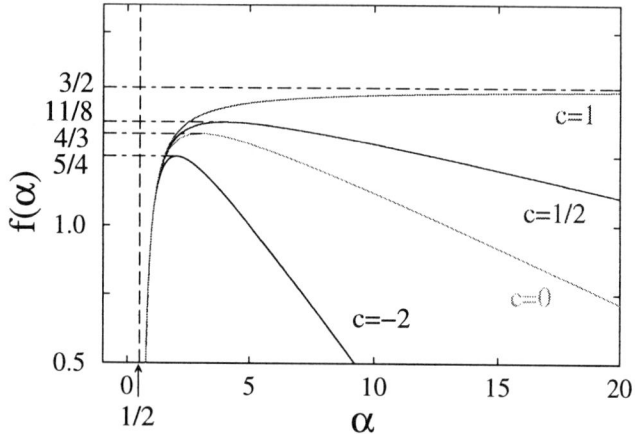

**Fig. 4.** Universal harmonic multifractal spectra $f(\alpha)$. The curves are indexed by the central charge $c$, and correspond respectively to: 2D spanning trees ($c = -2$); self-avoiding or random walks, and percolation ($c = 0$); Ising clusters or $Q = 2$ Potts clusters ($c = \frac{1}{2}$); $Q = 4$ Potts clusters ($c = 1$). The upper values $\max_\alpha f(\alpha)$ give the dimensions of the *accessible* frontiers. These harmonic multifractal functions $f$ satisfy Makarov's theorem: $f(1) = 1$. Note that the left branches of the various $f(\alpha)$ curves are indistinguishable, while their right branches split for large $\alpha$, corresponding to negative values of $n$.

Some particular cases are worth considering. An *Ising* cluster possesses a multifractal spectrum with respect to the harmonic measure ($c = \frac{1}{2}$):

$$\tau(n) = \frac{1}{2}(n-1) + \frac{7}{48}\left(\sqrt{48n+1} - 7\right), \tag{85}$$

$$f(\alpha) = \frac{49}{96}\left(3 - \frac{1}{2\alpha - 1}\right) - \frac{\alpha}{48}, \quad \alpha \in \left(\tfrac{1}{2}, +\infty\right), \tag{86}$$

with a dimension of the accessible perimeter

$$D_{\mathrm{EP}} = \max_\alpha f(\alpha, c = \tfrac{1}{2}) = \frac{11}{8}. \tag{87}$$

The $Q = 4$ Potts model provides an interesting example of a *left-handed* multifractal spectrum ($c = 1$)

$$\tau(n) = \frac{1}{2}(n-1) + \sqrt{n} - 1, \tag{88}$$

$$f(\alpha) = \frac{1}{2}\left(3 - \frac{1}{2\alpha - 1}\right), \quad \alpha \in \left(\tfrac{1}{2}, +\infty\right), \tag{89}$$

with accessible sites forming a set of Hausdorff dimenson

$$D_{\mathrm{EP}} = \max_\alpha f(\alpha, c = 1) = \frac{3}{2}, \tag{90}$$

which is also the *maximal* value common to all multifractal generalized dimensions $D(n) = \frac{1}{n-1}\tau(n)$.
Here finishes this study of the universal conformal multifractality of the harmonic measure of random fractals, by quantum gravity methods.

## References

1. K. Symanzyk (1972). In *Local Quantum Theory*, edited by R. Jost (Academic, London, New-York, 1969); P.G. de Gennes, Phys. Lett. **A38**, 339.
2. M. Aizenman(1981). Phys. Rev. Lett. **47**, 1.
   M. Aizenman(1982). Commun. Math. Phys. **86**, 1.
   D.C. Brydges, J. Fröhlich and T. Spencer (1982). Commun. Math. Phys. **83**, 123.
3. B. Mandelbrot (1982): *The fractal geometry of nature*. New-York, Freeman.
4. G.F. Lawler (1982). Commun. Math. Phys. **86**, 539.
   G.F. Lawler (1991): *Intersection of Random Walks*. Boston, Birkhäuser.
5. M. Aizenman (1985). Commun. Math. Phys. **97**, 91.
   G. Felder and J. Fröhlich. ibid., 111; G.F. Lawler, ibid., 583.
6. B. Duplantier (1987). Commun. Math. Phys. **117**, 279.
7. B. Duplantier and K.-H. Kwon (1988). Phys. Rev. Lett. **61**, 2514.
8. B. Li and A. D. Sokal (1990). J. Stat. Phys. **61**, 723.
   E. E. Puckette and W. Werner (1996). Elect. Comm. in Probab. **1**, 5.
9. K. Burdzy and G.F. Lawler(1990). Probab. Th. Rel. Fields **84**, 393.
   G.F. Lawler(1990). Ann. Probab. **18**, 981.
10. B.B Mandelbrot(1974). J. Fluid. Mech. **62**, 331.
11. H.G.E. Hentschel and I. Procaccia (1983). Physica (Amsterdam) **8D**, 835.
12. U. Frisch and G. Parisi (1985). In Proceedings of the International School of Physics "Enrico Fermi", course LXXXVIII, edited by M. Ghil (North-Holland, New York) p. 84.
13. T.C. Halsey, M.H. Jensen, L.P. Kadanoff, I. Procaccia, and B.I. Shraiman (1986). Phys. Rev. **A33**, 1141.
14. A.A. Belavin, A.M. Polyakov and A.B. Zamolodchikov (1984). Nucl. Phys. **B241**, 333.
15. D. Friedan, J. Qiu, and S. Shenker (1984). Phys. Rev. Lett. **52**, 1575.
16. B. Nienhuis (1987). In *Phase transition and critical phenomena*, vol. 11, ed. C. Domb and J.L. Lebowitz (Academic, London).
17. B. Duplantier (1986). Phys. Rev. Lett. **57**, 941.
    B. Duplantier (1989). J. Stat. Phys. **54**, 581.
18. B. Duplantier and H. Saleur (1986). Phys. Rev. Lett. **57**, 3179.
    H. Saleur (1986). J. Phys. **A19** L807.
19. M. Cates and J.M. Deutsch (1987). Phys. Rev. **A35**, 4907.
    B. Duplantier and A. Ludwig (1991). Phys. Rev. Lett. **66**, 247.
    C. von Ferber (1997). Nucl. Phys. **B490**, 511.
20. M.E. Cates and T.A. Witten (1986). Phys. Rev. Lett. **56**, 2497.
    M.E. Cates and T.A. Witten (1987). Phys. Rev. **A35**, 1809.
21. N.G. Makarov (1985). Proc. London Math. Soc. **51**, 369.
22. M. Aizenman and A. Burchard (to appear). math.FA/981027. In Duke Math. J..
23. R. Langlands, P. Pouliot and Y. Saint-Aubin (1994). Bull. AMS **30**, 1.
    J. L. Cardy (1992). J. Phys. **A25**, L201.

24. M. Aizenman (1998). In *Mathematics of Multiscale Materials*; the IMA Volumes in Mathematics and its Applications, K.M. Golden et al. eds, Springer-Verlag.
25. I. Benjamini and O. Schramm (1998). Commun. Math. Phys. **197**, 75.
26. B. Duplantier (1998). Phys. Rev. Lett. **81**, 5489.
27. B. Duplantier (1999). Phys. Rev. Lett. **82**, 880.
28. B. Duplantier (to be published). http://xxx.lanl.gov/cond-mat 9901008.
29. G.F. Lawler and W. Werner (to be published).
30. A.M. Polyakov (1987). Mod. Phys. Lett. **A2**, 893.
    Knizhnik, A.M. Polyakov and A.B. Zamolodchikov (1988). Mod. Phys. Lett. **A3**, 819.
31. A.M. Polyakov (1987): *Gauge fields and Strings*. **Harwood-Academic, Chur.**
32. D.V. Boulatov, et al. (1986). Nucl. Phys. **B275**, 641.
    F. David (1985). Nucl. Phys. **B257**, 45, 543.
    J. Ambjorn, B. Durhuus and J. Fröhlich. ibid., 433.
33. V.A. Kazakov (1986). Phys. Lett. **A119**, 140.
34. B. Duplantier and I.K. Kostov (1988). Phys. Rev. Lett. **61**, 1436.
    B. Duplantier and I.K. Kostov (1990). Nucl. Phys. **B 340**, 491.
35. F. David (1988). Mod. Phys. Lett. **A3**, 1651.
    J. Distler and H. Kawai (1988). Nucl. Phys. **B321** 509.
36. I.K. Kostov and M.L. Mehta (1987). Phys. Lett. **B189**, 118.
37. W. Werner (1997). Probab. Th. Rel. Fields **108**, 131.
38. G.F. Lawler (1996). Elect. Comm. in Probab. **1** (29).
39. J.F. Joanny, L. Leibler and R.C. Ball (1984). J. Chem. Phys. **81**, 4640.
40. C. von Ferber and Y. Holovatch (1997). Europhys. Lett. **39**, 31.
    C. von Ferber and Y. Holovatch (1997). Phys. Rev. E **56**, 6370.
41. J. Cardy (1984). Nucl. Phys. **B240** [FS12], 514.
42. B.B. Mandelbrot and C.J.G. Evertsz (1990). Nature **348**, 143.
43. G.F. Lawler (1998). To be published.
44. L. V. Ahlfors (1973): *Conformal Invariants. Topics in Geometric Function Theory.* McGraw-Hill, New York.
45. R.C. Ball and R. Blumenfeld (1991). Phys. Rev. A **44**, R828.
46. R. C. Ball, B. Duplantier and T. C. Halsey (1999). Unpublished.
47. B. Duplantier (1999). To be published, Isaac Newton Institute preprint.
48. M. Aizenman, B. Duplantier and A. Aharony (to be published). http://xxx.lanl.gov/cond-mat 9901018.
49. P. Meakin *et al.* (1986). Phys. Rev. A **34**, 3325.
    P. Meakin (1986). *ibid.* **33**, 1365.
    P. Meakin *et al.* (1988). In *Phase Transitions and Critical Phenomena*, vol. 12, edited by C. Domb and J.L. Lebowitz (Academic, London).
50. P. Meakin and B. Sapoval (1992). Phys. Rev. A**46**, 1022.
51. H. Saleur and B. Duplantier (1987). Phys. Rev. Lett. **58**, 2325.
52. T. Grossman and A. Aharony (1987). J. Phys. A **20**, L1193.

# Fractal Pores and Fractal Tunnels: Traps for "Particles" or "Sound Particles"

Jérôme Dorignac[1] and Bernard Sapoval[2]

[1] Centre de physique théorique
Ecole Polytechnique, 91128 Palaiseau Cedex - FRANCE
dorignac@cpht.polytechnique.fr
[2] Laboratoire de physique de la matière condensée
Ecole Polytechnique, 91128 Palaiseau Cedex - FRANCE
Bernard.Sapoval@polytechnique.fr

**Abstract.** The statistical behavior of ballistic trajectories in irregular pores with axial symmetry is investigated analytically and numerically. The geometrical irregularity is described by a specific quadratic Koch prefractal shape of order $\nu$ which is periodically translated along the symmetry axis. This generates an axi-symmetric prefractal surface which models an irregular media. The statistical properties of particles trajectories are studied in the limit case of uniform angular re-emission from the pore wall. The results could help to understand the effect of fractal macro-roughness on sound absorption when sound propagation can be approximated by sound particles uniformly scattered by a micro-roughness of the surface. The ballistic trajectories also mimic diffusion in porous material at very low pressure (Knudsen's diffusion) when tangential re-emission events are present. Several properties of the distribution of the free paths of these ballistic trajectories are discussed. They exhibit the particular role played by the irregular shape and the existence of sharp nooks or recoins which may act as dynamical traps. We also study the effect of increasing the system irregularity on the collision number and frequency and the dependence of the "sound particles" absorption on the irregularity in the limit of uniform re-emission.

Keywords: Knudsen, Ballistic trajectories, Sound absorption.

## 1 Introduction

The purpose of this paper is to study the statistical properties of ballistic trajectories in an irregular media in the particular situation where the re-emission of the particles hitting the walls is uniformly distributed in angle. Ballistic trajectories may represent the trajectories of atoms or molecules in porous media in the low pressure, so-called Knudsen, regime where the free paths are limited by collisions with the walls rather than collisions between molecules. Here we are interested in trying to predict the interaction of sound waves with fractal walls neglecting the wave propagation aspect. Such a simplified picture of the sound interaction with irregular geometries was first proposed by Kutruff to model the noise due to traffic in adjacent building areas. In that limiting picture, the sound waves are regarded as "sound particles" [1]. Here the hypothesis of uniformly distributed re-emission of sound particles is related to the results of Jaggard and Sun [2] on the diffraction of waves by fractally corrugated surfaces. When the characteristic size of the geometrical patch defect is of the order of the wavelength one finds that the diffraction pattern is widely distributed in angle. We consider here a deterministic fractal structure in which the smaller cutoff length is itself a fractally corrugated surface. For instance, for sound waves with wavelengths of the order of a few cm we consider that the building blocks of the prefractal deterministic structure have a size of a few cm. Below that size they are supposed to be corrugated and then give rise, as indicated by Jaggard, to a wide angular re-emission which here is postulated to be uniform.

In the case of atoms or molecules this law does not apply generally as previously discussed by Valleau and co-workers [3]. These authors have indicated that the Lambert's law re-emission ( *cosine law*) preserves the detailed balance in the flux of molecules at the wall while the uniform does not. Nevertheless, the interaction of atoms or molecules with surfaces is very complex. The real angular distributions generally depend on the particles incident angles and energies as much as the surface corrugation [4]. Moreover, experiments show that scattering of noble gas from Ag[111] (Ne, Ar, Kr, Xe) or Ag[100] may give rise to non-vanishing tangential events. The most important fact here is that the probability for a particle to restart from the wall with a trajectory almost parallel to it can be non zero and increases of course when the incidence angle tends to 90° . Here, only the limit case where the particles are re-emitted from the wall without privileged direction is considered. Even if this does not represents adequately the molecule wall interaction this allows us to derive some interesting results concerning the trapping role played by the recoins and to study a situation where the interaction of the particles with the surface is maximal.

This paper is organized as follows. In the first section the chord (or free path) probability distribution (or CPD) and the probability distribution of the projection of the chord (PPD) along the structure axis is discussed for a smooth cylinder and for irregular axi-symmetric pores. In the case of the irregular structure one finds a trapping phenomenon which is explained analytically in the appendices. In the second part, the "sound particles" absorption is computed and compared with the probability that a particle undergoes $N$ collisions either in a system of given length or in a given time. This permits us to find two important quantities: the absorption coefficient $A$ and the mean collision frequency $F$ versus the degree of irregularity of the structure.

## 2 Probability distribution of the chord length

The statistical properties of the trajectories are determined by two quantities. First the probability distribution of the chord length (CPD). In the case of uniform angular distribution that is considered here, the chords are classically called s-chords in the litterature [5]. The chord distribution gives information on the way the rays or the particles interact with the pore wall. The second quantity of interest is the probability distribution of the projections of the chords along the symmetry axis (PPD). This quantity gives information on the way particle would diffuse in the structure along its symmetry axis. In the case of a smooth cylindric pore, it is possible to derive analytic expressions for these quantities. This is done in detail in Appendix A. This allows us to compare the respective contributions of the global geometry from that of the irregularity to the mean CPD and PPD obtained numerically for the fractal system.

In this paper, we consider that the velocity $c$ of the particles (atoms, molecules or "sound particles") is equal to 1. The length of the free path $\lambda$ is related to the time of free flight by $\lambda = ct = t$. Its projection along the axis is denoted by $z_\lambda$. The results of the analytical calculation are given as a function of $\lambda$ or $z_\lambda$. They are represented on Fig. (1). The CPD $p(\lambda)$ is found to diverge when $\lambda \longrightarrow d$ whereas the PPD $p(z_\lambda)$ is found to diverge when $z_\lambda \longrightarrow 0$.

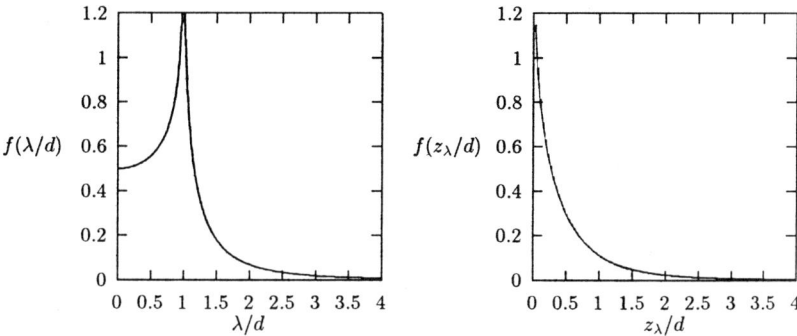

**Fig. 1.** Reduced probability distributions $f(\lambda/d)$ of the reduced chord lengths (or free paths) $\lambda/d$ and $f(z_\lambda/d)$ of the projections of the free paths in reduced units $z_\lambda/d$ where $d$ is the smooth cylinder diameter.

The exact expressions given in the Appendix A also gives the asymptotic behavior of the distributions and their moments. One finds

$$p(z_\lambda) \sim \frac{1}{\pi d} \ln\left(\frac{d}{z_\lambda}\right) \quad z_\lambda \to 0 \qquad (1)$$

$$p(z_\lambda) \sim \frac{1}{4} \frac{d^2}{z_\lambda^3} \quad z_\lambda \to \infty \qquad (2)$$

and

$$p(\lambda) \sim \frac{1}{2d} \quad \lambda \to 0 \tag{3}$$

$$p(\lambda) \sim \frac{1}{\pi d} \ln\left(\frac{16d}{(d^2 - \lambda^2)^{1/2}}\right) \quad \lambda \to d^- \tag{4}$$

$$p(\lambda) \sim \frac{1}{\pi d} \ln\left(\frac{16\lambda}{(\lambda^2 - d^2)^{1/2}}\right) \quad \lambda \to d^+ \tag{5}$$

$$p(\lambda) \sim \frac{1}{2} \frac{d^2}{\lambda^3} \quad \lambda \to \infty \tag{6}$$

Moreover, we can determine the mean chord length or first moment of $p(\lambda)$.

$$\langle \lambda \rangle = d \tag{7}$$

Thus, the mean time of flight or mean chord length of the smooth cylinder is equal to its diameter $d$ while, by symmetry,

$$\langle z_\lambda \rangle = 0 \tag{8}$$

Other moments of the distributions are at least logarithmically divergent. In particular,

$$\langle \lambda^2 \rangle \sim \frac{d^2}{2} \ln \lambda \quad \lambda \to \infty \quad ; \quad \langle z_\lambda^2 \rangle \sim \frac{d^2}{2} \ln z_\lambda \quad z_\lambda \to \infty \tag{9}$$

Note that, due to the particularly slow decay of $p(\lambda)$ and $p(z_\lambda)$ and to the cylindrical (straight) geometry, the behavior of particles is hyperdiffusive because the second moments of both distributions diverge. This is due to the fact that the system has infinite horizon and that tangential trajectories are permitted. This is one of the major difference with the so called $\cos\theta$ (Lambert's) law where the tangential events are forbidden. See for example [6] and [7]. These divergences do not play any role in our study because they do not intervene in the interaction of particles with the wall irregularity.

The geometrical irregularity is introduced by the mean of the prefractal shape shown in Fig. (2). The vertical amplitude is chosen to be relatively small as compared to the cylinder diameter $d$ (roughly 10%). The periodicity of the prefractal shape along the cylinder axis is equal to 16 computer units (CU) whereas the diameter is chosen to be $d = 110$ CU. The particle trajectories are studied numerically. They are made of successive free paths (or chords). For a given irregularity degree $\nu$, one can define a *mean* chord probability density along a trajectory of total length $L$, $\overline{p}_\nu^L(\lambda)$ by collecting all the chords along a trajectory and computing the associated probability. For sufficiently long trajectories ($L \to \infty$), $\overline{p}_\nu^L(\lambda)$ becomes independent of the initial conditions and of the specific random realization of a trajectory. That is, although all the trajectories are different, they possess the same $p_\nu(\lambda) \equiv \overline{p}_\nu^{L \to \infty}(\lambda)$. In other words the numerical trajectories possess an ergodic property.

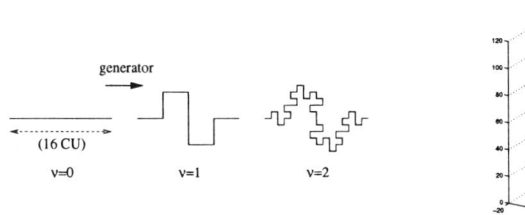
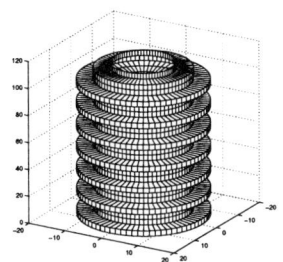

**Fig. 2.** Geometrical structure of the irregular system. Left: Iteration of the specific quadratic Koch motif used as a model of the longitudinal irregulatity. $\nu$ is the prefractal order. Right: A $3D$ perspective of the axisymmetric surface generated by the rotation of the generator at $\nu = 1$ and its translation along the rotation axis. In the computer simulations the diameter of the smooth cylinder is equal to 110 computer units (CU). For visibility the last figure is not at real scale.

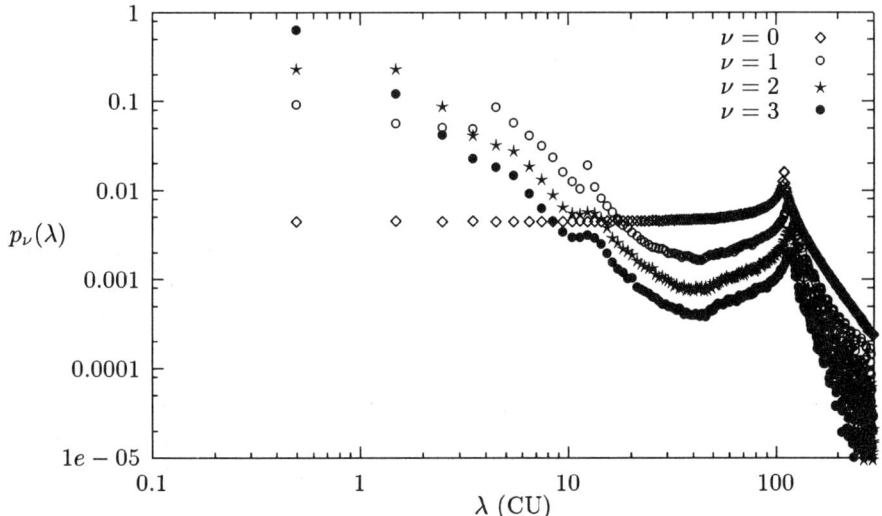

**Fig. 3.** Mean CPD $p_\nu(\lambda)$ for several irregularity orders ($\nu = 0, 1, 2, 3$).

Figure (3) gives the numerical CPD $p_\nu(\lambda)$ for increasing irregularity ($\nu = 0, 1, 2, 3$). These curves present several analogies. All of them reach a maximum[1] for $\lambda \sim d$. For very large values of $\lambda$ (not shown in the figure) $p_\nu(\lambda)$ varies like $1/\lambda^3$. A detailed graph of the behavior of $p_\nu(\lambda)$ for large $\lambda$ shows that this probability density decreases with oscillations. This is due to the particular shape of the system which possesses well defined geometrical oscillations. This creates some blind zones inaccessible to a particle situated far away and is responsible for gaps between accessible chord lengths from a given point. Apart from this detail, which is specific of our particular geometry and has no general significance, the tail of the distributions for $\nu = 1, 2, 3$ are very similar to that of the smooth cylinder.

We focus now on the specific effect of the irregularity which are found for small $\lambda$. First one observes several discontinuities arising at $\lambda = 4$ and 12 (CU) for $\nu = 1$. These discontinuities correspond to the shortest distance between two crowns facing each other (see Fig. (4,$a$)). Indeed, as shown in Fig. (4,$b$), if

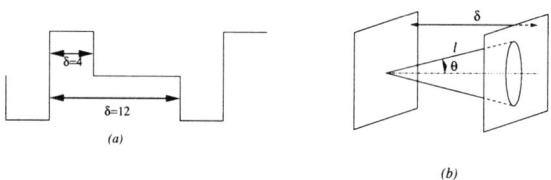

**Fig. 4.** Transverse view of the shape of prefractal order $\nu = 1$ ($a$). Shematic representation of two plane surfaces facing each other ($b$). The cone represents the solid angle such that any particle starting from its vertex executes a chord of length $\lambda \in [\delta, l]$.

the particle starts from one of the surface, the probability that it reaches the other with a chord length less than $l$ is $\Omega/2\pi = 1 - \delta/l$. Thus the associated CPD is $p(\lambda) = \delta/\lambda^2$ for $\lambda > \delta$ and 0 for $\lambda < \delta$. This discontinuity for $\lambda = \delta$ explains those observed in $p_\nu(\lambda)$ for $\lambda = 4$ and 12. We similarly observe such discontinuities for $\nu = 2, 3$ (for lengths $\delta = 1$ and 0.25 CU) although too low resolution of the plot (3) (1 CU) masks them. Here again, these discontinuities are specific of our particular geometry and have no general significance. They would not exist in a random geometry.

A more interesting behavior appears for very short chord lengths. Indeed, Fig. (5) shows a detailed plot of $p_\nu(\lambda)$ for $\lambda \in [0, 0.25]$ that is below the smallest slit length at $\nu = 3$ (0.25 CU). It is found that $p_\nu(\lambda)$ is not constant contrary to the case of the smooth cylinder. This behavior is different from the result obtained by Mering and Tchoubar in the case of Lambert re-emission [8]. Note that the transverse curvature responsible for short chords in the smooth cylinder, also exists here but gives, in the same manner, a constant density (see (3)) when

---

[1] In fact the real maximum is slightly greater than $d$ due to the radial enlargement caused by the irregularity but this has no interesting significance.

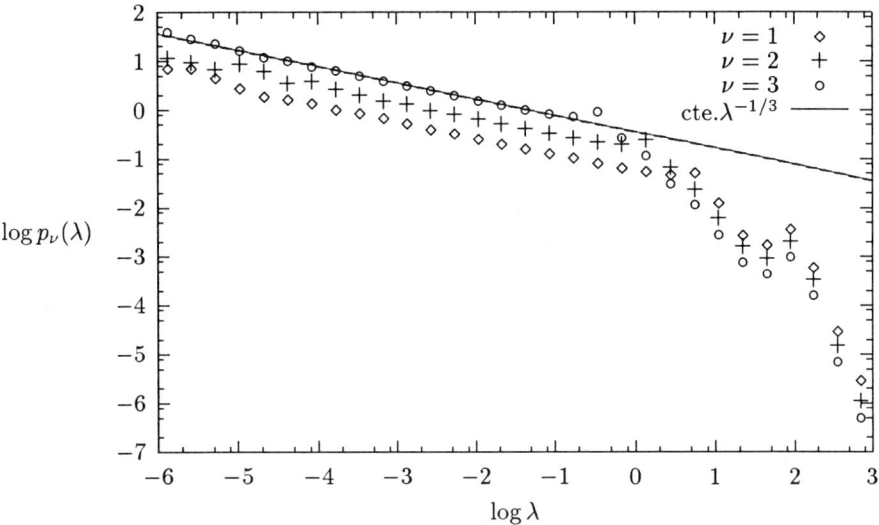

**Fig. 5.** Divergence of the mean CPD $p_\nu(\lambda)$ as $\lambda \to 0$ for several irregularity orders ($\nu = 1, 2, 3$) and comparison with analytical result.

$\lambda \to 0$.
In irregular structures, in which there exists necesserally recoins, there exists an additionnal mechanism which creates short chord lengths. Particles may jump from one face to adjacent one in the irregular geometry. These trajectories are responsible for the power law divergence observed for all irregularities in Fig. (5). This behavior can be understood analytically from the situation depicted in Fig. (6).

Consider a particle situated in $O$ at a distance $l$ from the edge. The probability, for that particle to reach the orthogonal plane with a chord smaller than $R$ is given by [2]

$$P(\lambda < R) = \frac{1}{2}\left(1 - \frac{l}{R}\right) \quad \text{for} \quad \lambda > l$$
$$P(\lambda < R) = 0 \quad \text{for} \quad \lambda < l$$

thus, the corresponding CPD is simply

$$p(\lambda; l) = \frac{l}{2\lambda^2} \quad \text{for} \quad \lambda > l$$
$$= 0 \quad \text{for} \quad \lambda < l \tag{10}$$

---

[2] As the cylinder diameter $d$ is very large as compared to the chords into consideration, one can locally assimilate the cylinder with its tangent plane and then its intersection with a perpendicular disk to the intersection of two planes.

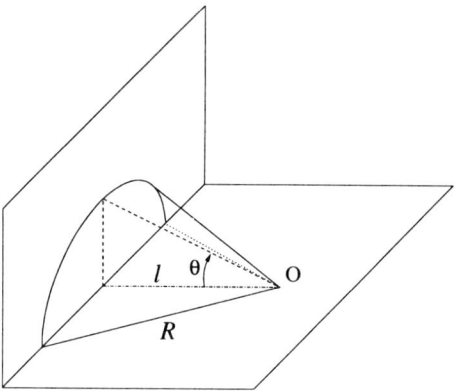

**Fig. 6.** Intersection of the sphere of radius $R$ centered in $O$ (starting point of the particle) with the orthogonal adjacent plane. The semi-cone of angle $\theta$ delimits the chord lengths $\lambda < R$.

Note that this probability density is not that of the chord length $p_\nu(\lambda)$ but the probability density that, starting at a distance $l$, the chord is equal to $\lambda$. To find $p_\nu(\lambda)$, one has to average this quantity by the probability that particles impacts are located at this particular distance $l$. It is shown in Appendix B that

The probability density $w(l, \alpha)$ of impacts at a distance $l$ of a corner of angle $\alpha$ behaves like

$$w(l, \alpha) \sim \frac{cst}{l^{\beta(\alpha)}} \quad \text{for} \quad l \to 0 \tag{11}$$

where the exponent $\beta$ is related to the angle $\alpha$ by the simple formula

$$\beta(\alpha) = \frac{\pi - \alpha}{2\pi - \alpha} \tag{12}$$

Note that this divergence is local and does not depend on the particular geometry outside the corner. Using (11) for an angle $\alpha = \pi/2$, one obtains

$$p_\nu(\lambda) \propto \int_0^\lambda dl\, w\left(l, \frac{\pi}{2}\right) p(\lambda; l) = \int_0^\lambda dl\, \frac{cst}{l^{1/3}} \frac{l}{2\lambda^2} \propto \frac{1}{\lambda^{1/3}} \quad \text{when} \quad \lambda \to 0 \tag{13}$$

The numerical simulations shown in Fig. (5) verify this theoretical prediction. Thus, in presence of uniform re-emission, sharp nooks or recoins of the geometrical structure are responsible for an increase of short chords by trapping the particle. This effect grows with increasing irregularity because of the multiplication of these structures.

It is also interesting to study the distribution functions $F_\nu(\lambda) = \int_0^\lambda p_\nu(u)du$. They are shown in Fig. (7). These curves indicate that irregularity increases drastically the probability of short chords and qualitatively that the collision number and the collision frequency are increasing functions of irregularity.

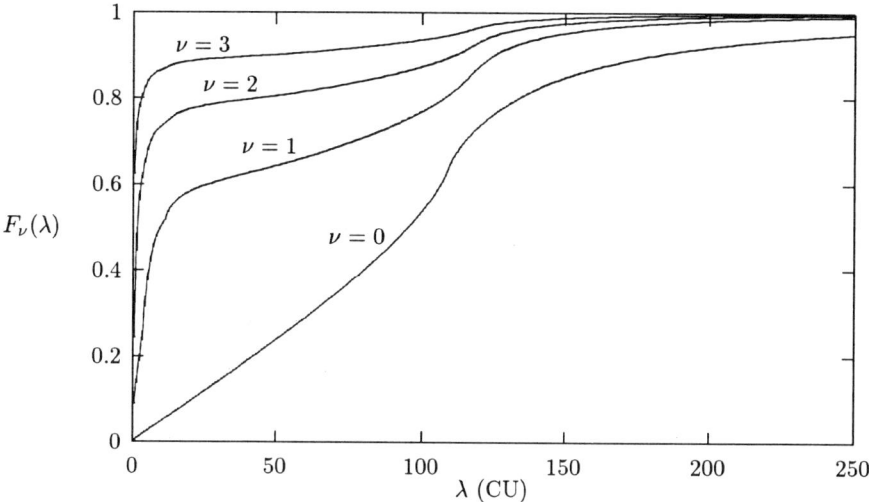

**Fig. 7.** Mean repartition function $F_\nu(\lambda)$ for several irregularity orders ($\nu = 0, 1, 2, 3$).

## 3 Sound particles absorption, collision number and collision frequencies

The physical properties of atoms or molecules or sound particles may depend on other probability distributions than those studied above. In particular here, we are interested in the absorption of sound particles. Suppose that, at each collision with the wall, only a fraction $r$ of the acoustic energy is reflected. For a trajectory in which there happens $N$ collisions the acoustical power absorbed by the walls is equal to $r^N$ and the absorption is given by $A = 1 - r^N$. For an ensemble of trajectories the absorption is $A = \langle 1 - r^N \rangle = 1 - \langle r^N \rangle$. Considering a delimited region of a media called $\mathcal{D}$ the average absorption will be given by

$$A_\mathcal{D} = 1 - \langle r^N \rangle = 1 - \sum_{N=0}^{\infty} r^N P(N; \mathcal{D}) \tag{14}$$

Here $P(N; \mathcal{D})$ is the probability for a particle to be submitted to $N$ collisions in the delimited region of a media $\mathcal{D}$ which represents a domain of interest. The probabilities $P(N; \mathcal{D})$ provides then a useful tool in understanding the absorbing power of the media. Figure (8) show a numerical evaluation of $P(N; \mathcal{D})$ for a cylinder section of longitudinal length $L_p = 2d$. The particle trajectories are initialized in such a way that the starting points are uniformly distributed in space and angle in a section $z = 0$. The trajectories end when the particle reaches the plane $z = L_p = 2d$ or returns to the origin $z = 0$. For measuring $P_\nu(N; 2d)$ we have studied $10^6$ trajectories ($\nu$ is the prefractal order of the shape or its irregularity degree). For large $N$ the decay of $P_\nu(N; L_p)$ is found to be exponential.

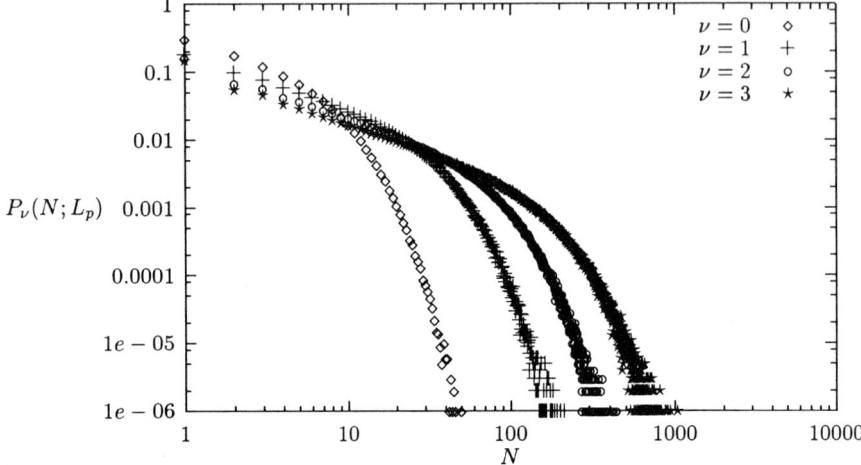

**Fig. 8.** Probability distribution of the collision number $N$ in a cylinder of increasing irregularity: $\nu = 0, 1, 2, 3$. The ratio $L_p/d$ of the longitudinal cylinder section $L_p$ to the cylinder diameter $d$ is equal to 2. Statistical datas have been calculated over $10^6$ particle trajectories.

This fact is not intrinsically related to the particular re-emission law that has been considered here. It occurs every time we deal with a Markovian re-emission probability. This is essentially different from the power law as encountered in the case of specular reflection (see for example (Santra and Sapoval [9])) which describes a correlated chaotic process. The exponential decay has been derived for a smooth cylinder in Appendix C.

The mean number of collisions has been calculated for a particular system length $L_p = 5d$. The following mean number of collisions $\langle N_\nu \rangle$ have been found: $\langle N_0 \rangle = 6.76$, $\langle N_1 \rangle = 23.10$, $\langle N_2 \rangle = 48.14$, $\langle N_3 \rangle = 96.02$. The mean collision number increases by a factor of order 3 from smooth to first irregularity. Above $\nu = 1$, the mean number of collisions increases roughly like $2^{\nu-1} \langle N \rangle_1$. If the quantity $A_{\text{est}} := 1 - r^{\langle N \rangle_\nu}$ would be a good estimate of the sound particles absorption, one could expect a rapid increase of absorption with irregularity. This increase exits but not to the extent corresponding to this estimate (see Fig. 9). This result is in fact a direct consequence of the Jensen's inequality which states that for any concave function $u(x)$, $\langle u(x) \rangle \leq u(\langle x \rangle)$ [10].

The other physical quantity of interest is the mean interaction between particles and irregular walls. This intervenes for instance in the average shift of the NMR (nuclear magnetic resonance) of atoms like $^{129}$Xe which are adsorbed at each collision for a short time $\tau$. During this short time, the NMR is submitted to a small shift in frequency. If $\tau$ is short enough [9], the averaged NMR shift is proportional to the collision frequency averaged over the time $T$ of observation of the phenomena ($T \gg \tau$). In this case, the mean frequency of collision along

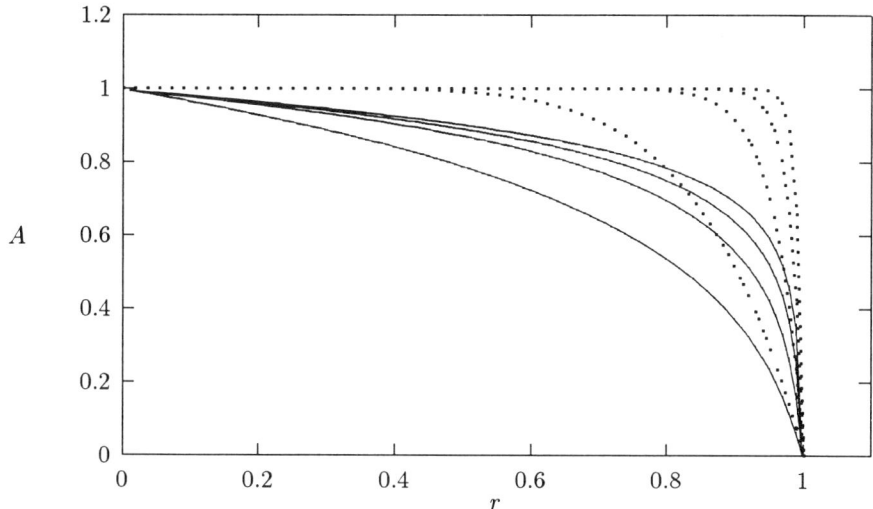

**Fig. 9.** Absorption coefficient $A$ (solid) and its estimate $A_{\text{est}}$ (dotted) versus the reflexion coefficient $r$ calculated with uniform re-emission for increasing irregularity $\nu = 0, 1, 2, 3$. The lowest curves (solid or dotted) correspond to the smooth ($\nu = 0$) cylinder. The others correspond to increasing $\nu$. The ratio $L_p/d$ of the cylinder length to its diameter is equal to 5. Statistics have been performed over $10^6$ trajectories.

a given trajectory $n$ is given by $F_n = N_n/T$. Thus, we just have to compute the probability the particle undergoes $N$ collisions during a time $T$ to retrieve the probability of $F$. This has been done numerically by taking into account the contributions of $10^6$ trajectories of total length (or duration) $T = 5d$ where $d$ is the cylinder diameter. The results are displayed in Fig. (10).

We can see that the irregularity does increase the frequency of collisions above the smooth cylinder, $\nu = 0$ value. For a fixed $\nu$, we have verified that the mean collision frequencies is independent of $T$ for enough large $T$. Indeed, the mean collision number $\langle N \rangle_\nu$ increases roughly proportionally to $T$ and thus, $\langle F \rangle_\nu = \langle N \rangle_\nu /T$ is stable. For $T = 5d$ and $T = 25d$, the results are reported in the array below.

| $\nu$ | 0 | 1 | 2 | 3 | |
|---|---|---|---|---|---|
| $\langle F \rangle_\nu$ (T=5d) | 1.01 | 2.06 | 3.67 | 6.98 | $(\times 10^{-2})$ |
| $\langle F \rangle_\nu$ (T=25d) | 0.95 | 2.08 | 3.81 | 7.02 | $(\times 10^{-2})$ |

(For the sake of clarity $P_\nu(N, T)$ have not been represented for $T = 25d$ in Fig. (10)). Statistics for determining $\langle F \rangle_\nu$ have been made over $10^6$ trajectories initialized at random inside the cylinder. We see that, although the irregularity is only a surface irregularity (its transverse amplitude is about 10% of the diameter) its effect is to increase by a factor close to 2 the mean collision frequency each time $\nu$ is incremented. Similar results (to appear in a following paper) for the

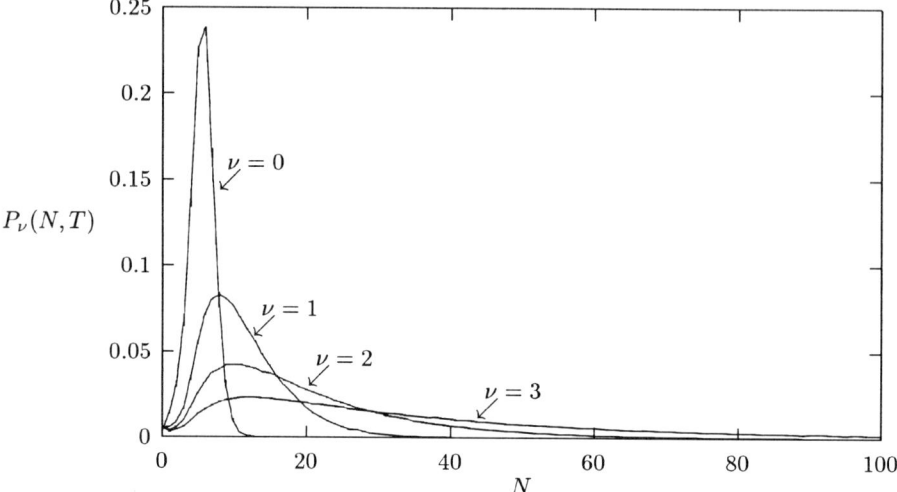

**Fig. 10.** Probability $P_\nu(N,T)$ the particle undergoes $N$ collisions in a time $T = 5d$ in a cylinder of increasing irregularity ($\nu = 0, 1, 2, 3$).

usual Lambert's law (cosine law) show that the particles have less interaction with the wall. Indeed, the absorption coefficient and the mean collision number are smaller in this case.

## 4 Conclusions

The goal of this study was to find how geometrical irregularity may, or may not, influence properties of atoms or "sound particles" traveling in irregular structures. Considering a particular case of deterministic geometry, several statistical properties of ballistic trajectories in an irregular tunnel or pore in 3D have been studied numerically and analytically whenever possible. The global geometry is axi-symmetric and the irregularity is modeled by a fractal, more precisely prefractal, unit translated along the symmetry axis. The re-emission law which has been used is that of uniform angular probability distribution.

The probability distribution of the free paths or chords is found to decrease like the inverse cube of the chord length for the large quasi-axial paths. The role of the geometrical irregularity is to induce a power law divergence of the probability of small paths. This divergence has been shown to arise from successive impacts of particles between adjacent faces in the irregular structure. This divergence should then be a general property of irregular structures which presents sharp angles between adjacent surfaces.

The probability distribution of the number of collisions inside a finite section of the irregular pore decreases exponentially for large collision numbers. The mean collision number increases with the irregularity of the structure. For the geometries that have been studied, the mean collision frequency increases approximately as the total area of the pore. In the heuristic picture of sound attenuation described by partial sound particle absorption the attenuation of sound is found to increase more rapidly with irregularity than in the case of specular reflections in 2d pores.

As such, these results can be considered as a limit case because of the uniform re-emission law that has been used. This limit can only be considered as a rough approximation for the interaction of a sound wave with a corrugated surface. In that limit, they may be used to guess the role of irregularity for the diffusion and absorption of sound waves by corrugated surfaces elements, themselves being the building blocks of a fractal tunnel. In the case of the interaction of atoms or molecules in irregular pores in the Knudsen limit, the above results can help to understand the role of tangential re-emission events after interaction with an irregular pore wall.

These studies will be extended in the future to the case of a re-emission probability distribution obeying the $\cos\theta$ Lambert's law. This should permit to better understand the role of the geometrical irregularity in 3D irregular porous material as those widely used in heterogeneous catalysis.

## Acknowledgments

The authors thank M.O. Coppens, A. de Martino and particularly P. Lévitz for several illuminating discussions.

## Appendix A: Derivation of the CPD and PPD

In this appendix, the probability distribution of chord lengths $\lambda$ CPD ($p(\lambda)$) and the PPD ($p(z)$) are calculated analytically for the case of a smooth cylinder. Consider a cylinder whose equation is $(x - d/2)^2 + y^2 = (d/2)^2$. Because of translational and rotational invariance, all the points of the cylinder are equivalent for determining the distributions $p(z)$ and $p(\lambda)$. Thus, we derive them from the origin $O$ ($x = y = z = 0$). In $O$, we consider the usual spherical coordinates. $\theta$ denotes the angle of the unitary vector $u$ with the $z$-axis while $\phi$ denotes the angle between the projection of $u$ on the plane $(O, x, y)$ with the $x$-axis. Any ray emitted from $O$ is characterized by its unitary vector $u = [\sin\theta\cos(\phi), \sin\theta\sin(\phi), \cos(\theta)]$. For $\phi \in [-\pi/2; \pi/2]$ and $\theta \in [0, \pi]$, the ray is inside the cylinder. Its equation reads

$$x = \lambda \sin\theta \cos\phi$$
$$y = \lambda \sin\theta \sin\phi$$
$$z = \lambda \cos\theta \qquad (15)$$

where $\lambda$ represents the distance to the origin, that is, the time (recall that $c = 1$). The ray intercepts the cylinder in

$$\lambda = \frac{d\cos\phi}{\sin\theta} \qquad (16)$$

and

$$z = \frac{d\cos\phi}{\tan\theta} \tag{17}$$

where $d$ is the cylinder diameter. Knowing the probability density of $(\theta, \phi)$: $p_{\theta,\phi} = \sin\theta/2\pi$ which corresponds to a uniform emission inside the cylinder, we look for the probability density of the variables $\lambda$ and $z$.

We first derive $p(z)$. Because of symmetry, we just deal with $z > 0$ ($p(z) = p(-z)$). The probability the ray intercepts the cylinder wall after a given $Z$ is

$$P(z \geq Z) = \frac{1}{2\pi} \iint_{\mathcal{D}_Z} \sin\theta \, d\theta \, d\phi \tag{18}$$

where

$$\mathcal{D}_Z = \left( \phi \in \left[ -\frac{\pi}{2}; \frac{\pi}{2} \right] \right) \times \left( \theta \in \left[ 0; \arctan\left( \frac{d}{Z} \cos\phi \right) \right] \right) \tag{19}$$

thus

$$P(z \geq Z) = \frac{1}{2\pi} \cdot 2 \int_0^{\frac{\pi}{2}} d\phi \int_0^{\arctan\left(\frac{d}{Z}\cos\phi\right)} d\theta \sin\theta$$

$$= \frac{1}{2} - \frac{Z}{\pi (Z^2 + d^2)^{1/2}} K\left( \frac{d}{(Z^2 + d^2)^{1/2}} \right) \tag{20}$$

By derivation one obtains

$$p(z) = -\left. \frac{\partial P(z \geq Z)}{\partial Z} \right|_{Z=z} = \frac{1}{\pi d} q(z) \left[ K\left( q(z) \right) - E\left( q(z) \right) \right] \tag{21}$$

where

$$q(z) = \frac{d}{(d^2 + z^2)^{1/2}}$$

and where we have used the complete elliptic integrals of first and second kind $K(k)$ and $E(k)$ defined as follows [14]

$$K(k) = \int_0^{\frac{\pi}{2}} \frac{d\theta}{(1 - k^2 \sin^2\theta)^{1/2}}$$

$$E(k) = \int_0^{\frac{\pi}{2}} d\theta \, (1 - k^2 \sin^2\theta)^{1/2}$$

We now derive $p(\lambda)$ the probability density of chords. For a fixed $R$, the set of chords of length less than $R$ relies on the cylinder part delimited by the intersection of a sphere of radius $R$ and the cylinder itself. If $R$ is less than $d$, this curve is closed but it splits up in two equivalent parts as soon as $R > d$. See Fig. (11).

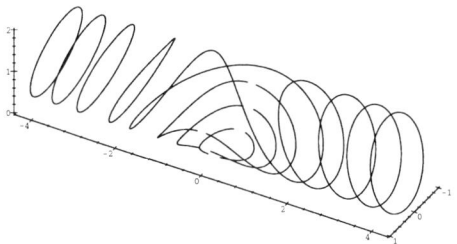

**Fig. 11.** Perspective view of lines iso-$R$: intersection of the cylinder ($d = 2$) with the sphere of radius $R$. The central crossed curve represents the limit case $R = d$.

Thus, for $R \leq d$,

$$P(\lambda \leq R) = \frac{1}{2\pi} \iint_{\mathcal{D}_R} \sin\theta \, d\theta \, d\phi \tag{22}$$

where

$$\mathcal{D}_R = \left(\phi \in \left[\arccos\left(\frac{R}{d}\sin\theta\right); \frac{\pi}{2}\right]\right) \times \left(\theta \in \left[0; \frac{\pi}{2}\right]\right) \tag{23}$$

This yields

$$P(\lambda \leq R) = \frac{2}{\pi} \int_0^{\frac{\pi}{2}} d\theta \sin\theta \arcsin\left(\frac{R\sin\theta}{d}\right) \tag{24}$$

and by derivating

$$p(\lambda) = \left.\frac{\partial P(\lambda \leq R)}{\partial R}\right|_{R=\lambda} = \frac{2d}{\pi\lambda^2}\left[K\left(\frac{\lambda}{d}\right) - E\left(\frac{\lambda}{d}\right)\right] \quad (\lambda < d) \tag{25}$$

whereas for $R \geq d$, we change the domain of integration

$$\mathcal{D}'_R = \left(\phi \in \left[-\frac{\pi}{2}; \frac{\pi}{2}\right]\right) \times \left(\theta \in \left[0; \arcsin\left(\frac{d}{R}\cos\phi\right)\right]\right) \tag{26}$$

$$P(\lambda \geq R) = 2 \cdot \frac{1}{2\pi} \int_{-\pi/2}^{\pi/2} d\phi \int_0^{\arcsin\left(\frac{d}{R}\cos\phi\right)} d\theta \sin\theta$$

$$= 1 - \frac{2}{\pi} E\left(\frac{d}{R}\right) \tag{27}$$

Thus we arrive at

$$p(\lambda) = -\frac{\partial P(\lambda \geq R)}{\partial R}\bigg|_{R=\lambda} = \frac{2}{\pi\lambda}\left[K\left(\frac{d}{\lambda}\right) - E\left(\frac{d}{\lambda}\right)\right] \quad (\lambda > d) \quad (28)$$

Using the developments of the exact analytical expressions

$$K(k) \sim E(k) + \frac{\pi}{4}k^2 \quad k \to 0 \quad (29)$$

$$K(k) \sim \frac{1}{2}\ln\left(\frac{16}{(1-k^2)^{1/2}}\right) \quad k \to 1 \quad (30)$$

$$E(k) \sim 1 \quad k \to 1 \quad (31)$$

one can derive the asymptotic expressions given in Section 2.

The mean chord lenght can be caculated by inserting (16) in the definition of the probability in spherical coordinates, for the domain $\mathcal{D} = (\phi \in [-\pi/2; \pi/2]) \times (\theta \in [0; \pi])$, one obtains

$$\langle \lambda \rangle = \frac{1}{2\pi}\iint_{\mathcal{D}} \sin\theta \ \lambda(\theta,\phi)\, d\theta\, d\phi = d \quad (32)$$

## Appendix B: Localization in a corner

### Introduction

We consider, in the following, a random walk which takes place in an isosceles triangle of angle $\alpha$. The random walk has the following properties: each times the mobile encounters a line of the triangle, it is reemited uniformly inside the triangle. We are interested here in the location of points of impact of the mobile on the two segments contiguous to the angle $\alpha$. More precisely, we want to determine the behavior of their probability density (PD) $p(\alpha, s)$ when the distance $s$ to the vertex of the angle $\alpha$ tends to 0. The solution of this problem is exactly the same for a three dimensional corner of angle $\alpha$ (intersection of two planes) with a 3D uniform re-emission law: $p(\theta,\phi) = d\Omega/2\pi$ where $\Omega$ represents the solid angle. In the following, we present analytical results and comparisons with simulations.

### Autocoherence equation for the probability density $p(\alpha, s)$

Let $\mathcal{C}$ be a closed contour and $s$ be the curvilinear abscissa along $\mathcal{C}$. Let now $\rho(s, s')$ be the PD to reach the point $s'$ from the point situated at $s$. Then $p(s)$, the PD of being at $s$ obeys the following autocoherence equation:

$$p(s) = \oint_{\mathcal{C}} p(s')\rho(s, s')ds' \quad (33)$$

Taking into account the specific geometry of the problem, we determine $\rho(s, s')$

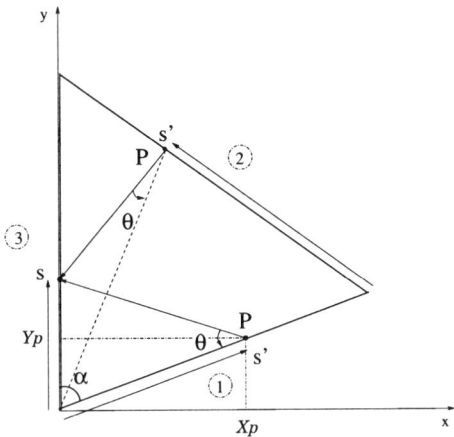

**Fig. 12.** Isosceles triangle of angle $\alpha$.

for an isosceles triangle of angle $\alpha$. We choose $s$ to belong to the third segment of the triangle as shown in the Fig. (12) and calculate $\rho(s, s')$ for $s'$ belonging to the first then to the second one. There is, obviously, no possibility for the mobile to reach a point of the segment it already belongs. We denote by $\theta$ the angle under which the segment of length $s$ is seen from the point $P$ (see figure 12). Because of the uniformity of emission, starting from $P$, the probability the mobile reaches a point of the segment is $\theta/\pi$. Then its PD is just

$$\rho(s, P) = \frac{1}{\pi}\frac{d\theta}{ds} = \frac{1}{\pi}\frac{x_p}{x_p^2 + (y_p - s)^2} \tag{34}$$

Applying this formula to the triangle, we find:

$$\rho_1(s, s') = \frac{1}{\pi} \frac{s'.\sin(\alpha)}{(s'^2 - 2ss'\cos(\alpha) + s^2)} \quad \text{for} \quad s' \in 1 \tag{35}$$

$$\rho_2(s, s') = \frac{1}{\pi} \frac{(a - s')\cos(\frac{\alpha}{2})}{(a - s')^2 \cos^2\frac{\alpha}{2} + \left(\frac{a}{2\sin\frac{\alpha}{2}} - (a - s')\sin\frac{\alpha}{2} - s\right)^2}$$

$$\text{for} \quad s' \in 2 \tag{36}$$

$$\rho_3(s, s') = 0 \quad \text{for} \quad s' \in 3 \tag{37}$$

where $a$ is the length of the segment 2. Using these results, (33) reads

$$p_3(\alpha, s) = \int_0^{a/2\sin(\frac{\alpha}{2})} \frac{ds'}{\pi} \frac{s'.\sin(\alpha)}{(s'^2 - 2ss'\cos(\alpha) + s^2)} p_1(\alpha, s')$$

$$+ \int_0^a \frac{ds'}{\pi} \frac{(a - s')\cos(\frac{\alpha}{2})}{(a - s')^2 \cos^2\frac{\alpha}{2} + \left(\frac{a}{2\sin\frac{\alpha}{2}} - (a - s')\sin\frac{\alpha}{2} - s\right)^2} p_2(\alpha, s') \tag{38}$$

Where we have denoted by $p_j(\alpha, s)$ the restriction of the whole function $p(\alpha, s)$ on the segment $j$. Remark that, because of symmetry, $p_3(\alpha, s) = p_1(\alpha, s)$.

### Behavior of $p(\alpha, s)$ when $s \longrightarrow 0$

It seems not very easy to calculate exactly the function $p(\alpha, s)$ which satisfies the functional equation (38). Nevertheless, it is possible to determine its asymptotic behavior when $s \longrightarrow 0$. Let us first have a look on some simulations. Fig. (13) shows two simulations for two different triangles of angle $\alpha = \pi/4$ and $\alpha = \pi/3$ respectively. The function $p(\alpha, s)$ is shown to be divergent when $s \longrightarrow 0$ with a power law of exponent $\beta(\alpha)$ i.e.

$$p(\alpha, s) \sim \frac{D(\alpha)}{s^{\beta(\alpha)}} \quad \text{when} \quad s \to 0 \tag{39}$$

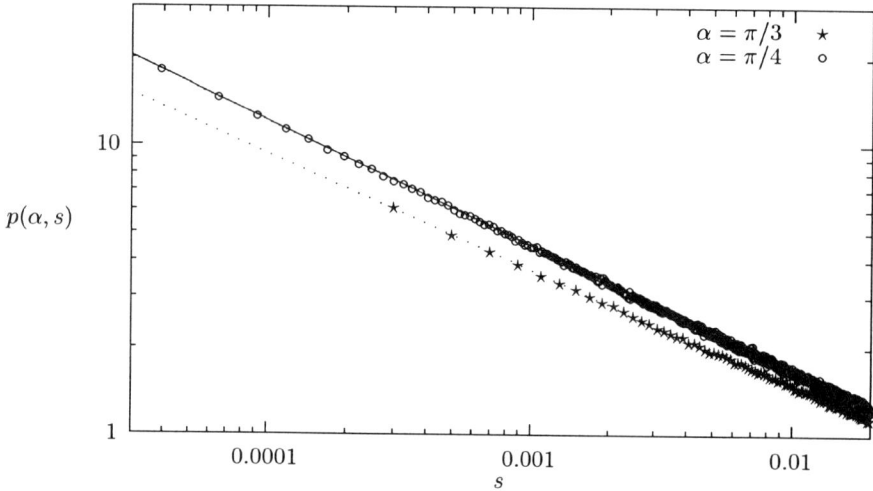

**Fig. 13.** $p(\alpha, s)$ for $\alpha = \pi/3$ and $\pi/4$. The lines represent the asymptotic behaviors deduced from formula (48). $p(\pi/3, s) \propto s^{2/5}$ (dotted) and $p(\pi/4, s) \propto s^{3/7}$ (solid).

Because of integrability of $p(\alpha, s)$, ( $\int ds\, p(\alpha, s) = 1$), the exponent $\beta(\alpha)$ is certainly strictly less than 1. So, we have

$$0 < \beta(\alpha) < 1 \tag{40}$$

Before to derive an analytical result for $\beta(\alpha)$, let us describe briefly why $p(\alpha, s)$ diverges when $s \longrightarrow 0$. When the mobile reach a point very close to the corner of the triangle, the probability it is reemited towards a point of the adjacent segment which is also close to the corner increases drastically. The result of this is a kind of trapping which leads to a localization of the impacts near the corner and explains the behavior of $p(\alpha, s)$ shown in the graphic.

Starting from equation (38) we have

$$\lim_{s \to 0} p_1(\alpha, s) = \tag{41}$$

$$\lim_{s \to 0} \int_0^{a/2\sin(\frac{\alpha}{2})} \frac{ds'}{\pi} \frac{s'.\sin(\alpha)}{(s'^2 - 2ss'\cos(\alpha) + s^2)} p_1(\alpha, s')$$

$$+ \lim_{s \to 0} \int_0^a \frac{ds'}{\pi} \frac{(a-s')\cos(\frac{\alpha}{2})}{(a-s')^2 \cos^2\frac{\alpha}{2} + \left(\frac{a}{2\sin\frac{\alpha}{2}} - (a-s')\sin\frac{\alpha}{2} - s\right)^2} p_2(\alpha, s') \tag{42}$$

At this stage, it is not difficult to see that we can directly take the limit in the second term of (41). Indeed, when $s \to 0$ and $s' \to 0$ (or $a$) together, the integrand tends to a constant times $p_2(\alpha, s')$. Of course, $p_2(\alpha, s')$ diverges when $s' \to 0$ (or $a$) but remains integrable because it represents a probability density. However, it is not the same for the first term for wich, taking directly the limit $s \to 0$ is not possible because the integrand behaves like $D(\alpha)/s^{1+\beta(\alpha)}$ which is not integrable due to the range of $\beta(\alpha)$. Thus, inserting (39) in (41), we arrive at

$$\frac{D(\alpha)}{s^{\beta(\alpha)}} \sim \int_0^{a/2\sin(\frac{\alpha}{2})} \frac{ds'}{\pi} \frac{s'.\sin(\alpha)}{(s'^2 - 2ss'\cos(\alpha) + s^2)} p_1(\alpha, s') \quad \text{when} \quad s \to 0 \tag{43}$$

Because of the corner of angle $\gamma = (\pi - \alpha)/2$ the function $p_1(\alpha, s')$ also diverges when $s' \to a/2\sin(\frac{\alpha}{2})$ but this gives no contribution to the divergence of the integral because when $s \to 0$, the integrand behaves like a constant times $p_1(\alpha, s')$ which is integrable. So, we split up the integral in two terms like $\int_0^{a/2\sin(\frac{\alpha}{2})} = \int_0^{a/4\sin(\frac{\alpha}{2})} + \int_{a/4\sin(\frac{\alpha}{2})}^{a/2\sin(\frac{\alpha}{2})}$ and forget the second term wich does not contribute to the divergence. Thus, (43) becomes

$$\frac{D(\alpha)}{s^{\beta(\alpha)}} \sim \int_0^{a/4\sin(\frac{\alpha}{2})} \frac{ds'}{\pi} \frac{s'.\sin(\alpha)}{(s'^2 - 2ss'\cos(\alpha) + s^2)} p_1(\alpha, s') \quad \text{when} \quad s \to 0 \tag{44}$$

If, furthermore, we assume that $p_1(\alpha, s')$ is a Laurent series in $]0, a/4\sin(\frac{\alpha}{2})]$ i.e.

$$p_1(\alpha, s) \sim \frac{D(\alpha)}{s^{\beta(\alpha)}} + \mathcal{O}(s^{1-\beta(\alpha)}) \quad \text{when} \quad s \to 0 \tag{45}$$

it is not difficult to see that only the first term of this expansion makes the integral diverge. Finally, we obtain

$$\frac{D(\alpha)}{s^{\beta(\alpha)}} \sim \int_0^{a/4\sin(\frac{\alpha}{2})} \frac{ds'}{\pi} \frac{s'.\sin(\alpha)}{(s'^2 - 2ss'\cos(\alpha) + s^2)} \frac{D(\alpha)}{s'^{\beta(\alpha)}} \quad \text{when} \quad s \to 0 \tag{46}$$

Changing of variable and puting $u = s'/s$ then taking the limit $s \to 0$ yields to the following condition for $\beta(\alpha)$

$$\int_0^\infty \frac{du}{\pi} \frac{u^{1-\beta(\alpha)} \sin(\alpha)}{(u^2 - 2u\cos(\alpha) + 1)} = 1 \tag{47}$$

which gives finally by integration

$$\frac{\sin(\beta\pi + \alpha - \alpha\beta)}{\sin(\beta\pi)} = 1$$

$$\beta\pi + \alpha - \alpha\beta = \pi - \beta\pi$$

because of the range of $\beta$.
We thus obtain the divergence exponent

$$\beta(\alpha) = \frac{\pi - \alpha}{2\pi - \alpha} \qquad (48)$$

## Appendix C: Exponential decay of $P(N; L_p)$

Using the probability density PDP found in (21), we may write

$$P_0(n \geq N) = \int_0^{L_p} d\zeta_1 \, p^+(\zeta_1) \int_0^{L_p} d\zeta_2 \, p_z(\zeta_2 - \zeta_1) \int_0^{L_p} d\zeta_3 \, p_z(\zeta_3 - \zeta_2)$$

$$\ldots \int_0^{L_p} d\zeta_N \, p_z(\zeta_N - \zeta_{N-1})$$

$$= \int_{[0,L_p]^N} \left(\prod_{i=1}^N d\zeta_i\right) p^+(\zeta_1) \prod_{j=2}^N p_z(\zeta_j - \zeta_{j-1}) \qquad (49)$$

where $p^+(z)$ is the probability density that the particle, starting from a given point $P$ of the disk situated at $z = 0$, reaches a point of coordinate $z$ on the cylinder, averaged uniformly over the points $P$. Thus, we find

$$P_0(N; L_p) = \int_{[0,L_p]^N} \left(\prod_{i=1}^N d\zeta_i\right) p^+(\zeta_1) \prod_{j=2}^N p_z(\zeta_j - \zeta_{j-1}) \Big\{ 1$$

$$- \int_0^{L_p} d\zeta_{N+1} \, p_z(\zeta_{N+1} - \zeta_N) \Big\} \qquad (50)$$

Although we have not calculated the exact result of this complicated integral, the above formula gives the way to deduce simple boundaries indicating its exponential decay. Indeed, denoting by

$$M = \max\nolimits_{z_2} \int_0^{L_p} dz_1 \, p_z(z_1 - z_2) \,, \quad m = \min\nolimits_{z_2} \int_0^{L_p} dz_1 \, p_z(z_1 - z_2) \,,$$

and
$$k = \int_0^{L_p} dz \, p^+(z) \qquad (51)$$

the following inequalities hold

$$km^{N-1}(1 - M) < P_0(N; L_p) < kM^{N-1} \qquad N \geq 1 \qquad (52)$$

Thus, the decay of $P_0(N; L_p)$ is exponential.
Fig. (8) shows that this behavior remains true in presence of irregularity.

# References

1. H. Kutruff (1982). *J. Sound Vib.*, **85**, p. 115.
2. D. L. Jaggard and X. Sun (1990). *J. Optic. Soc. Am. A*, **7** (6), p. 1131-9.
3. J.P. Valleau, D.J. Diestler, J.H. Cushman, M. Schoen, A.W. Hertzner and M.E. Riley (1991). *J.Chem. Phys.*, **95** (8), p. 6194.
4. M.B. Andersson and J.B.C. Pettersson (1995). *J. Chem. Phys.*, **102** (10), p. 4239.
5. Coleman (1969). *J. Appl. Prob.*, **6**, p. 430–441.
6. J. Klafter and G. Zumofen G. (1993). *Physica A*, **196**, p. 102.
7. G. Zumofen and J. Klafter (1993). *Physica D*, **69**, p. 436.
8. J. Mering and D. Tchoubar (1968). *J. Appl. Cryst.*, **1**, p. 153.
9. S. B. Santra and B. Sapoval (1998). *Phys. Rev. E*, **57**, p. 6888.
10. W. Feller (1966). *An introduction to probability theory and its applications*, Ed. Wiley.
11. H. Kutruff (1995). *J. Acoust. Soc. Am.*, **98**, p. 288.
12. P. Levitz (1993). *J. Phys. Chem.*, **97**, p. 3813.
13. P. Levitz (1997). *Europhys. Lett.*, **39**, p. 593–598.
14. M. Abramowitz and I. Stegun (1970). *Handbook of mathematical functions*, Ed. Dover.
15. M.-O. Coppens and G.F. Froment (1995). *Fractals*, **3** (4), p. 807–20.

# Fractal Pores and the Degradation of Shales

Luis E. Vallejo[1] and Ann Stewart Murphy[2]

[1] Department of Civil and Environmental Engineering, University of Pittsburgh,
Pittsburgh PA 15261 - U.S.A.
vallejo@engrng.pitt.edu
[2] Office of Surface Mining, 3 Parkway Center, Pittsburgh PA 15220 - U.S.A.

**Abstract.** This study presents an application of fractal theory to analyze the degradation behavior of shales when in contact with water. The durability to water of sixty-eight shale samples from the Appalachian region of the United states was evaluated using the Jar Slake (Soak) Test. Of the sixty-eight samples tested, fourteen degraded into a pile of flakes or mud, four developed small fractures, and fifty experienced no degradation at all. X-ray diffraction analysis identified kaolinite as the predominant clay mineral present in the shales. Shales composed primarily of kaolinite degrade (slake) as a result of capillarity induced-pore air compression. Pore air compression is favored by small pore radii. Petrographic analysis of the samples revealed that on average, the fourteen samples that slaked had a system of macropores with a diameter equal to 0.06 mm, the four samples that developed small fractures had macropores with a diamater equal to 0.07 mm, and the fifty samples that did not slake had a macopore diameter equal to 0.092 mm. However, there were some exceptions. Many shale samples with a diameter equal to or smaller than 0.06 mm did not slake. A fractal analysis of the pore boundaries of the these shales found them to be very rough. Capillary pores with rough boundaries absorb water not through their whole cross sectional area, but through their corners and crannies. This partial filling of the pore cross sectional area will prevent the development of pore air compression that is the necessary cause for the slaking of shales. Thus, the roughness of the pore boundaries (measured by the fractal dimension) has a significant influence on the slaking of shales and needs to be considered in any evaluation of their durability to water. Also, the use of the fractal dimension to measure the degree of roughness of the pore boundaries proved to be a simple but powerful tool to analyze the degradation behavior of the shales.

## 1 Introduction

Shale is a sedimentary argillaceous rock which is the most common group of rock material found in the Earth's crust. Because of their abundance, shales have been used in the construction of earth embankments for highways as well as earth dams for the retention of water. Shales also form part of the roof system in underground coal mines and of natural or engineered slopes. Many types of shales degrade (slake) when in contact with water. Shale degradation has been associated with many infrastructure problems such as: (a) settlement and instability of highway embankments, (b) subsidence due to roof collapse in coal mines, and (c) slope instability [19, 14, 3]. To date, our knowledge of the mechanisms of shale degradation is incomplete [15]. The purpose of this study is to present a microstructural analysis that investigates the pore geometry in shales and its influence on their degradation when in contact with water. The pore geometry will be evaluated using fractal geometry [9].

## 2 Durability Tests to Measure the Slaking Potential of Shales

### 2.1 Shale Samples

In order to understand the mechanisms involved in the slaking of shales, sixty-eight shale samples were collected at recently blasted highwalls at surface mines in Kentucky, Tennessee, Virginia and West Virginia. The samples were subjected in the laboratory to a combination of slake durability tests and petrographic analyses using thin sections of the shales. Image analysis of the photographs of the thin sections provided information about the pore geometry in the shale samples.

### 2.2 Durability Tests

Several types of tests have been developed or modified to provide qualitative and/or quantitative assessments of the slake potential of geologic materials [1]. The durability tests that have been suggested include the following: (a) Slake Durability Test, and (b) Jar Slake (Soak) Test. In the Slake Durability Test, oven-dried samples of shale are placed in a 2 mm mesh-faced drum partially immersed in water. The drum is then rotated at 20 rpm for about 10 minutes; the samples and drum are then removed, dried, and weighed. This process is done twice. The ratio of the final to the initial dry samples weight provides an index, the Slake Durability Index (SDI), as a measure of the durability of the shales [5]. In the Jar Slake Test, the oven dried samples are subjected only to immersion in water without the rotation component involved in the Slake Durability Test. In the present study, the Jar Slake (Soak) Test was used to measure the durability of the shales. This test provides a qualitative measure of rock behavior after its immersion in water for a 24-hour period. The Jar Slake Test uses a ranking system [the Jar Slake Index $(I_j)$] which is based on the appearance of the sample after soaking to measure the durability or slaking potential of the samples. This ranking system was established by [8]. The Jar Slake Index $(I_j)$ is ranked on a scale from one to six (Table 1).

The Lutton (1977) (Table 1) ranking system was applied to the sixty-eight shale samples forming part of this study. Irregularly-shaped dry samples weighing at least 100 grams were soaked in distilled water at $20°C$ for 24 hours, and then photographed. Pre- and post-test photographs were compared to assess the mode

Table 1. Jar Slake Ranking (Lutton 1977)

| Jar Slake Index $I_j$ | Behavior |
|---|---|
| 1 | Degrades to a pile of flakes or mud |
| 2 | Breaks rapidly and/or forms many chips |
| 3 | Breaks slowly and/or forms few chips |
| 4 | Breaks rapidly and/or develops several fractures |
| 5 | Breaks slowly and/or develops few fracture |
| 6 | No change |

and degree of breakdown. Figures 1 and 2 show the before and after results of the Jar Slake Test conducted on two shale samples, TN-5 and TN-9. The TN-5 sample Fig. 1 had an $I_j$ value equal to 1 and the TN-9 sample Fig. 2 had an $I_j$ value equal to 6.

The full range of slaking behavior, from $I_j=1$ to $I_j=6$, was observed in the tested rock. Of the sixty-eight samples tested by the jar slake method, fourteen samples were rated at an $I_j$ of either 1 or 2, four samples were rated at an $I_j$ equal to 3, and fifty samples were rated at an $I_j$ of either 5 or 6. The samples with an $I_j$ of either 1 or 2 disintegrated in the form of many flakes Fig. 1. In order to understand the mechanisms involved in the disintegration or non-disintegration of the shales when immersed in water, the sixty-eight samples were subjected to petrographic analyses that involved both x-ray diffraction analysis and thin section examination.

## 3 X-Ray Diffraction and Thin Section Analyses

Complete description of shales involves both x-ray diffraction analysis and thin section examination [13]. X-ray diffraction identifies clay and carbonate minerals, and thin section analysis provides information about the fabric of the shales. The x-ray diffraction analysis of the shale samples was made using a Phillips XRG 3000 Diffractometer. The x-ray scan of the samples produces prints called diffractograms. These diffractograms consist of a combination of relatively smooth and peak intensity patterns. The recorded patterns are compared with standard patterns produced by known pure substances in order to obtain the quantities and types of clays and non-clay materials present in. In addition, to detect the presence of montmorillonite, x-ray diffraction analysis was undertaken in powder samples exposed to ethylene glycol vapors. Using this method, the presence of expansive clay minerals in the samples is detected by identifying characteristic peak intensity levels and their location in the diffractograms [13].

Table 2 shows the results from the x-ray diffraction analysis for the shales from Kentucky, Tennessee, Virginia and West Virginia.

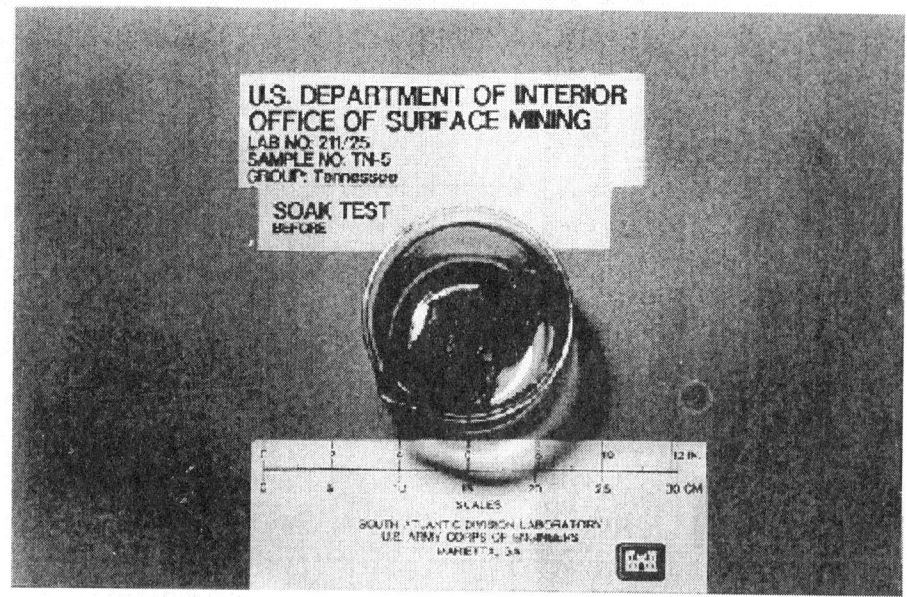
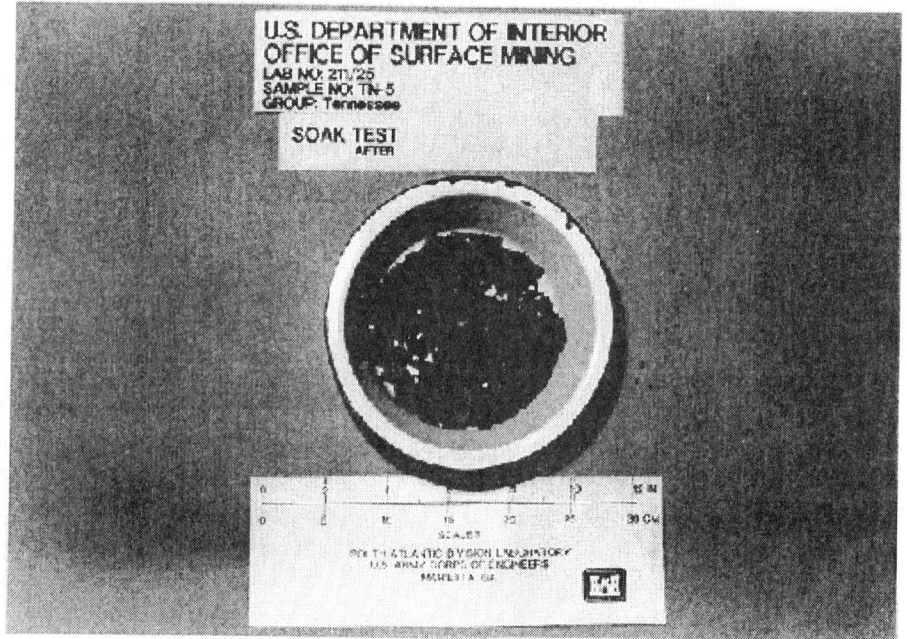

Fig. 1. Before and after Jar Slake Test-Sample TN-5 ($I_j=1$, degrades into mud)

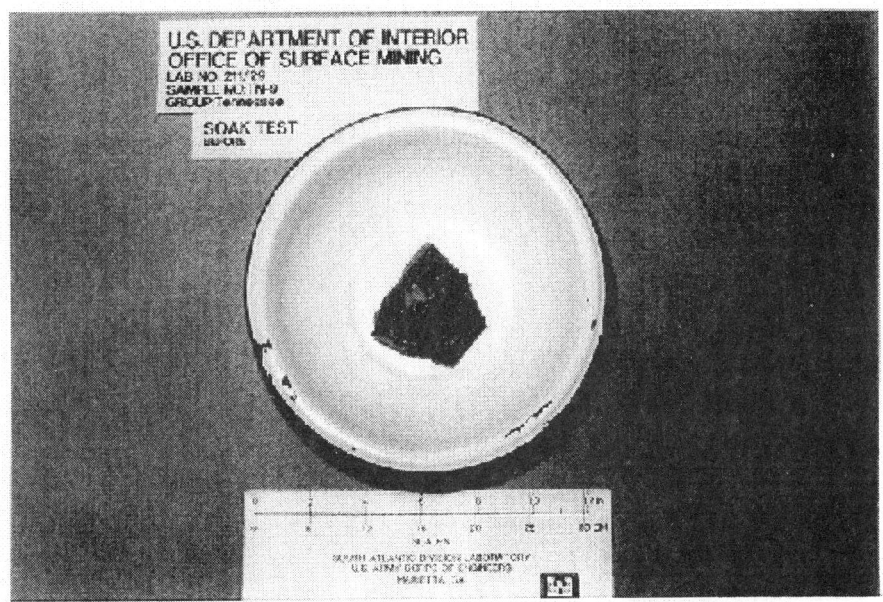

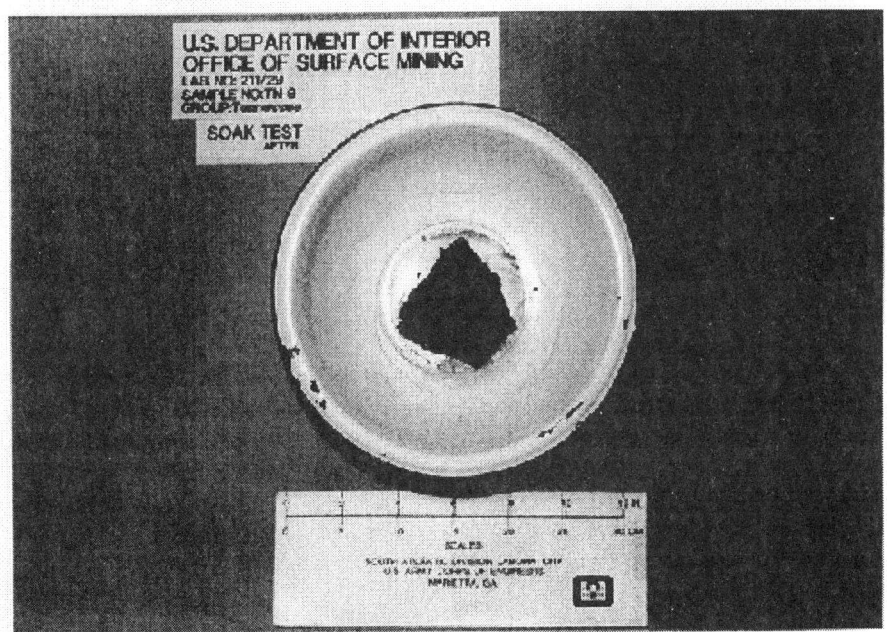

**Fig. 2.** Before and after Jar Slake Test-Sample TN-9 ($I_j=6$, no degradation)

Table 2. X-Ray Diffraction Analysis of Shales

| Shales Origin (# of samples) | Kaolinite (%) | Quartz (%) | Mica (%) | Feldespar (%) | Other* (%) |
|---|---|---|---|---|---|
| Kentucky (30) | 18 | 26 | 9 | 17 | 30 |
| Tennessee (12) | 6 | 42 | 26 | 5 | 21 |
| Virginia (11) | 8 | 30 | 37 | 6 | 19 |
| West Virginia (15) | 23 | 26 | 3 | 25 | 23 |

*Other includes: Chlorite, Ankerite, Calcite, Siderites and Opaques.

An analysis of Table 2 indicates that the most prevalent clay mineral present in the shales was kaolinite. No expansive clay minerals were present in the shales. As a complement to the x-ray diffraction analysis, a petrographic analysis of the shales was performed. This latter analysis involved the use of thin sections (30 mm in thickness), that were first examined optically with a polarizing microscope, and then photographed at two different magnifications (25X and 63X).

# 4 Slaking Mechanisms

There are different mechanisms discussed in the geotechnical literature which explain the slaking of shales when immersed in water. One slaking phenomenon is attributed to the compression of entrapped air in the pores of the shale when water enters the shales as a result of capillary suction [12]. This entrapped air in the pores exerts tension on the solid skeleton, causing the material to fail in tension. According to [12], pore air compression is the predominant slaking mechanism in shales composed primarily of non-expansive clay minerals such as kaolinite. Clay surface hydration by ion adsorption has been suggested as another mechanism that causes slaking through the swelling of montmorillonite clays in the shales [1]. The removal of cementing agents from the shales by groundwater dissolution is also considered to be a mechanism that causes slaking

Since the sixty-eight shale samples tested in the Jar Slake (Soak) Test have kaolinite as the primary clay mineral in the structure, pore air compression should be [12] the primary mechanism causing the slaking of the shales. The Jar Slake Tests resulted in fourteen samples completely disintegrating when immersed in water; four samples developing small fractures; and fifty samples experiencing little or no change after soaking in water. In order to understand the reasons for the slaking and non-slaking behavior of the shales, an analysis of their pore structure was undertaken using the photographs of the thin sections.

# 5 Pore Structure

## 5.1 Pore Pressures

Shales with non-expansive clays (i.e. kaolinite clay) slake because of pore air compression [12]. Pore air compression is exhibited when water is drawn into the macropore system of the shales during the immersion process of the Jar Slake Test. The suction of water by the shales is the result of capillary forces. This suction process is illustrated by Fig. 3, which represents a shale sample with a system of cylindical macropores that run continuously through it (Fig. 3(A)). These macropores, which are assumed not to interconnect, resemble small tubes inside the shale. When the sample is immersed in water, water will be pulled into the individual macropores, as a result of capillary forces, and the air that originally filled the macropores will be subjected to compression (Fig. 3(B)). The system of forces acting at the interface between the air and the water in the macropores are modeled in Fig. 3(C) [11]. According to [11], the following equation applies at equilibrium conditions:

$$\pi d T_s - p(\pi d^2)/4 + u(\pi d^2)/4 = 0 \tag{1}$$

where d = the diameter of the cylindrical macropore, Ts = the surface tension of water acting on the meniscus, p = the air pressure, u = pore water pressure. From Eq. 1 the following relationship can be obtained

$$p = u + \left(\frac{4T_s}{d}\right) \tag{2}$$

An analysis of Eq. 2 indicates that the pore pressure, p , in the portion of the macropore filled with air (Fig. 3(B)) increases as the diameter, d , of the cylindrical macropore decreases. Thus, the smaller the diameter of the macropore, the larger the air pressure will be. Since pore air compression is favored by small pore radii, slaking of shales by air compression will be more pronounced in those shales containing small diameter macropores. In addition, small diameter macropores will more readily confine the air pressure developed during the suction process. That is, diffusion of the air pressure will decrease with the decrease in surface area (which is a function of the diameter of the macropore) of the pore that is in contact with the air. From the previous discussion it can be concluded that the diameter of the macropore system of the shales has a marked influence on their slaking in water.

## 5.2 Pore Diameter

In order to study the pore geometry of the sixty-eight shale samples, thin sections (30 mm in thickness) were prepared and photographs of the sections were made using a polarizing microscope. The study of pore space geometry of rocks using photographs of thin sections is a standard procedure in petrographic analysis [4, 17]. The photographs of the thin sections were made at two different magnifications (25X and 63X).

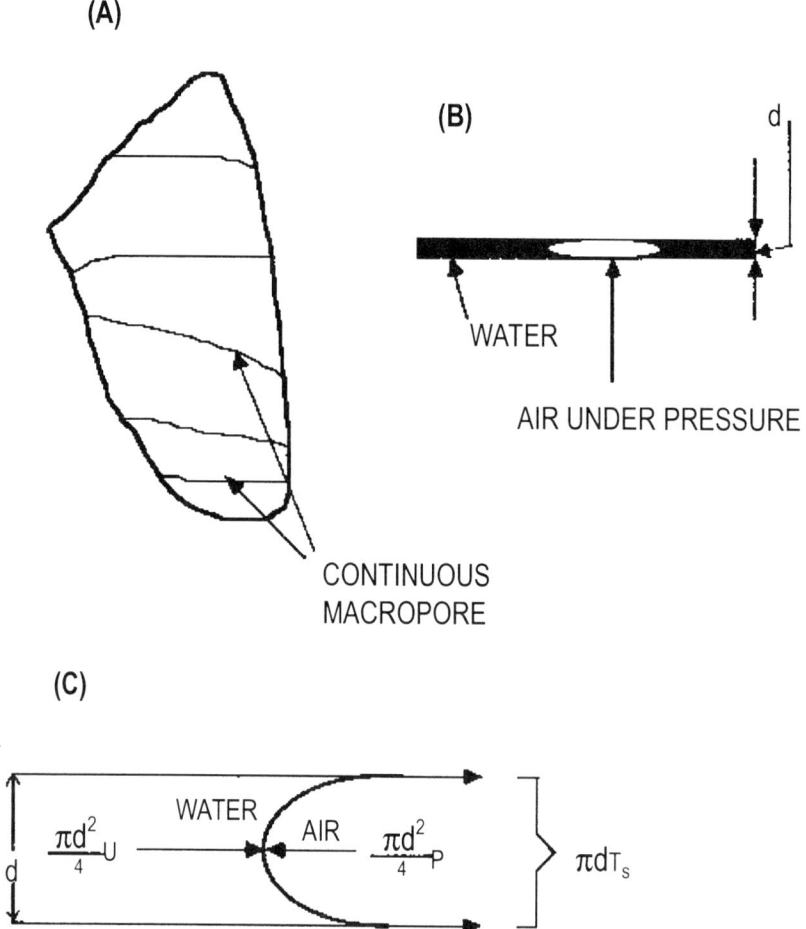

Fig. 3. (A) Shale sample, (B) Macropore with water and air pressure, (C) Air and water forces at the air-water interface in a macropore

From the photographs of the thin sections, the cross-sectional areas and the shapes of the perimeters of the macropores in the shales were obtained. To obtain the areas of the macropores as well as a plot of their perimeters, the macropores were determined by standard digitizing procedures [18]. Digitizing records the x-y coordinates of the perimeter of each macropore. This information is analyzed by a microcomputer with software necessary to calculate the areas of the macropores and to plot their perimeter. Figure 4 shows the typical results of the process for shale samples TN-5 and TN-9.

Using the area of each macropore, the equivalent diameter of the macropore was obtained. This equivalent diameter corresponds to the diameter of a circular area that has the same area as the macropore. From each photograph, forty representative macropores were chosen to calculate the overall average "equivalent diameter" for each shale sample. For example, the average equivalent diameter for the forty macropores of sample TN-5 (Fig. 4(A)) is equal to 0.053 mm. In the case of sample TN-9 (Fig. 4(B)), this overall average equivalent diameter is equal to 0.107 mm.

### 5.3 Fractal Evaluation of the Roughness of the Pore Boundaries

The roughness of the pore boundaries was evaluated using fractal theory. Fractal theory makes use of a number called the fractal dimension, D, to evaluate the irregularity of objects in nature [9]. In this study, the fractal dimension will be used to measure the degree of roughness of the pore boundaries in the 68 shale samples. The fractal dimension of the pore boundaries was calculated using the area-perimeter method [10, 6, 2, 7, 17].

The fractal dimension, D, is obtained from the slope, m, of the best fit line that connects the values of the area and perimeter for each of the pores in a shale sample. Fig. 5 shows this plot for the case of the pores forming part of the shale TN-5 and TN-9 (Fig. 4). Once the slope, m, is obtained, the fractal dimension is calculated from the ratio between 2 and m (D= 2/m). The fractal dimension, D, measures the average roughness of the pore boundaries in the shale samples. The higher the the value of D, the rougher are the pore boundaries in the shales. The fractal dimension, D, for the pore boundaries in sample TN-5, was equal to 1.1989 (Fig. 5(A)). The fractal dimension, D, for the pore boundaries of sample TN-9 was measured to be 1.4587 (Fig. 5(B)).

## 6 Relationship between Pore Diameter and Degree of Slaking

As previously mentioned, from the Jar Slake Test on the sixty-eight shale samples, fourteen of them completely disintegrated during this test and had a Jar Slake Index ($I_j$) of 1 or 2 (Table 1); four samples developed small fractures and had an $I_j$ equal to 3; and fifty shale samples experienced no degradation at all and had an $I_j$ value of either 5 or 6. Thus, three groups of shales were established with respect to their durability. One group of fourteen samples disintegrated into

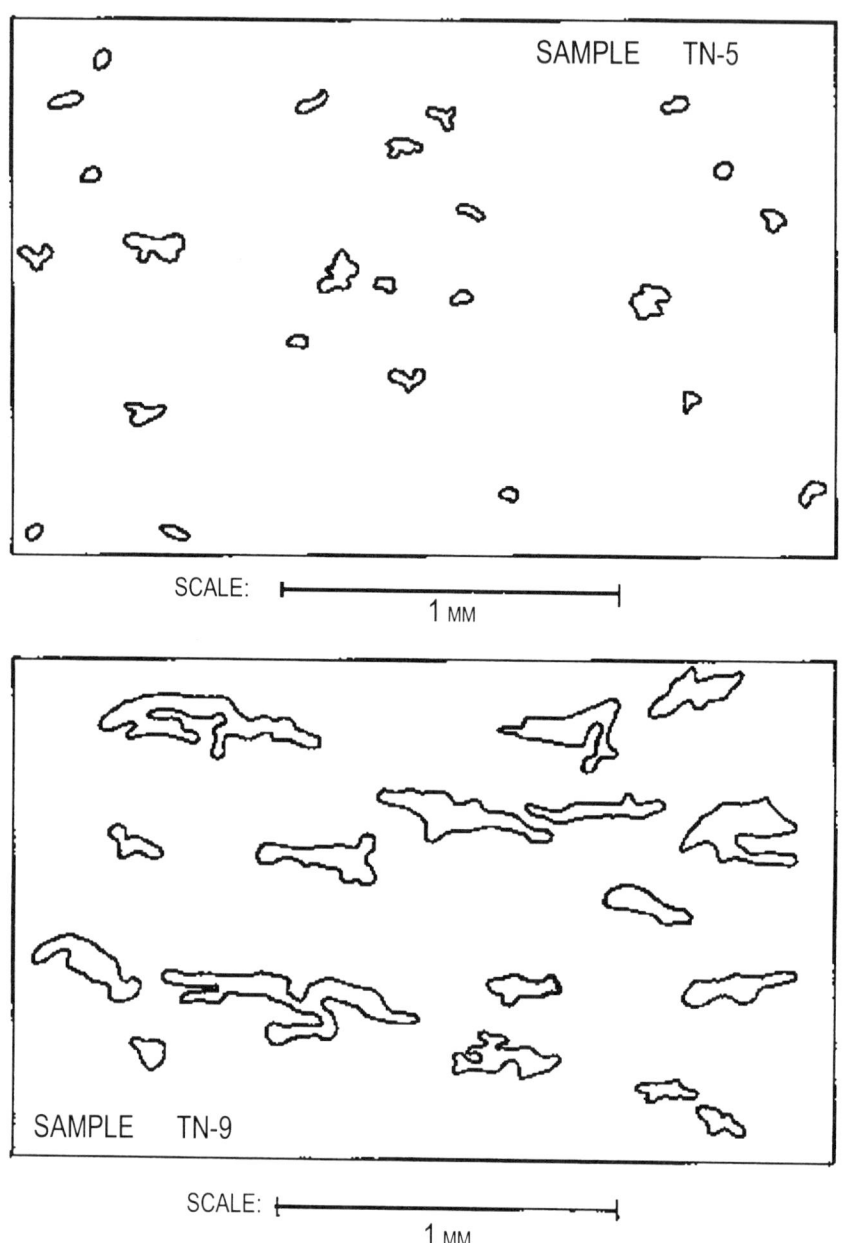

Fig. 4. Typical pore geometry in samples TN-5 and TN-9

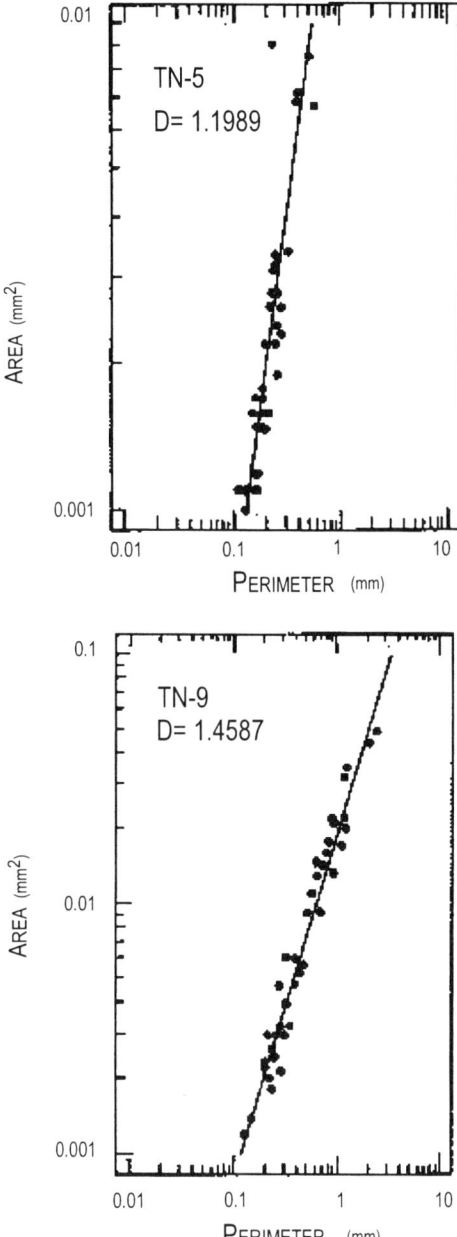

**Fig. 5.** Area-perimeter method to obtain the fractal dimension, TN-5 and, TN-9.

soil-size particles, a second group of four samples experienced very small changes, and a third group of fifty samples suffered no change when immersed in water.

Next, the mean macropore diameters of each of these three groups were calculated and compared with their slaking behavior. This was done in order to determine if there is a correlation between macropore diameter and their degree of slaking in water. The results of these calculations are shown in Table 3.

An analysis of Table 3 indicates that the fourteen samples that completely disintegrated during the Jar Slake Test had a system of macropores with an average diameter equal to 0.06 mm, the four samples that developed small fractures had macropores with an average diameter equal to 0.07 mm, and the fifty samples that experienced no degradation al all had macropores with an average diameter equal to 0.092 mm. Thus, there appears to be an optimum macropore size at which air pressures are effective in breaking the shales. However, there are exceptions to this behavior that seem to be explained by the roughness of the boundaries of the pores.

**Table 3.** Slaking Behavior and Pore Diameter

| Number of Samples | Slake Index $I_j$ | Maximum Diameter mm | Minimum Diameter mm | Standard Deviation | Mean Diameter mm |
|---|---|---|---|---|---|
| 14 | 1 or 2 | 0.081 | 0.043 | 0.0125 | 0.060 |
| 4 | 3 | 0.082 | 0.050 | 0.0136 | 0.070 |
| 50 | 5 or 6 | 0.180 | 0.050 | 0.030 | 0.092 |

# 7 Relationship between Pore Roughness and Degree of Slaking

Even though the size of the pores in shales is a good indicator for their slaking susceptibility, it was not a parameter that indicated without doubt the slaking behavior of shales. Table 4 shows the results of the Jar Slake Tests on 4 shale samples from Tennessee. The pore size in these shales was found to be less than 0.060 mm. However, it was determined that the shales that slaked had very smooth boundaries as compared to the shales that did not slake. The degree of roughness of the pores of the samples TN-5, TN-7, TN-15 and TN-16 was evaluated using the area-perimeter method [6, 7, 17].

An explanation for the lack of slaking of the shales with the small, rough pores seems to rest on the roughness of the pore boundaries. According to [16], when the tip of a capillary tube with either a square, triangular or an irregular cross section as shown in Fig. 6 is immersed in water, the water does not advance in the capillary tube following the whole cross sectional area, but advances in

Table 4. Slaking behavior of Tennessee shales

$I_j = 1$ or 2 slakes when immersed in water; $I_j = 6$ does not slake when immersed in water. The Fractal Dimension, D, of the pore boundaries was measured using the Area-Perimeter Method.

| Sample | Pore diameter (mm) | Jar Slake Index $I_j$ | Degree of Roughness | Fractal Dimension, D |
|---|---|---|---|---|
| TN-5  | 0.053 | 1 | smooth     | 1.1989 |
| TN-7  | 0.056 | 6 | very rough | 1.7767 |
| TN-15 | 0.043 | 2 | smooth     | 1.2263 |
| TN-16 | 0.045 | 6 | rough      | 1.4246 |

the tube following the corners and crannies of the tube. This partial filling of the tube cross sectional area will prevent the development of air pressure that is the necessary cause for the slaking of shales (Fig. 3). Thus, the roughness of the boundary of the pores (measured by the fractal dimension) appears to have a significant influence on the slaking of shales and needs to be considered in any evaluation of the durability of shales to water.

## 8 Conclusions

The degradation behavior of shale samples when in contact with water was analyzed using a combination of slake durability tests, petrographic analysis, and fractal analysis of the pore structure of the shales. From this combination of tests, the following determinations were made:

1. The predominant degradation mechanism of the shales was the result of pore air compression that takes place when the shales are immersed in water and is the result of capillary suction pressures.
2. The degradation (slaking) by pore air compression was directly related to the average pore diameter and the roughness of the pore boundaries of the macropore system in the shales. The smaller the diameter and the smoother the boundaries of the pores were, the more pronounced appears to be the slaking of the shales by air compression.
3. The roughness of the pore boundaries in the shales was measured using the fractal dimension concept from fractal theory. The use of the fractal dimension to measure the degree of roughness of the pore boundaries proved to be a simple but powerful tool to analyze the degradation behavior of the shales.

## References

1. D.E. Andrews, J.L. Withiam, E.F. Perry and H.L. Crouse (1980): Environmental Effects of Slaking of Surface Mine Spoils: Eastern and Central United States. Re-

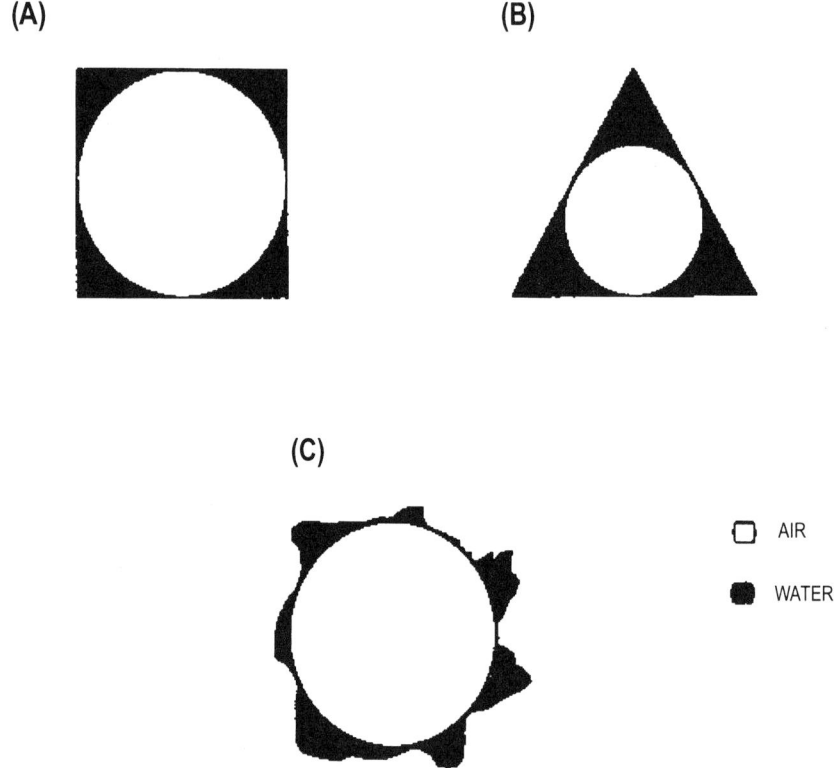

**Fig. 6.** Water advance in capillary tubes with irregular or rough cross sectional areas (after Ransohoff and Radke, 1988)

port, Contract No.J02855024, U.S. Bureau of Mines, U.S. Dept. of the Interior, Denver, Colorado, 247 p.
2. B.L. Cox and Y. Wang (1993): "Fractal surfaces: measurement and applications in the earth sciences". *Fractals*, Vol. 1, No. 1, pp. 87–115.
3. R.A. Cummings, M.M. Singh and N.N. Moebs (1983): "Effect of atmospheric moisture on the deterioration of coal mine roof shales". *Mining Engineering*, Vol. 35, No. 3, pp. 243–245.
4. F.A.L. Dullien (1979): Porous Media: Fluid Transport and Pore Structure. Academic Press, New York.
5. J.A. Franklin and R. Chandra (1972): "The slake durability test". *Int. J. of Rock Mech. and Mining Sciences*, Vol. 9, pp. 325–341.
6. G. Korvin (1992): Fractal Models in the Earth Sciences. Elsevier, Amsterdam.
7. J.P. Hyslip and L.E. Vallejo (1997): "Fractal analysis of the roughness and size distribution of granular materials". *Engineering Geology*, Vol. 48, No. 3-4, pp. 231–244.
8. R.J. Lutton (1977): "Design and Construction of Compacted Shale Embankments: Slaking Indeces for Design". Report FHWA-RD-1, Federal Highways Administration, Washington, D.C., 88 p.
9. B.B. Mandelbrot (1977): Fractals: Forms, Chance and Dimension. W.H. Freeman, San Francisco.
10. B.B. Mandelbrot, D.E. Passoja and A.J. Paullay (1984): "Fractal character of fracture surface of metals". *Nature*, Vol. 308, pp. 721–722.
11. R.E. Means and J.V. Parcher (1963): Physical Properties of Soils. Charles E. Merrill Books Inc., Columbus, Ohio.
12. Y. Moriwaki (1974): Causes of Slaking of Argillaceous Materials. Ph.D. Dissertation, Dept. of Civil Engineering, University of California, Berkeley, CA, 291 p.
13. G. Muller and H. Schmincke (1967): Methods in Sedimentary Petrology. Hafner Publishing Co., New York.
14. H.J. Olivier (1979): "A new engineering-geological rock durability classification". *Engineering Geology*, Vol. 14, pp. 255–279.
15. D. Pappas and L.E. Vallejo (1997): "The settlement and degradation of non-durable shales associated with coal mine waste embankments". *Int. J. of Rock Mech. and Mining Sciences*, Vol. 34, No. 3-4, Paper 241, pp. 779–789.
16. T.C. Ransohof and C.J. Radke (1988): Laminar flow of a wetting liquid along corners of predominantly gas-occupied non-circular pores. *J. of Colloudal and Interface Science*, Vol. 121, No. 2, pp. 391–401.
17. E.M. Schlueter, R.W. Zimmerman, N.G.W. Cook and P.A. Witherspoon (1997): "The fractal dimension of pores in sedimentary rocks and its influence on permeability". *Engineering Geology*, Vol. 49, No. 3-4, pp. 199–215.
18. T. Stanton (1987): "Tablets for precision graphics". *PC Magazine*, Vol. 6, No. 14, pp. 159–181.
19. W.E. Strom, G.H. Bragg and T.W. Ziegler (1978): Design and Construction of Compacted Shale Embankments. Report HWA-RE-78-141, Federal Highway Adminsitration, Washington, D.C., 207 p.

# Continuous Wavelet Transform Analysis of Fractal Superlattices

Hervé Aubert[1]* and Dwight L. Jaggard[2]

[1] Ecole Nationale Superieure d'Electrotechnique,
d'Electronique, d'Informatique et d'Hydraulique,
Institut National Polytechnique, 31071 Toulouse - FRANCE
aubert@len7.enseeiht.fr
[2] Complex Media Laboratory, Moore School of Electrical Engineering,
University of Pennsylvania, Philadelphia, PA 19104-6390 - U.S.A.
jaggard@seas.upenn.edu

**Abstract.** Fractal superlattices, or multilayer structures designed by alternating dielectric layers according to an iterative process, are interrogated by an electromagnetic pulse. We show that the remote extraction of their scaling properties can be achieved by applying the continuous wavelet transform to the rapidly fluctuating reflected signal by using the *wavelet skeleton*. For a sufficiently narrow pulse a set of particular identifiable lines of the skeleton exhibit a hierarchical structure in the time-scale domain. Such well-ordered structure reveals that some detectable singularities in the impulse response are located on the governing fractal set. Finally a wavelet-based dimension allows us to extract the similarity dimension of these Cantor superlattices from their time-domain reflection data.

## 1 Introduction

The remote detection of fractal features of many self-similar objects can be achieved from interrogation by an incident electromagnetic wave [1]- [2],[14],[15]. The problem consists of extracting the primary fractal descriptors, such as fractal dimension, stage of growth and lacunarity, characteristic of interrogated fractal objects from the observed scattering data [1]-[13]. This is one of the canonical problems in *fractal electrodynamics* [1],[3],[14],[15].

Fractal superlattices are designed by alternating dielectric layers according to an iterative process. The modeling of wave interactions with such multi-layered structures has recently been investigated [14]- [23],[16],[25]. However, to the authors' knowledge, only a limited amount of work has been reported on the time-domain reflection of waves from finely divided stratified media and the remote

---

* This research work was performed while the author was with the Complex Media Laboratory, Moore School of Electrical Engineering, University of Pensylvania, Philadelphia, PA 19104-6390 - U.S.A.

description of their fractal characteristics [11]. Here we explore the scaling properties of fractal superlattices from interrogation by an incident electromagnetic pulse.

Typically, the impulse response of a superlattice exhibits a wildly irregular structure that originates in its multiple reflections. We interpret this erratic reflected signal as a complex arrangement of abrupt changes -or singularities- which can be related, more or less explicitly, to the spatial distribution of discontinuities in the refractive index profile. The *wavelet transform* [28]-[31] is particularly well adapted for detecting singularities in signals [32]. From the local maxima of the wavelet-transform modulus, Arneodo *et al.* [33]- [36] have recently proven the ability of such transform to explore the distribution of singularities in fractal signals such as those from fully developed turbulence data, Brownian motion or DNA sequences. We use here a similar approach for investigating the distribution of singularities in the rapidly fluctuating reflected signal of fractal superlattices.

We focus on the wavelet analysis of the impulse response of one-dimensional Cantor superlattices [37]-[39]. In the time-scale domain the skeleton of the wavelet-transform modulus-maxima is computed. When a hierarchical structure emerges at large scales, we show that such structure provides enough information on the scaling properties of interrogated superlattices to characterize the underlying fractal set. In particular it reveals the existence in the impulse response of singularities that are distributed on a Cantor set, that is, on the same fractal set which governs the distribution of layers. Finally the remote extraction of the similarity dimension may be achieved from reflection data.

## 2 Impulse Response of Fractal Superlattices

A Cantor superlattice is a multi-layered dielectric structure genererated as follows. Start from a nondispersive layer of length $L$ and refractive index $n_1$ immersed in the nondispersive host medium of refractive index $n_0$. The stage of growth $S = 1$ is obtained by proceeding to the following operation : first, the division of the layer of index $n_1$ into $N$ non overlapping segments and then, the replacement of the middle $N_m$ segments by the medium of index $n_0$. The stage $S = 2$ is obtained by applying the same operation to each remaining layers of index $n_1$. Fig. 1 displays the refractive index profile $n(x)$ of the resulting multilayered structures stage by stage. The stage $S$ is generated by dividing all layers of index $n_1$ which exists at stage $S - 1$ into $N$ non overlapping segments and then, replacing the middle $N_m$ segments of such layers by the medium of index $n_0$.

As stage of growth $S \to \infty$ the discontinuities in the refractive index profile are located upon a discrete scale invariant set called *Cantor set*. Since each stage of growth consists of $\eta = 2$ scaled replicas of the previous stage, with a scaling factor $\rho = (N - N_m)/2N$, the similarity dimension $D_S$ of the superlattice is

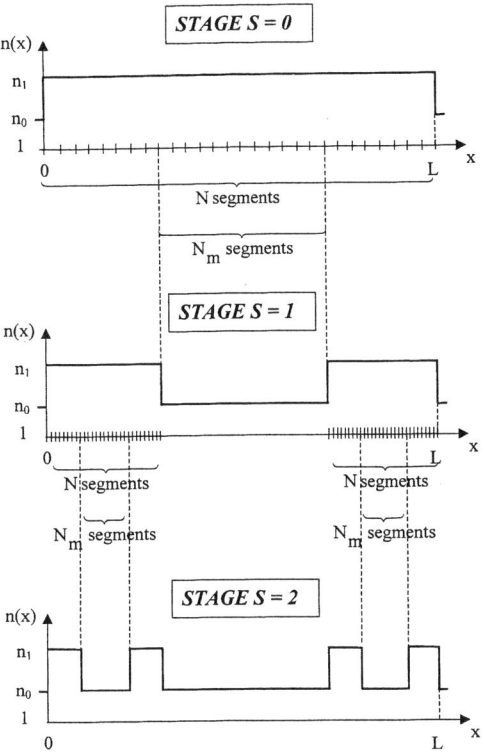

**Fig. 1.** Schematic of the iterative process which governs the construction rule of Cantor superlattices
As stage of growth $S \to \infty$, the discontinuities in the refractive index profile $n(x)$ are located on a Cantor set.

found to be

$$D_S = \frac{\ln \eta}{\ln \frac{1}{\rho}} = \frac{\ln 2}{\ln \frac{2N}{N-N_m}} \quad (1)$$

Now consider a normally incident pulse with Gaussian profile given by :

$$d_\sigma(t) = \frac{1}{2\sigma\sqrt{\pi}} e^{-\left(\frac{t}{2\sigma}\right)^2} \quad (2)$$

where $\sigma$ denotes the radius, or root mean square duration, of the pulse. The impulse response $r_\sigma(t)$ of the fractal superlattice is then given by the following inverse Fourier transform :

$$r_\sigma(t) = \frac{1}{2\pi} \int_{-\infty}^{+\infty} R_{S,L}(\omega) e^{-(\sigma\omega)^2} e^{-j\omega t} d\omega \quad (3)$$

where $R_{S,L}$ denotes the normal incident reflection coefficient of the multi-layered structure of length $L$ at stage of growth $S$. This reflection coefficient is computed on the basis of an efficient recursive computational technique which takes advantage of the iterative process governing the construction rule of the fractal superlattice [14]-[17],[20], [25]-[27]. This approach, called the *self-similarity method of computation*, reduces dramatically the amount of calculation required to compute the reflection coefficient of discrete self-similar structures compared to the conventional chain matrix method (an extension to this iterative process for variable angle of incidence and lacunarity has been recently proposed [26] and formulated in detail [27]). First, it consists in defining two generating functions $\mathbf{F}[x|y|d|\omega]$ and $\mathbf{G}[x|y|d|\omega]$ as

$$\mathbf{F}[x,y,d,\omega] = x + \frac{xy^2 \exp\left(-j2\frac{\omega}{c}n_0 d\right)}{1 - x^2 \exp\left(-j2\frac{\omega}{c}n_0 d\right)} \quad (4)$$

$$\mathbf{G}[x,y,d,\omega] = \frac{y^2 \exp\left(-j\frac{\omega}{c}n_0 d\right)}{1 - x^2 \exp\left(-j2\frac{\omega}{c}n_0 d\right)} \quad (5)$$

where $x, y, d$ are (positive) real numbers and $c$ designates the vacuum speed of light. Then, the reflection and transmission coefficients of the superlattice of longitudinal length $L$ at stage of growth $S$ are respectively obtained from the following recursive relationships

$$R_{S,L} = \mathbf{F}\left[R_{S-1,\rho L}(\omega), T_{S-1,\rho L}(\omega), qL, \omega\right] \quad (6)$$

and

$$T_{S,L} = \mathbf{G}\left[R_{S-1,\rho L}(\omega), T_{S-1,\rho L}(\omega), qL, \omega\right] \quad (7)$$

where $q = \frac{N_m}{N}$. The initialization ($S = 0$) is given by

$$R_{0,\rho^S L}(\omega) = -r + \frac{rtt' \exp\left(-j2\frac{\omega}{c}n_1\rho^S L\right)}{1 - r^2 \exp\left(-j2\frac{\omega}{c}n_1\rho^S L\right)} \tag{8}$$

and

$$T_{0,\rho^S L}(\omega) = \frac{tt' \exp\left(-j\frac{\omega}{c}n_1\rho^S L\right)}{1 - r^2 \exp\left(-j2\frac{\omega}{c}n_1\rho^S L\right)} \tag{9}$$

with

$$r = \frac{n_1 - n_0}{n_1 + n_0}, \, t = \frac{2n_0}{n_1 + n_0} \text{ et } t' = \frac{2n_1}{n_1 + n_0} \tag{10}$$

Figure 2 displays the impulse response of a triadic Cantor superlattice ($N = 3$ and $N_m = 1$) at the sixth stage of growth, $S = 6$. This response is computed from formulas (3)-(10) with refractive indices $n_0 = 1$ and $n_1 = 1.5$. Here the spectral radius of the pulse is chosen equal to the highest spatial frequency of the multi-layered structure, that is, $1/2\sigma = c/n_0\rho^{S-1}qL$. Because of the multiple reflections, this reflected signal presents what appear to be noise-like variations. Meanwhile, because of the scaling features of the interrogated superlattice, such fluctuations may be viewed as a succession of abrupt changes, the origin of which are not readily apparent. However, we will next show how the wavelet analysis places in evidence the underlying order of these fluctuations.

## 3 Remote Extraction of Fractal Descriptors

### 3.1 The Wavelet Transform

The *wavelet transform* (WT) consists of expanding signals in a basis of wavelets that are constructed from a single function, the *mother wavelet* $\Psi(t)$ by means of dilations and translations [28]-[31]. The WT of the signal $r_\sigma(t)$ is defined as

$$W_\Psi[r_\sigma](a,b) = \frac{1}{a}\int_{-\infty}^{+\infty} r_\sigma(t)\Psi\left(\frac{t-b}{a}\right)dt \tag{11}$$

where $b$ is a time parameter and, $a$ ($a > 0$) is a scale parameter. In formula (11) $\Psi(t)$ is assumed to be a real valued function. In the present work, we use the second derivative of the Gaussian function as mother wavelet, so that, $\Psi(t) =$

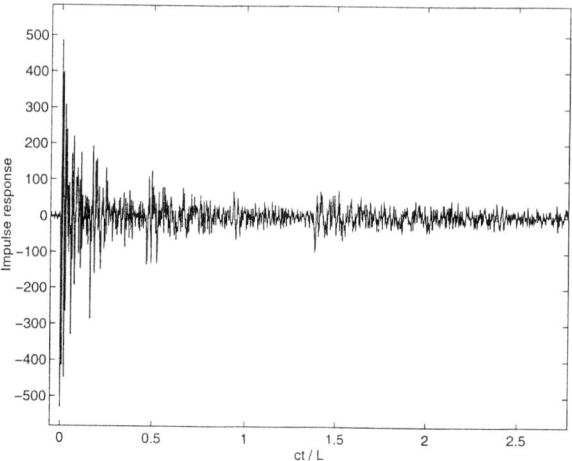

**Fig. 2.** Impulse response of the triadic Cantor superlattice [$N = 3$ and $N_m = 1$] at sixth stage of growth.
[$n_0 = 1$, $n_1 = 1.5$ and $L/\sigma c = 1460$.]
(Figure from Aubert, H., Jaggard, D.L. (1998): Fractal Superlattices and Their Wavelet Analyses. *Opt. Comm.* **149**, 207-212. ©1998 Elsevier Science B.V.)

$(1 - t^2) \exp\left(-\frac{t^2}{2}\right)$ and we proceed to the computation of the transform (11) using available subroutines [40].

In practice the reflected signal $r_\sigma(t)$ is approximated at a finite resolution, that is, by a set of $2^K$ samples ($K = 12$ in case of the impulse response displayed in Fig. 2). We associate to this resolution a characteristic scale length $a_0$ given by $a_0 = 2^{-K}$. Figure 3 represents in the time-scale domain the WT modulus of the impulse response given in Fig. 2. (In order to improve the images, the WT are displayed from the fine scale $2^3 a_0 = 2^{-9}$ to the large scale $2^{10} a_0 = 2^{-2}$.)

A well-ordered architecture appears. The hierarchy underlying such architecture is clearly revealed on the skeleton of Fig. 4 that displays the local maxima - at a given scale - of Fig. 3. In the context of signal processing using wavelets, it has been demonstrated that WT maxima, denoted the *wavelet-transform modulus-maxima* (WTMM), detect singularities analyzed signals [32] and are able to reveal, if it exists, the iterative process underlying their structures [33]-[36]. Here we take advantage of these properties to explore the spatial distribution of singularities in $r_\sigma(t)$.

## 3.2 Remote Extraction of the Construction Rule

In Fig. 4 we observe that *some* maxima are clearly located on a *hierarchical structure* which emerges at large scales and can be followed among other maxima when decreasing the scale. Such arch-like structures in time-scale domain

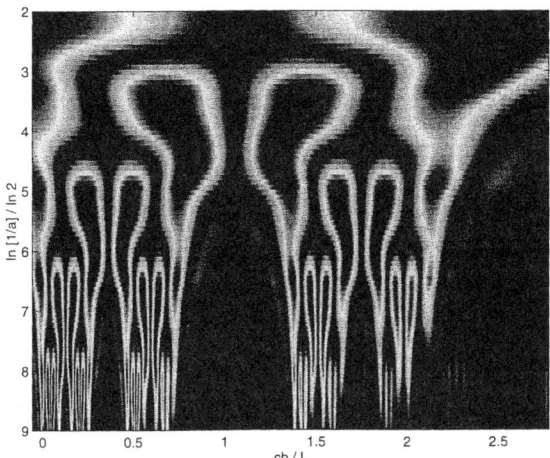

**Fig. 3.** Modulus of the wavelet transform (black is zero) of the impulse response of Fig. 2.
The mother wavelet is the second derivative of the Gaussian function.
(Figure from Aubert, H., Jaggard, D.L. (1998): Fractal Superlattices and Their Wavelet Analyses. *Opt. Comm. 149*, 207-212. ©1998 Elsevier Science B.V.)

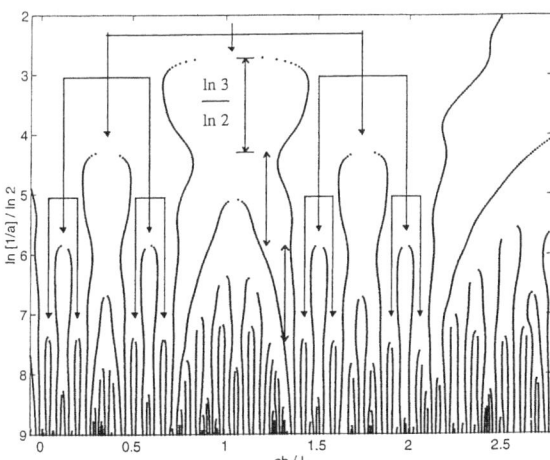

**Fig. 4.** Skeleton showing the location of wavelet-transform modulus-maxima (each point indicates the location in time-scale domain of a maximum) of Fig. 3. A well-ordered architecture (pointed out by a set of arrows) appears in time-scale domain and greatly contrasts with the apparent erratic fluctuations of the analyzed signal (see Fig. 2).
The scale factor between two successive bifurcations is constant and very close to 3.
(Figure from Aubert, H., Jaggard, D.L. (1998): Fractal Superlattices and Their Wavelet Analyses. *Opt. Comm. 149*, 207-212.©1998 Elsevier Science B.V.)

reveal the iteration process that governs the construction rule of the interrogated Cantor superlattice. As a matter of fact the number $Z_{hier.}$ of maxima belonging to it follows a characteristic scaling law: $Z_{hier.}$ increases by $2^m$ when decreasing the scale from $k/\alpha^{m-1}$ to $k/\alpha^m$ where $\alpha$ denotes the *constant* scale factor between two successive bifurcations ( $\alpha \simeq 3$ in Fig. 4). An analogous scaling law governs the number of discontinuities in the refractive index profile: at stage of growth $S$ ($S \geq 1$), after $S$ successive applications of the reduction factor $\rho$ ( $\rho = 1/3$ in case of the triadic Cantor superlattice), the number of discontinuities is increased by $2^S$ compared to one at the previous stage. Such scaling law originates in the application of the iterative process described in the previous section. The magnification factor $1/\rho$ and the scale factor $\alpha$ play clearly an analogous role. We deduce that the arch-like structures in time-scale domain allows one to identify singularities in the impulse response that are located on a Cantor set, that is, the same fractal set which determines the spatial distribution of discontinuities in the refractive index profile. Thus, the discrete scale-invariance property of the fractal superlattice imprints in its impulse response an analogous fractal distribution of easily identifiable singularities.

For sufficiently narrow pulses and as far as the hierarchical structure is apparent in time-scale domain, the above results can be extended to Cantor superlattices of arbitrary order $N$ and degree $N_m$. As a matter of fact the arch-like structure is always found to be analogous to that of Fig. 4. For example, Fig. 5 and Fig. 6 display the WTMM for the cases $(N, N_m) = (5, 3)$ [i.e., $\rho = 1/5$] and $(N, N_m) = (11, 9)$ [i.e., $\rho = 1/11$] respectively. In these cases, the scale factor between two successive bifurcations is found to be very close to 5 and 11 respectively, and the above discussion is still valid.

Thus, although the amount of data is dramatically reduced by discarding the maxima of wavelet-tranform modulus which are not located on an identifiable hierarchical structure in the time-scale domain, direct information about the fractal distribution of layers is obtained. We now proceed to the remote extraction of the similarity dimension of fractal superlattices.

### 3.3 The Remote Extraction of the Similarity Dimension

Following section 3.3 the number $Z_{hier.}$ of WTMM at scale $a = k/\alpha^n$ that belongs to the arch-like structure is systematically given by

$$Z_{hier.}\left(a = \frac{k}{\alpha^n}\right) = 2 + 2^2 + 2^3 + ... + 2^n = 2(2^n - 1) \qquad (12)$$

where $\alpha$ ($\alpha > 1$) is the constant scale factor between two successive bifurcations. From this scaling law a wavelet-based dimension $D_W$ is now introduced by the following relationship

253

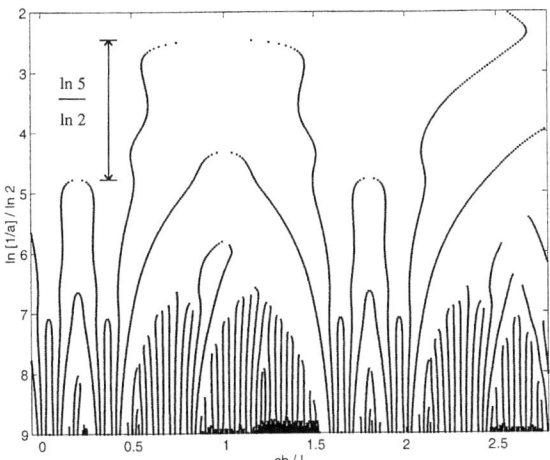

**Fig. 5.** Skeleton of the impulse response of a Cantor superlattice at $6^{th}$ stage of growth with $N = 5$ and $N_m = 3$
[i.e., $D_S = \ln 2 / \ln 5$]. $n_0, n_1$ and $L/\sigma c$ are given in Fig. 2.
The scale factor between two successive bifurcations is very close to 5.
(Aubert, H., Jaggard, D.L. (1998): Wavelet Analysis of Transients in Fractal Superlattices. Submitted to *IEEE Trans. Antennas Propagat.*. ©1998 IEEE)

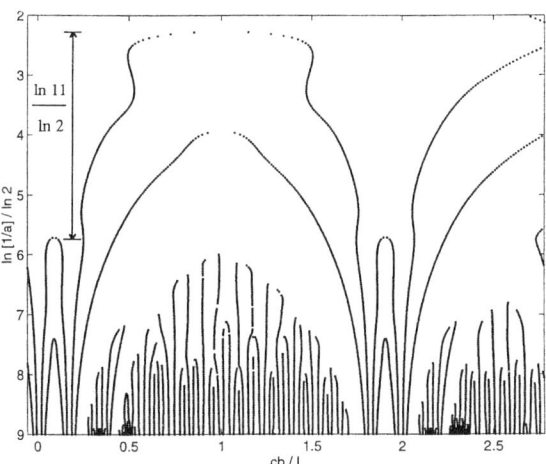

**Fig. 6.** Skeleton of the impulse response of a Cantor superlattice at $6^{th}$ stage of growth with $N = 11$ and $N_m = 9$
[i.e., $D_S = \ln 2 / \ln 11$]. $n_0$, $n_1$ and $L/\sigma c$ are given in Fig. 2.
The scale factor between two successive bifurcations is very close to 11.
(Aubert, H., Jaggard, D.L. (1998): Wavelet Analysis of Transients in Fractal Superlattices. Submitted to *IEEE Trans. Antennas Propagat.*. ©1998 IEEE)

$$D_W = \lim_{a \to 0^+} -\frac{\ln Z_{hier.}(a)}{\ln a} \qquad (13)$$

Following formula (12) we obtain

$$D_W = \lim_{n \to +\infty} \frac{\ln[2(2^n - 1)]}{\ln \alpha^n} = \frac{\ln 2}{\ln \alpha} \qquad (14)$$

This formula is similar to the expression of the fractal dimension given in formula (1) and reinforces the above noted analogy between the magnification factor $1/\rho$ and the scale factor $\alpha$. If we had considered *all* WT maxima lines of skeletons in the computation of formula (13) instead of *only ones* that belongs to the arch-like structure we would have recovered the definition of the wavelet-based dimension given in [33] and [35].

As shown in Fig. 7, where a large range of $D_S$ is investigated by appropriate choices of $N$ and $N_m$, the wavelet-based dimension $D_W$ and the scale factor $\alpha$ still provide a very good approximation of the similarity dimension $D_S$ and the magnification factor $1/\rho$, respectively, for many Cantor superlattices. Note that few scales are required in practice to estimate the similarity dimension. As a matter of fact the scale factor $\alpha$ [and consequently $D_W$ from formula (14)] may be determined from the two first bifurcations at large scales and retrieved between other bifurcations at fine scales. This result accounts for the well-known fact that, since the construction rule of fractal superlattices is governed by an iterative process the knowledge of few steps of refinement of such structures are sufficient for carrying on the refinement for any fine resolution. Moreover the remarkable ability of the wavelet analysis to give a remote accurate approximation of the similarity dimension of superlattices from the behavior at *large scales* of reflected signals is consistent with the ability of frequency-domain techniques [16],[20],[25] to extract such dimension from *low frequency* components of the reflection coefficient.

We have observed a graceful degradation of the hierarchy as the refractive indices increase: this degradation primarily affects the late-time part of the skeleton and gradually reaches the early-time part. As $D_S$ increases the maximal value of $n_1$ which allows the detection of a hierarchical structure in time-scale domain decreases (see Fig. 8).

### 3.4 Role of the Pulse Radius in the Estimation of the Stage of Growth

Fractal superlattices which are interrogated by wide pulses do not reveal their scaling properties. Figs. 9 displays the skeletons for various pulse radii.

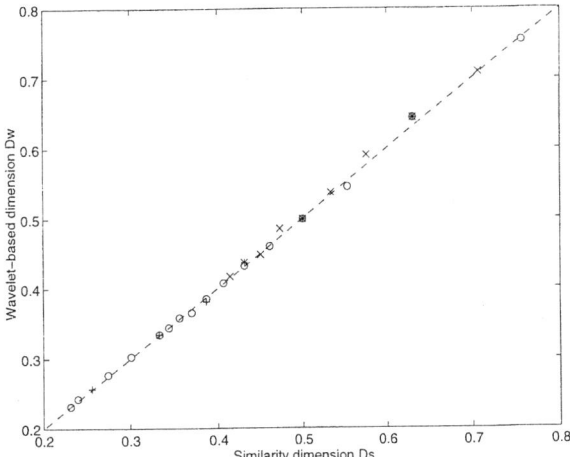

**Fig. 7.** Wavelet-based dimension $D_W$ as a function of the similarity dimension $D_S$ for Cantor superlattices:

$(+)N - N_m = 2$, $(\bigcirc)N - N_m = 4$ $and$ $(\times)N - N_m = 6$
$[S = 6, n_0 = 1, n_1 = 1.5$ and $L/\sigma c = 1460]$.

All the symbols are very close to a straight line (dashed line) so that $D_W \simeq D_S$.
(Figure from Aubert, H., Jaggard, D.L. (1998): Fractal Superlattices and Their Wavelet Analyses. *Opt. Comm.* **149**, 207-212. ©1998 Elsevier Science B.V.)

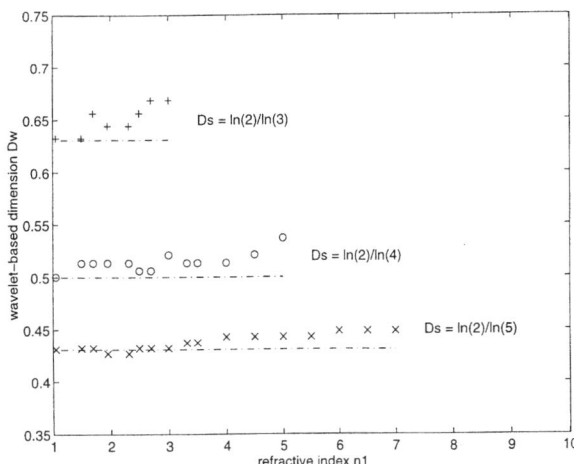

**Fig. 8.** Wavelet-based dimension $D_W$ as a function of the refractive index $n_1$ for various similarity dimensions $D_S$

$[S = 6, n_0 = 1$ and $L/\sigma c = 1460]$.

(Aubert, H., Jaggard, D.L. (1998): Wavelet Analysis of Transients in Fractal Superlattices. Submitted to *IEEE Trans. Antennas Propagat.*. ©1998 IEEE)

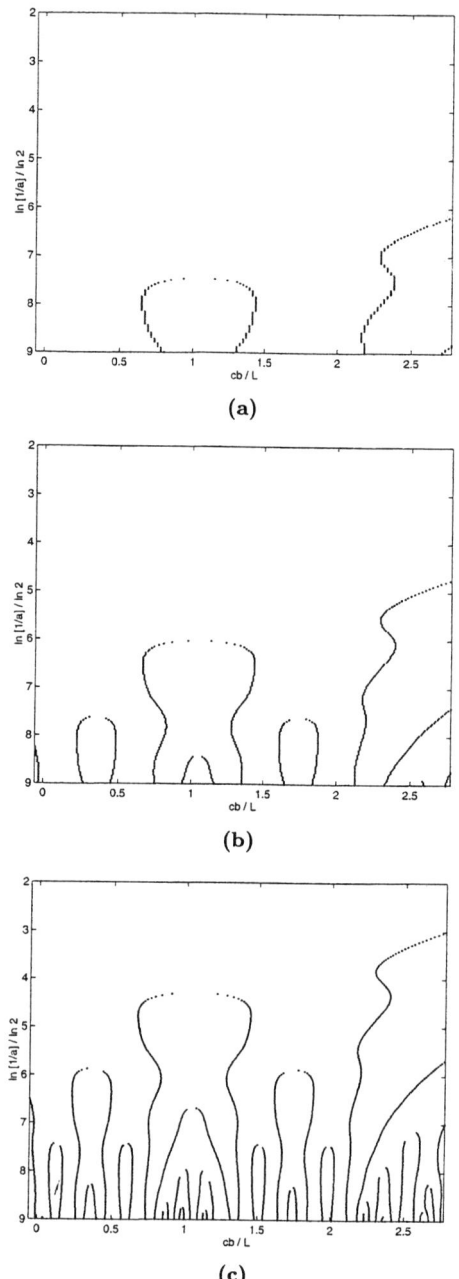

**Fig. 9.** Skeletons of the impulse responses of a triadic Cantor superlattice at $6^{th}$ stage as a function of the incident pulse radius for :
(a) $L/\sigma c = 1460/\rho^3$, (b) $L/\sigma c = 1460/\rho^2$ and (c) $L/\sigma c = 1460/\rho$ with $\rho = 1/3$ (for the case $L/\sigma c = 1460$ see Fig. 4).
As $\sigma$ decreases a well-ordered structure rises in the time-scale domain and reveals the scaling properties of the interrogated superlattice.

As $\sigma$ decreases the arch-like structure rises in time-scale domain. The repetitive application of the reduction factor $\rho$ to the pulse radius allows a gradually increasing number of bifurcations in time-scale domain ($\rho = 1/3$ in case of Figs. 9. The radius of the pulse is then viewed as an adjustable parameter which allows the remote determination of the stage of growth from the following iterative process: at $k^{th}$ iteration the superlattice is interrogated by a pulse of radius $\sigma_k = \rho\sigma_{k-1}$ and the number of bifurcations is calculated. If this number has changed compared to one at the $(k-1)^{th}$ iteration, the superlattice is interrogated at the following iteration by a new pulse of radius $\sigma_{k+1} = \rho\sigma_k$ and then, the number of bifurcations is calculated again. When it remains constant the iterative process is stopped and the stage of growth is equal to the number of bifurcations. The initialization of this process may be achieved by taken $\sigma_0 = L/2c$.

## 4 Conclusion

In summary, the wavelet analysis of impulse responses of discrete self-similar superlattices allows us to easily specify singularities in reflected signals that are located on the same fractal set which underlies the spatial distribution of discontinuities in the refractive index profile. For sufficiently narrow pulses, the construction rule, the similarity dimension, and the stage of growth of superlattices can be remotely extracted.

The present approach allows us to extract information concerning the multi-scale fluctuation of the refractive index and characterize these fluctuation without *in situ* measurements. Moreover, it does not require the reconstruction of the index profile via an inversion technique. From our numerical work to date applications for impulse responses corrupted by noise looks very promising and consequently, application to experimental scattering data obtained from the interaction of a superlattice with a laser would be a useful extension of this work

**Acknowledgement** : H. Aubert acknowledges the Complex Media Laboratory, University of Pennsylvania for its hospitality and the INPT-ENSEEIHT, France, for support. D. L. Jaggard acknowledges partial support through NATO Grant 930923 and the Complex Media Laboratory, University of Pennsylvania.

## References

1. D.L. Jaggard (1990): On Fractal Electrodynamics. In *Recent Advances in Electromagnetic Theory*, H. N. Kritikos and D. L. Jaggard, editors, Springer-Verlag, New York, 183-224.
2. M.V. Berry (1979): Diffractals. *J. Phys. A: Math. Gen.* **12**, 781-797.

3. D.L. Jaggard (1991): Fractal Electrodynamics and Modeling. In *Directions in Electromagnetic Wave Modeling*, H. L. Bertoni and L. B. Felsen, editors,.Plenum Publishing Co., New York, 435-446.
4. A. Le Méhauté (1990): Les Géometries Fractales : l'Espace-Temps Brisé. Hermès, Paris.
5. E. Jakeman (1982): Scattering by a Corrugated Random Surface with Fractal Slope. *J. Phys. A: Math. Gen. 15*, L55-L59.
6. H.D. Bale and P.W. Schmidt (1984): Small-Angle X-Ray-Scattering Investigation of Submicroscopic Porosity with Fractal Properties. *Phys. Rev. Lett. 53*, 596-599.
7. C. Allain and M. Cloitre (1986): Optical Diffraction on Fractals. *Phys. Rev. B 33*, 3566-3569.
8. D.L. Jaggard and X. Sun (1989): Scattering from Band-Limited Fractal Fiber Surfaces. *IEEE Trans. Antennas Propagat. 37*, 1591-1597.
9. S.K. Sinha (1989): Scattering from Fractal Structures. *Physica D 38*, 310-314.
10. M.V. Berry and T.M. Blackwell (1981): Diffractal Echoes. *J. Phys. A: Math. Gen. 14*, 3101-3110.
11. Y. Kim and D.L. Jaggard (1991): Wave Interactions with Continuous Fractal Layers. *Proc. SPIE 1558*, 113-119.
12. T. Lo, H. Leung, J. Litva and S. Haykin (1993): Fractal Characteristisation of Sea-Scattered Signals and Detection of Surface Targets. *IEE Proc. F 140*, 243-250.
13. S. Rouvier, P. Borderies and I. Chenerie (1997): Ultra Wide Band Electromagnetic Scattering of a Fractal Profile. *Radio Science. 32* , 285-293.
14. D.L. Jaggard (1995): Fractal Electrodynamics: Wave Interactions with Discretely Self-Similar Structures. In *Electromagnetic Symmetry*, C. Baum and H. Kritikos, editors, Taylor and Francis Publishers, Washington, D.C., 231-281.
15. D.L. Jaggard (1997): Fractal Electrodynamics: From Super Antennas to Superlattices. In *Fractals in Engineering*, J. L. Vehel, E. Lutton and C. Tricot, editors, Springer-Verlag, Berlin, 204-221.
16. A.D. Jaggard and D.L. Jaggard (July 13-17, 1998): Fractal Superlattices: a Frequency Domain Approach. Presented at the *1998 Progress in Electromagnetics Research Symposium*, Nantes, France.
17. D.L. Jaggard and X. Sun (1990): Reflection from Fractal Multilayers. *Opt. Lett. 15*, 1428-1430.
18. V.V. Konotop, O.I. Yordanov and I.V. Yurkevich (1990): Wave Transmission through a One-Dimensional Cantor-Like Fractal Medium. *Europhys. Lett. 12*, 481-485.
19. T. Megademini (1990): Principles and Possibilities of Interferential Multilayer Mirrors of Non Integral Dimensionality. *Phys. Rev. B 41*, 4693-4699.
20. X. Sun and D.L. Jaggard (1991): Wave Interaction with Generalized Cantor Bar Multilayers. *J. Appl. Phys. 70*, 2500-2507.
21. T. Megademini, B. Pardo and R. Jullien (1991): Fourier Transform and Theory of Fractal Multilayer Mirrors. *Opt. Comm. 80*, 312-316.
22. M. Bertolotti, P. Masciulli and C. Sibilia (1994): Spectral Transmission Properties of a Self-Similar Optical Fabry-Perot Resonator. *Opt. Lett. 19*, 777-779.
23. S. De Nicola (1994): Reflection and Transmission of Cantor Fractal Layers. *Opt. Comm. 111*, 11-17.
24. S.A. Bulgakov, V.V. Konotop and V. Vasquez (1995): Wave Interaction with a Random Fat Fractal: Dimension of the Reflection Coefficient. *Waves in Random Media 5*, 9-18.

25. A.D. Jaggard and D.L. Jaggard (June 1997): Fractal Superlattices and Scattering: Lacunarity, Fractal Dimension, and Stage of Growth. Presented at the *1997 AP-S/URSI Meeting*, Montréal, Canada.
26. D.L. Jaggard and A.D. Jaggard (1997): Polyadic Cantor Superlattices with Variable Lacunarity. *Opt. Lett. 22*, 145-147.
27. A.D. Jaggard and D.L. Jaggard (1998): Scattering from Fractal Superlattices with Variable Lacunarity. *J. Opt. Soc. Am. A154*, 1626-1635.
28. P. Goupillaud, A. Grossmann and J. Morlet (1984-1985): Cycle-Octave and Related Transforms in Seismic Signal Analysis. *Geoexploration 23*, 85-102.
29. Y. Meyer (1990): Ondelettes. Hermann, Paris.
30. I. Daubechies (1992): Ten Lectures on Wavelets. S.I.A.M, Philadelphia, PA.
31. M. Holschneider (1995): Wavelets: An Analysis Tool. Oxford University Press, New York.
32. S. Mallat and W.L. Hwang (1992): Singularity Detection and Processing with Wavelets. *IEEE Trans. Inform. Theory 38*, 617-643.
33. A. Arneodo (1996): Wavelet Analysis of Fractals: from the Mathematical Concepts to Experimental Reality. In *Wavelets: Theory and Applications*, G. Erlebacher, M. Y. Hussaini and L. M. Jameson, editors, Oxford University Press, New York, 349-502.
34. A. Arneodo, G. Grasseau and M. Holschneider (1988): Wavelet Transform of Multifractals. *Phys. Rev. Lett. 61*, 2281-2284.
35. J.F. Muzy, E. Bacry and A. Arneodo (1991): Wavelets and Multifractal Formalism for Singular Signals: Application to Turbulence Data. *Phys. Rev. Lett. 67*, 3515-3518.
36. A. Arneodo, E. Bacry, P.V. Graves and J.F. Muzy (1995): Characterizing Long-Range Correlations in DNA Sequences from Wavelet Analysis. *Phys. Rev. Lett. 74*, 3293-3296.
37. H. Aubert and D.L. Jaggard (1998): Fractal Superlattices and Their Wavelet Analyses. *Opt. Comm. 149*, 207-212.
38. H. Aubert and D.L. Jaggard (July 13-17, 1998): Remote Characterization of Fractal Superlattices using Wavelets. Presented at the *1998 Progress in Electromagnetics Research Symposium*, Nantes, France.
39. H. Aubert and D.L. Jaggard (1998): Wavelet Analysis of Transients in Fractal Superlattices. Submitted to *IEEE Trans. Antennas Propagat.*.
40. J. Buckheit, S. Chen, J. Crutchfield, D. Donoho, H.-Y. Gao, I. Johnstone, E. Kolaczyk, J. Scargle and K. Young (1996): WaveLab .701. In http://playfair.stanford.edu/~ wavelet/, Stanford Univ., Berkeley Univ. and NASA-Ames.

# CHEMICAL ENGINEERING

# Mixing in Laminar Chaotic Flows: Differentiable Structures and Multifractal Features

Massimiliano Giona

Dipartimento di Ingegneria Chimica, Universitá di Roma "La Sapienza",
via Eudossiana 18, 00184 Roma - ITALY
max@giona.ing.uniroma1.it

**Abstract.** This article discusses some recent advances in the theory of laminar incompressible 2D chaotic flows. Attention is focused on the geometrical properties characterizing the asymptotic evolution of material lines and on the statistical measure-theoretical properties associated with the resulting invariant structures. The basic quantity describing the lamellar structure produced by chaotic advection is the intermaterial contact-area density, which possesses highly singular multifractal features in the case of generic laminar chaotic flows.

## 1 Introduction

There is a consolidated history of common research bridging fluid dynamics and fractal analysis [1]. Most of this research refers to the statistical characterization of turbulent eddies and to their multifractal features [2–4]. The application of fractal concepts in the description of turbulent structures is not particularly surprising, due to the stochastic nature of turbulence. Less obvious is the occurrence of multifractal features in the description of coherent structures arising in laminar flows under chaotic mixing conditions [5, 6]. The global geometry of laminar chaotic flows has attracted great interest in recent years [7–9, 9–13] due to its implications in dynamical system theory and its practical consequences in the optimization of mixing performances and mixing-controlled phenomena such as chemical reactions in industrial equipments. The formation of invariant coherent structures in laminar chaotic flows is an interesting physical and mathematical problem of diffeomorphic dynamical systems, giving rise to differentiable spatial patterns whose measure-theoretical properties are highly singular and amenable to multifractal description.

Laminar mixing corresponds to fluid element motion under the condition that the nonlinear inertial term in the Navier-Stokes equation is negligible, so that the velocity field $\mathbf{v}(\mathbf{x}, t)$ is a solution of a linear partial differential equation. If diffusion is negligible compared to the convective term, as occurs for highly viscous fluids at short/intermediate time-scales, the kinematics of a passive tracer particle is described by means of an ordinary differential equation

$$\frac{d\mathbf{x}(t)}{dt} = \mathbf{v}(\mathbf{x}(t), t) , \quad (1)$$

where **v** is a time-dependent velocity field and $\mathbf{x} = (x, y)^t$ the position vector of the advected particle. Throughout this article 2D time-periodic incompressible velocity fields are considered. This means that Eq. (1) is defined on a 2D manifold $\mathcal{M}$ representing the physical space, $\mathbf{v}(\mathbf{x}, t + T) = \mathbf{v}(\mathbf{x}, t)$, where $T$ is the period, and the velocity field **v** is solenoidal, i.e. $\nabla \cdot \mathbf{v} = 0$.

In the physics of mixing, a central issue concerns the properties of vectors and curve arcs advected by a given velocity field. It has been observed both numerically and experimentally [7, 8, 14, 15] that incompressible time-periodic flows give rise to invariant patterns in the evolution of curve arcs, when observed at periodic instants of time, under the condition that the kinematics of particle motion is chaotic (in point of fact efficient mixing performance and the chaotic nature of Eq. (1) are two complementary sides of the same coin). The geometrical origin of this phenomenon lies in the fact that the restriction of a Poincaré section of a time-periodic 2D differentiable area-preserving flow within a chaotic region $\mathcal{C} \subseteq \mathcal{M}$ defines a hyperbolic transformation upon it. This means that all the global properties characterizing the spatial distribution of the coherent structures generated by chaotic advection can be derived from the geometry of the asymptotic unstable vector sub-bundle $\{\mathcal{E}_\mathbf{x}^u\}_{\mathbf{x} \in \mathcal{C}}$ along which tangent vectors are stretched.

The hyperbolic nature of a chaotic differentiable area-preserving dynamics not only determines the geometry and topology of partially mixed structures but also makes possible a quantitative framing of the measure-theoretical properties involved in the evolution of material interfaces.

The relation between the geometry of asymptotic invariant properties and the statistical characterization of the intermaterial area density has recently been analyzed in detail [16–19]. This article summarizes the main results and focuses on the open problems, in the solution of which fractal and multifractal analysis may contribute significantly towards achieving a unified picture. This article is organized as follows. Section 2 briefly discusses the physical phenomenology involved in the formation of invariant structures in laminar chaotic flows and their geometrical properties. Section 3 connects this to the structure of the invariant unstable foliation characterizing differentiable area-preserving hyperbolic diffeomorphisms. Section 4 addresses the multifractal characterization of the probability measure associated with the length content of a generic leaf of the unstable foliation. Finally, section 5 briefly discusses some problems remaining open for further analysis.

## 2 Invariant geometry in laminar chaotic flows

Under the condition of time-periodicity for the velocity field **v**, we may associate with Eq. (1) a Poincaré map $\Phi$ obtained by sampling the trajectories modulo the period $T$. In this way, particle motion is described by means of an autonomous dynamical system

$$\mathbf{x}_{n+1} = \Phi(\mathbf{x}_n), \qquad (2)$$

where $\mathbf{x}_n = \mathbf{x}(nT)$, defined on a two-dimensional manifold $\mathcal{M}$. Attention is restricted to differentiable dynamics. It is therefore assumed that $\Phi$ is at least a $C^2$-diffeomorphism i.e. *a fortiori* that both $\Phi$ and its inverse $\Phi^{-1}$ are differentiable and their Jacobian matrices continuous. We use the symbol $\Phi^*(\mathbf{x}) = \partial \Phi(\mathbf{y})/\partial \mathbf{y}|_{\mathbf{y}=\mathbf{x}}$ to indicate the differential of $\Phi$ at $\mathbf{x}$, i.e. its Jacobian matrix. $\Phi^*(\mathbf{x})$ is a mapping of the tangent space $T\mathcal{M}_\mathbf{x}$ at $\mathbf{x}$ onto the tangent space $T\mathcal{M}_{\Phi(\mathbf{x})}$ at the image point $\Phi(\mathbf{x})$. The differential $\Phi^*(\mathbf{x})$ is the basic mapping needed in order to analyze the first-order properties of the Poincaré map $\Phi$, and in particular the evolution of vectors tangent to curves (representing the boundary of a fluid/phase-space element) advected by the velocity field $\mathbf{v}$. The condition of area preservation means that the determinant of the differential is equal to 1 in absolute value for all $\mathbf{x} \in \mathcal{M}$. In the case of the dynamical system considered, the more restrictive condition $\det(\Phi^*(\mathbf{x})) = 1$ holds, thus implying that the differentials $\Phi^*(\mathbf{x})$ for $\mathbf{x} \in \mathcal{M}$ belong to the group of unimodular matrices $SL(2, \Re)$. By definition, the differential $\Phi^{n*}(\mathbf{x})$ of the $n$-th iterative of $\Phi$ is given by $\Phi^{n*}(\mathbf{x}) = \Phi^*(\Phi^{n-1}(\mathbf{x})) \cdots \Phi^*(\Phi(\mathbf{x}))\Phi^*(\mathbf{x}) = \prod_{j=0}^{n-1} \Phi^*(\Phi^j(\mathbf{x}))$.

We assume a fairly general definition of chaos for area-preserving mappings. A map $\Phi$ is said to be *chaotic* within an invariant submanifold $\mathcal{C} \subseteq \mathcal{M}$ if, for each $\mathbf{x} \in \mathcal{C}$ and for any vector $\mathbf{v} \in T\mathcal{C}_\mathbf{x}$, the sequence of vectors $\Phi^{n*}(\mathbf{x})\mathbf{v}$ is unbounded in norm, either forward ($n \to \infty$) or backward ($n \to -\infty$) in time [20], i.e.

$$\sup_{-\infty < n < \infty} ||\Phi^{n*}(\mathbf{x})\mathbf{v}|| = \infty \,. \tag{3}$$

This definition is related to the sensitive dependence with respect to the initial conditions typical of chaotic dynamical systems since the distance between the evolution of nearby points $||\Phi^n(\mathbf{x} + \varepsilon) - \Phi^n(\mathbf{x})||$ for small $\varepsilon$ can be expanded to the first order to give $||\Phi^{n*}(\mathbf{x})\varepsilon|| \sim \exp(n\lambda)||\varepsilon||$ with $\lambda > 0$.

The standard map introduced by Chirikov [21] is analyzed as a model system representative of the generic behaviour of Poincaré maps of area-preserving differentiable dynamics of the form given by Eq. (1). This map is defined on the two-dimensional torus and expressed by

$$\begin{cases} x_{n+1} = x_n - (\kappa/2\pi) \sin(2\pi y_n) \bmod. 1 \\ y_{n+1} = y_n + x_{n+1} \qquad \qquad \bmod. 1 \,, \end{cases} \tag{4}$$

For $\kappa > \kappa^* \simeq 0.97$ (value of the parameter $\kappa$ corresponding to the break-up of the last KAM torus [22]) there exists one region (at least) in the phase space within which the map is chaotic.

In fluid mixing, a central problem is the understanding of the asymptotic geometrical structures arising as a result of the advection of a generic material line $\gamma$ contained within a chaotic region C. This entails analysis of the asymptotic evolution of the sequence $\{\Phi^n(\gamma)\}$ of images through $\Phi^n$ of $\gamma$, with $\gamma \cap \mathcal{C} \neq 0$. It has been shown in [14, 15, 8, 9] that for $n \to \infty$ the evolution of a generic material line lying within a chaotic region attains an invariant spatio-temporal pattern formed by thousands of striations, which progressively fills the whole chaotic region $\mathcal{C}$. This means that $\Phi^n(\gamma)$ becomes dense on $\mathcal{C}$. The origin of this phenomenon comes lies in the fact that the restriction of a diffeomorphic area-preserving map $\Phi$ within a chaotic submanifold $\mathcal{C}$ defines a hyperbolic structure on it. This means that $\Phi$ restricted to $\mathcal{C}$ induces a splitting of the tangent space point belonging to $\mathcal{C}$ into two vector subspaces $\mathcal{E}_\mathbf{x}^u$ (the unstable or dilating subspace) and $\mathcal{E}_\mathbf{x}^s$ (the stable or contracting subspace):

$$T\mathcal{C}_\mathbf{x} = \mathcal{E}_\mathbf{x}^u \oplus \mathcal{E}_\mathbf{x}^s, \qquad (5)$$

where $\oplus$ indicates direct sum. The unstable and stable vector sub-bundles are invariant under the differential, and respectively possess the dynamic properties of expanding and contracting vectors. If $\mathbf{v}_u \in \mathcal{E}_\mathbf{x}^u$ and $\mathbf{v}_s \in \mathcal{E}_\mathbf{x}^s$, it follows that

$$\lim_{n\to\infty} ||\Phi^{n*}(\mathbf{x})\mathbf{v}_u|| = \infty, \quad \lim_{n\to\infty} ||\Phi^{-n*}(\mathbf{x})\mathbf{v}_u|| = 0,$$
$$\lim_{n\to\infty} ||\Phi^{-n*}(\mathbf{x})\mathbf{v}_s|| = \infty, \quad \lim_{n\to\infty} ||\Phi^{n*}(\mathbf{x})\mathbf{v}_s|| = 0. \qquad (6)$$

In particular, the spatial patterns formed by $\Phi^n(\gamma)$ for a generic initial curve arc $\gamma$ attain an invariant orientation at any point $\mathbf{x} \in \mathcal{C}$: the tangent vector at a point $\mathbf{x} \in \Phi^n(\gamma)$ becomes parallel to the unstable direction spanned by $\mathcal{E}_\mathbf{x}^u$.

In 2D systems both $\mathcal{E}_\mathbf{x}^u$ and $\mathcal{E}_\mathbf{x}^s$ are one-dimensional vector subspaces. The numerical evaluation of a vector basis for the unstable vector subspace $\mathcal{E}_\mathbf{x}^u$ can be obtained if we enforce the concept of asymptotic directionality [18] by considering the sequence of unstable eigenvectors $\mathbf{e}_n^u(\mathbf{x})$ of the differential $\Phi^{n*}(\Phi^{-n}(\mathbf{x}))$:

$$\Phi^{n*}(\Phi^{-n}(\mathbf{x}))\mathbf{e}_n^u(\mathbf{x}) = \omega_n(\mathbf{x})\mathbf{e}_n^u(\mathbf{x}). \qquad (7)$$

Indeed, if a diffeomorphism is hyperbolic in $\mathcal{C}$, the sequence of eigenvectors $\mathbf{e}_n^u(\mathbf{x})$ eventually exists with $|\omega_n(\mathbf{x})| \to \infty$ and converges (in direction) to a limit eigenvector $\mathbf{e}^u(\mathbf{x})$ spanning $\mathcal{E}_\mathbf{x}^u$.

The connection between invariant patterns in the evolution of material lines and hyperbolicity makes it possible to obtain a quantitative description of the space-filling properties of the sequence $\{\Phi^n(\gamma)\}$ and to obtain numerical approximations for the intermaterial contact area. This issue is explored in the next section.

## 3 Unstable foliation and associated invariant measures

Given a normalized vector basis for the unstable vector sub-bundle $\{\mathbf{e}^u(\mathbf{x})\}_{\mathbf{x}\in\mathcal{C}}$ $||\mathbf{e}^u(\mathbf{x})|| = 1$, it is possible to define the dynamical system generated by $\mathbf{e}^u(\mathbf{x})$,

$$\frac{d\mathbf{x}_w(p)}{dp} = \mathbf{e}^u(\mathbf{x}_w(p)), \qquad (8)$$

starting from $\mathbf{x}_w(p = 0) = \mathbf{x}_o \in \mathcal{C}$. In Eq. (8) the parameter $p$ plays the role of a generic parametrization of the trajectories generated by the unstable sub-bundle. Eq. (8) makes sense since by hypothesis $\mathbf{e}^u(\mathbf{x}_w(p))$ is continuous along the trajectories, and the corresponding dynamical system is referred to as the $w$-system associated with the Poincaré map $\Phi$ [16, 18].

Let us consider a generic point, $\mathbf{x}_o \in \mathcal{C}$, and the solution $\mathbf{x}_w(p; \mathbf{x}_o)$ of Eq. (8) for $p \in [0, p_1]$ passing as the initial condition $(p = 0)$ through $\mathbf{x}_o$. This solution defines a curve arc $\gamma_{\mathbf{x}_o}^+(p_1)$ on $\mathcal{C}$. Similarly, by replacing $p$ with $-p$ and integrating from 0 to $-p_1$, it is possible to define a new curve arc $\gamma_{\mathbf{x}_o}^-(p_1)$. These two curve arcs can be joined together to form a single curve arc $w_{\mathbf{x}_o}^u(p_1) = \gamma_{\mathbf{x}_o}^-(p_1) \cup \gamma_{\mathbf{x}_o}^+(p_1)$ passing through $\mathbf{x}_o$. By definition $w_{\mathbf{x}_o}^u(p_1)$ is continuous and differentiable at $\mathbf{x}_o$ and may be referred to as a *local unstable leaf* at $\mathbf{x}_o$. An important property of the local unstable leaves of the $w$-system is their invariance with respect to $\Phi$: the image $\Phi(w_{\mathbf{x}_o}(p_1))$ of $w_{\mathbf{x}_o}(p_1)$ through $\Phi$ is a local unstable leaf, the solution of Eq. (8) passing through $\Phi(\mathbf{x}_o)$. This result stems from the invariance of the unstable sub-bundle with respect to the differential $\Phi^*$. Because of the expanding properties of $\Phi^*$ along the unstable sub-bundle, a local unstable leaf $w_{\mathbf{x}_o}^u(\mathbf{x}_o)$ for $p_1 \to \infty$ becomes unbounded in length and space-filling on $\mathcal{C}$ in the sense that it is dense on $\mathcal{C}$. As a result, it is possible to define a global unstable leaf at a generic point $\mathbf{x}_o \in \mathcal{C}$ as the limit of $w_{\mathbf{x}_o}(p_1)$ for $p_1 \to \infty$, or equivalently as the limit for $n \to \infty$ of the sequence of integral manifolds $\{\Phi^n(w_{\Phi^{-n}(\mathbf{x}_o)}^u(p_1))\}$. We shall use the notation $w_{\mathbf{x}_o}^u$ to indicate such a global unstable leaf passing through $\mathbf{x}_o$. The family of all the distinct global unstable leaves for $\mathbf{x}_o \in \mathcal{C}$ is referred to as the *invariant unstable foliation* $\mathcal{F}_u = \{w_{\mathbf{x}_o}^u\}_{\mathbf{x}_o \in \mathcal{C}}$ of $\Phi$ in $\mathcal{C}$.

As a numerical exemplification, Fig. 1 shows a portion of a leaf of $\mathcal{F}_u$ for the standard map at $\kappa = 2$.

Although $\mathcal{F}_u$ is transitive on $\mathcal{C}$ and the invariant ergodic measure associated with $\Phi$ restricted to $\mathcal{C}$ is uniform, the spatial structure of any leaf of $\mathcal{F}_u$ is nonuniform. To clarify the meaning of this statement, let us consider the portion $w_{\mathbf{x}}^u(p_1)$. For finite $p_1$, this curve arc possesses a length equal to $2p_1$ (because the vector field in the equation for the $w$-system is normalized to 1). It is thus possible to define a probability measure associated with the length content of $w_{\mathbf{x}}^u(p_1)$. Let us define the length function $L(\gamma, B)$, where $\gamma$ is a curve arc and $B$ any measurable set of $\mathcal{C}$, returning the length of $\gamma \cap B$, i.e. the length of the subarcs of $\gamma$ falling within $B$. A probability measure $\tilde{\mu}_w(B, \mathbf{x}, p_1)$ can be defined as

$$\tilde{\mu}_w(B, \mathbf{x}, p_1) = \frac{L(w_{\mathbf{x}}^u(p_1), B)}{2p_1} . \quad (9)$$

In the limit for $p_1$ tending to infinity, the measure $\mu_w$ can be defined as

$$\mu_w(B) = \lim_{p_1 \to \infty} \tilde{\mu}_w(B, \mathbf{x}, p_1) \quad (10)$$

The measure defined by Eq. (10) exists, is independent of $\mathbf{x}$, i.e. of the particular leaf of foliation $\mathcal{F}_u$ considered, and corresponds to the probability measure associated with the pointwise length distribution of any manifold of unstable foliation. The probability measure $\mu_w$ is referred to as the $w$-measure associated with a hyperbolic diffeomorphism $\Phi$ [17, 18]. The geometrical meaning of the $w$-measure follows from Eqs. (9)-(10): $\mu_w(B)$ represents the fraction of length

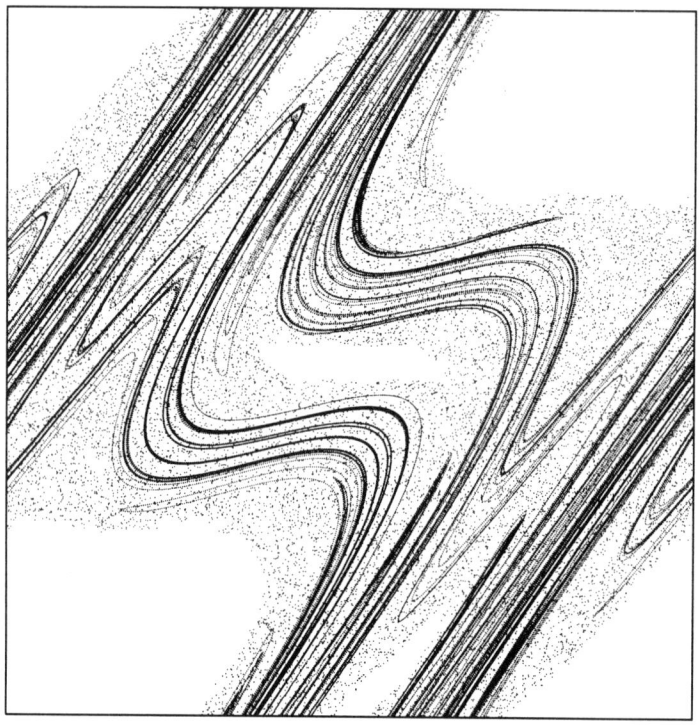

**Fig. 1.** Portion of an integral manifold (continuous line) obtained by integrating Eq. (8) (standard map with $\kappa = 2$) starting from a point belonging to $\mathcal{C}$. The chaotic region $\mathcal{C}$ is dotted.

of a generic leaf of $\mathcal{F}_u$ falling within the set $B$. The physical meaning of the $w$-measure arises from the equivalence between the leaves of $\mathcal{F}_u$ and the images of a material line $\gamma$ through $\Phi^n$ for large enough $n$. For hyperbolic area-preserving diffeomorphisms, the sequence $\Phi^n(\gamma)$ converges towards a leaf of $\mathcal{F}_u$, and the $w$-measure therefore corresponds to the stationary pointwise length distribution of a generic material line advected by $\Phi^n$.

In order to obtain an analytical approximation for $\mu_w$, it is convenient to analyze the measure-theoretical properties of Eq. (8) by enforcing asymptotic directionality, Eq. (7). Let us set $\mathbf{A}^{(n)} = \Phi^{n*}(\Phi^{-n}(\mathbf{x})) = (A_{ij}^{(n)})$. For sufficiently large $n$, within the chaotic region $\mathcal{C}$

$$\operatorname{trace}[\Phi^{n*}(\Phi^{-n}(\mathbf{x}))] = A_{11}^{(n)} + A_{22}^{(n)} = \lambda_n^u + 1/\lambda_n^u = \lambda_n^u + o(n) , \quad \mathbf{x} \in \mathcal{C} , \quad (11)$$

where $o(n)$ is a quantity tending (exponentially) to zero for $n \to \infty$. Within this approximation, the eigenvector $\mathbf{e}_n^u(\mathbf{x}) = (e_{n,1}^u, e_{n,2}^u)^t$ can be obtained from the solution of the equation

$$-A_{22}^{(n)} e_{n,1}^u + A_{12}^{(n)} e_{n,2}^u = 0 , \quad (12)$$

or equivalently

$$A_{21}^{(n)} e_{n,1}^u - A_{11}^{(n)} e_{n,2}^u = 0 . \quad (13)$$

A non-normalized basis for $\mathcal{E}_\mathbf{x}^u$ can therefore be approximated by

$$\widehat{\mathbf{e}}_n^u(\mathbf{x}) = (A_{12}^{(n)}, A_{22}^{(n)})^t , \quad (14)$$

or equivalently by

$$\widehat{\mathbf{e}}_n^u(\mathbf{x}) = (A_{11}^{(n)}, A_{21}^{(n)})^t . \quad (15)$$

By differentiating the identity $\Phi^n(\Phi^{-n}(\mathbf{x})) = \mathbf{x}$, it follows that

$$\mathbf{A}^{(n)} = \Phi^{n*}(\Phi^{-n}(\mathbf{x})) = [(\Phi^{-n}(\mathbf{x}))^*]^{-1} , \quad (16)$$

and the vector $\widehat{\mathbf{e}}_n^u(\mathbf{x})$ can therefore be expressed as

$$\widehat{\mathbf{e}}_n^u(\mathbf{x}) = \left( -\frac{\partial \Phi_i^{-n}(\mathbf{x})}{\partial x_2}, \frac{\partial \Phi_i^{-n}(\mathbf{x})}{\partial x_1} \right)^t \quad i = 1, 2 , \quad (17)$$

where $\mathbf{x} = (x_1, x_2)^t$ is a Cartesian coordinate system for the phase manifold and $\Phi_i^{-n}(\mathbf{x})$ the $i$-th component of the map $\Phi^{-n}$. As can be observed from Eq. (17), the functions $\Phi_i^{-n}$ ($i = 1, 2$) play the role of asymptotic invariant "stream functions", determining through Eq. (8) the invariant properties of the unstable manifolds of $\Phi$ in the limit for $n \to \infty$. The choice of $i = 1, 2$ in Eq. (17) determines two different approximations for $\widehat{\mathbf{e}}_n^u(\mathbf{x})$ which return a vector co-linear to $\mathbf{e}^u(\mathbf{x})$ in the limit of $n \to \infty$ and are therefore equivalent.

Eq. (17) is the starting point needed in order to determine a sequence of analytic approximations $\mu_w^{(n)}$,

$$d\mu_w^{(n)}(\mathbf{x}) = \rho_w^{(n)}(\mathbf{x})\, d\mathbf{x}\,, \tag{18}$$

for the $w$-measure $\mu_w$. From Eq. (17) it follows that

$$\nabla \cdot \widehat{\mathbf{e}}_n^u(\mathbf{x}) = -\frac{\partial^2 \Phi_i^{-n}(\mathbf{x})}{\partial x_1\, \partial x_2} + \frac{\partial^2 \Phi_i^{-n}(\mathbf{x})}{\partial x_2\, \partial x_1} = 0\,. \tag{19}$$

By definition, the $n$-th approximation $\rho_w^{(n)}(\mathbf{x})$ fulfils the continuity equation

$$\nabla \cdot (\rho_w^{(n)}(\mathbf{x})\, \mathbf{e}_n^u(\mathbf{x})) = 0\,, \tag{20}$$

where $\mathbf{e}_n^u(\mathbf{x})$ are co-linear with $\widehat{\mathbf{e}}_n^u(\mathbf{x})$ and possess unit norm. Eq. (20) is a continuity equation defining $\rho_w^{(n)}(\mathbf{x})$. Since $\mathbf{e}_n^u(\mathbf{x}) = \widehat{\mathbf{e}}_n^u(\mathbf{x})/\|\widehat{\mathbf{e}}_n^u(\mathbf{x})\|$, it follows from Eq. (20) that $\nabla \cdot (\|\widehat{\mathbf{e}}_n^u(\mathbf{x})\|\, \mathbf{e}_n^u(\mathbf{x})) = 0$, and an analytical expression from $\rho_w^{(n)}(\mathbf{x})$ is therefore given by

$$\rho_w^{(n)}(\mathbf{x}) = \text{const}\, \|\widehat{\mathbf{e}}_n^u(\mathbf{x})\| + o(n) = \rho_w^{(n)}(\mathbf{x}) = \text{const}\, \|\nabla \Phi_i^{-n}(\mathbf{x})\| + o(n)\quad i=1,2\,, \tag{21}$$

or equivalently

$$\rho_w^{(n)}(\mathbf{x}) = \text{const}\, \widehat{\rho}_w^{(n)}(\mathbf{x}) + o(n)\,,$$

$$\widehat{\rho}_w^{(n)}(\mathbf{x}) = \left[\left(\frac{\partial \Phi_i^n}{\partial x_1}\right)^2_{\Phi^{-n}(\mathbf{x})} + \left(\frac{\partial \Phi_i^n}{\partial x_2}\right)^2_{\Phi^{-n}(\mathbf{x})}\right]^{1/2}\quad i=1,2\,. \tag{22}$$

Eq. (21)-(22) provides a sequence of analytical approximations for $\rho_w(\mathbf{x})$ from which the statistical properties of the measures $\mu_w^{(n)}$ generated from these densities, and ultimately of $\mu_w$, can be addressed. For each $n$, the densities $\rho_w^{(n)}(\mathbf{x})$ are differentiable despite the fact that in the limit for $n \to \infty$ the sequence of measures $\mu_w^{(n)}$ converges towards a stationary singular measure. A convenient way to depict the structure of the $w$-measure is to consider the sectional box-measures $\widehat{\mu}_w^{(n)}(x_i, \varepsilon)$

$$\widehat{\mu}_w^{(n)}(x_i, \varepsilon) = C^{(n)} \int_{x_i}^{x_i+\varepsilon} \rho_w^{(n)}(\xi, y_c)\, d\xi\,, \tag{23}$$

where $C^{(n)}$ is a normalization constant such that $\sum_i \widehat{\mu}_w^{(n)}(x_i, \varepsilon)\varepsilon = 1$, and $\rho_w^{(n)}$ is given by Eq. (21). Fig. 3 shows the behaviour of the sectional box-measure for the standard map at $\kappa = 2$, $n = 11$ along a circumference at $y_c$, indicating its high singular structure. The analysis of the properties of $\mu_w$ discussed in [19] indicates that: **(A)** the sequence of approximants $\mu_w^{(n)}$ obtained through Eq. (21),

$$\mu_w^{(n)}(\Delta) = \int_\Delta \rho_w^{(n)}(\mathbf{x})\, d\mathbf{x}\,, \tag{24}$$

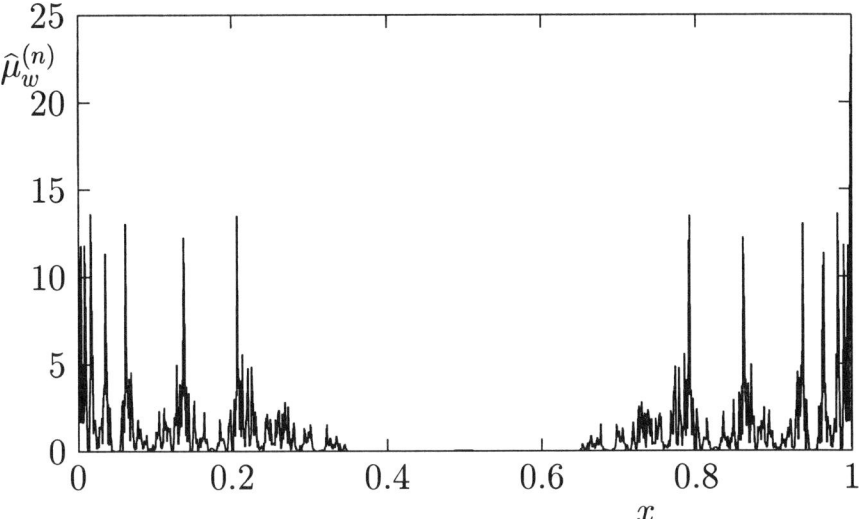

**Fig. 2.** Sectional box measure $\widehat{\mu}_w^{(n)}(x,\varepsilon)$ vx $x$ along $y_c = 1/2$ for the standard map, $\kappa = 2$, $n = 11$, $\varepsilon = 5 \cdot 10^{-4}$.

where $\Delta$ is a Borel set, converges with sufficient rapidity towards a spatially non-uniform stationary measure; **(B)** this measure coincides with the box measure obtained from the box-counting of the length of a generic material line evolved for a sufficiently long time, or equivalently from the box-counting of the pointwise length distribution of a leaf of $\mathcal{F}_u$. The measure $\mu_w^{(n)}$ yields an approximate expression for the intermaterial contact area (which in 2D corresponds to an intermaterial contact length) entering into mean-field models of interfacial phenomena controlled by the formation of the lamellar structure in chaotic advection [23–25].

## 4 Multifractal properties

The singularity properties of the $w$-measure can be investigated by means of the standard multifractal approach [26] by analyzing the scaling properties of the moments of the normalized sectional box-measures

$$\sum_{i=1}^{N_k} (p_i^{(n)}(\varepsilon_k))^q \sim \varepsilon_k^{(q-1)\widehat{D}(q)}, \qquad (25)$$

where $p_i^{(n)}(\varepsilon_k)$ is the $n$-th order approximation of the sectional box-measure of the $i$-th interval of the partition of size $\varepsilon_k$. The spectrum of generalized dimensions collapses for sufficiently large $n$ onto an invariant curve, thus giving further quantitative confirmation that $n$-th-order sectional box-measures converge towards an invariant measure.

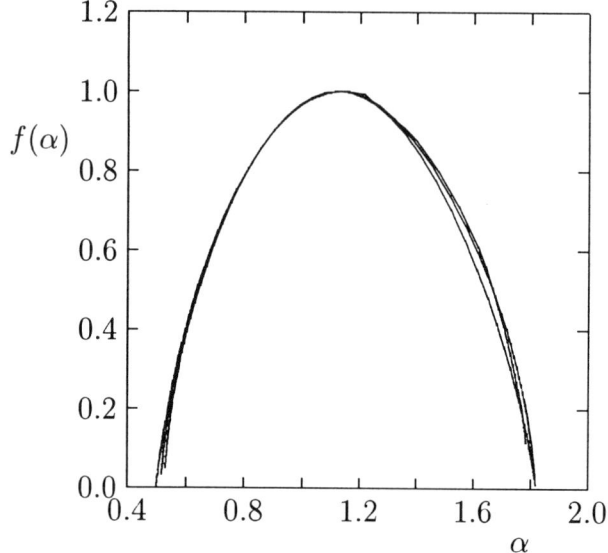

**Fig. 3.** $f(\alpha)$-spectrum of the sectional box-measures for the standard map ($\kappa = 2$) at $y_c = 1/2$, $n = 10, 11, 12$.

Moreover different sectional box-measures (obtained for different $y_c$) return the same spectrum of generalized dimensions, thus confirming that the simpler analysis of the multifractal properties of sectional box-measures provides an insight into the singularity structure of the $w$-measure as a whole. The $f(\alpha)$-spectra obtained through a Legendre transform of the $\widehat{D}(q)$-spectrum of sectional box measures is shown in Fig. 4 for different values of $n$.

The observation that the singularity properties of the $w$-invariant measure can be inferred from the analysis of sectional box-measures is further supported by the numerical observation that there exists a simple relationship between the spectrum of generalized dimensions $\widehat{D}(q)$ evaluated from sectional box-measures and the spectrum $D(q)$ evaluated from the $w$-invariant measure defined on $\mathcal{C}$ as a whole:

$$D(q) = \widehat{D}(q) + 1 , \qquad (26)$$

as was to be expected given the meaning of sectional box-measures [19].

Another interesting feature associated with the complex structure of the $w$-measure and related to its physical meaning in advection problems is its sign-singularity. In 1992, Ott et al. [27] introduced the so-called cancellation exponent in order to characterize the sign-singular properties of the signed measure [29] associated with the magnetic field in fast magnetic dynamos. If $\nu$ is a signed measure (a signed measure of a set can take either positive or negative values), $\nu$ is said to be sign-singular if it changes sign almost everywhere on arbitrarily small lengthscales. For a signed measure, the cancellation exponent may be defined as follows:

$$\kappa_\nu = \lim_{\varepsilon \to 0} \sup \frac{\log \sum_i |\nu(I_i)|}{\log(1/\varepsilon)}, \qquad (27)$$

where $I_i$ denotes the $i$-th interval of a $\varepsilon$-partition. For a probability measure and for a signed measure with a smooth density, $\kappa_\nu = 0$, while $\kappa_\nu > 0$ indicates an oscillation in sign on arbitrarily small lengthscales, i.e. the sign-singularity of the measure. The natural signed measure associated with the $w$-measure may be defined by considering the normal component of the $n$-th order unstable eigenvectors $\hat{\mathbf{e}}_n^u(\mathbf{x})$ (defining through Eq. (17), the $w$-density) along a circumference of the torus.

The geometrical meaning of this signed measure is related to the folding of the invariant unstable manifolds and consequently to the folding dynamics involved in the evolution of partially mixed structures.

The log-log plot $\sum_i |\nu(I_i)|$ vs $\varepsilon$, is shown in Fig. 4 for different values of $n$. The corresponding slope (in a log-log plot) yields the cancellation exponent $\kappa_\nu^{(n)}$. The sequence $\kappa_\nu^{(n)}$ quickly moves towards a constant value equal to $\kappa_\nu = 0.83$, and the same numerical result was obtained by analyzing different sections, i.e. different values of $y_c$.

## 5 Concluding remarks

This article has shown numerically that the Poincaré maps of incompressible 2D chaotic flows give rise within a chaotic region to a stationary measure (referred to as the $w$-measure) possessing multifractal properties. This result appears to be valid for generic area-preserving chaotic diffeomorphisms possessing bounded dynamics that cannot be reduced upon conjugation to a linear map, as in the case of uniform hyperbolic maps of the 2D torus conjugate to the linear toral automorphism.

The results obtained for laminar flows could be used as a starting point to tackle the more complex problem of the geometry of coherent structures in turbulent conditions. The latter issue is strongly related to the fundamental question of whether turbulent flows admit a dynamical explanation in terms of hyperbolic models.

As regards the physical theory of laminar chaotic flows, it remains an open problem whether interfacial phenomena such as fast chemical reactions are sensitive to the multifractal singular structure observed for the intermaterial contact-area density. In any case, the existence of a highly singular measure associated with the spatial distribution of the length content of a generic leaf belonging

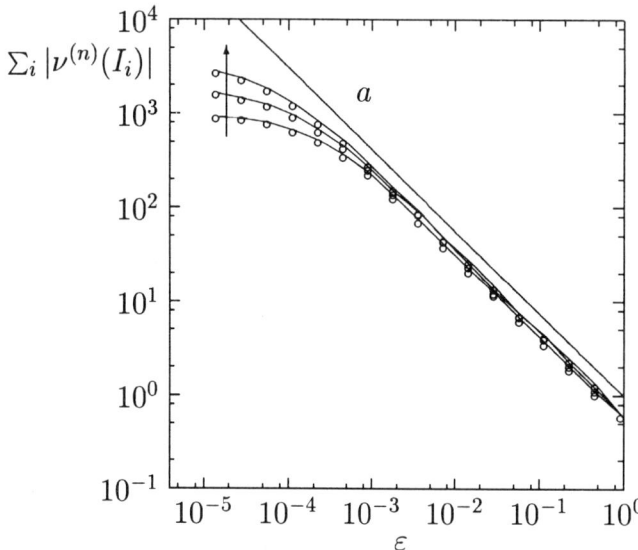

**Fig. 4.** Log-log plot of $\sum_i |\nu^{(n)}(I_i)|$ vs $\varepsilon$ for the signed sectional box-measure ($y_c = 0.4$) of the standard map at $\kappa = 2$, $n = 12, 13, 14$. The arrow indicates increasing values of $n$. Line a) is $\sum_i |\nu^{(n)}(I_i)| \sim \varepsilon^{-\kappa_\nu}$ with $\kappa_\nu = 0.83$.

the unstable foliation is an important result *per se* in the global analysis of differentiable incompressible flows.

To conclude, many results obtained for 2D systems can be extended with minor modifications to 3D systems of physical and engineering interest [30].

## References

1. U. Frisch (1995): Turbulence. Cambridge Univ. Press, Cambridge.
2. K. R. Sreenivasan (1991). Annu. Rev. Fluid Mech. **23**, 539.
3. J. Jimenez and C. Martel (1991). Phys. Fluids A **3**, 1261.
4. J.C. Vassilicos and J.C.H. Fung (1995). Phys. Fluids **8**, 1970.
5. H. Aref. (1984). J. Fluid Mech. **143**, 1.
6. J. Ottino (1989): The Kinematics of Mixing: Stretching, Chaos and Transport. Cambridge Univ. Press, Cambridge.
7. J.M. Lopez and A. D. Perry (1992). J. Fluid Mech. **143**, 1.
8. V. Rom-Kedar, A. Leonard and Wiggins S. (1990). J. Fluid Mech. **214**, 347.
9. R. Camassa and S. Wiggins (1991). Phys. Rev. A **43**, 774.
10. A. Pentek, T. Tel and Z. Toroczkai (1995). J. Phys. A **28**, 2191.
11. G. Karolyi and T. Tel. (1997). Phys. Rep. **290**, 125.
12. M.A.F. Sanjuan, J. Kennedy, E. Ott and J. A. Yorke (1997). Phys. Rev. Lett. **78**, 1892.
13. D. Beigie, A. Leonard, S. Wiggins (1994): Chaos, Solitons & Fractals **4**, 749.

14. M.M. Alvarez, F. J. Muzzio, S. Cerbelli and A. Adrover (1997): in Fractals in Engineering, Lévy Véhel J., Lutton E., Tricot C. (Eds.), p. 323, Springer Verlag, Berlin.
15. M.M. Alvarez, F.J. Muzzio, S. Cerbelli, A. Adrover and M. Giona (1998). Phys. Rev. Lett. **81**, 3395.
16. M. Giona, S. Cerbelli, F.J. Muzzio and A. Adrover (1998). Physica A **254**, 251.
17. M. Giona, A. Adrover, S. Cerbelli, F.J. Muzzio and M.M. Alvarez (1999). Physica D (in press).
18. M. Giona and A. Adrover (1998). Phys. Rev. Lett. **81**, 3864.
19. A. Adrover and M. Giona (1999). Phys. Rev. E (in press).
20. R. Mañé (1977). Trans. Am. Math. Soc. **229**, 351.
21. B.V. Chirikov (1979). Phys. Rep. **52**, 265.
22. J.M. Greene (1979). J. Math. Phys. **20**, 1183.
23. F.J. Muzzio and J. Ottino (1989). Phys. Rev. Lett. **63**, 47.
24. I.M. Sokolov and A. Blumen (1991): Phys. Rev. A **43**, 6545.
25. I.M. Sokolov and A. Blumen (1991): Phys. Rev. Lett. **66**, 1942.
26. D. Auerbach, P. Cvitanovic, J.-P. Eckmann, G. Gunaratne and I. Procaccia (1987). Phys. Rev. Lett. **58**, 2387.
27. E. Ott, Y. Du, K.R. Sreenivasan, A. Juneja and A.K. Suri (1992). Phys. Rev. Lett. **69**, 2654.
28. Y. Du and E. Ott (1993). Physica D **67**, 387.
29. P. R. Halmos (1950): Measure Theory, Van Nostrand, New York.
30. M. Giona and A. Adrover (1999). Physica D (submitted).

# Adhesion AFM Applied to Lipid Monolayers. A Fractal Analysis.

Gianina Dobrescu[1], Camelia Obreja[2], and Mircea Rusu[2]

[1] Romanian Academy, Institute of Physical Chemistry,
spl. Independentei, 202, 77208, Bucharest - ROMANIA
gdobresc@pcnet.pcnet.ro
[2] University of Bucharest, Faculty of Physics,
Magurele - ROMANIA
mrusu@meganet.ro

**Abstract.** Adhesion Atomic Force Microscopy images of lipid monolayers on mica are studied using methods of image processing and fractal analysis. We show that liquid-condensed domains are fractal. Free energy computation indicates that fractal domains have an advantage over circular domains as far as free energy is concerned. This may provide an answer to the question why LC domains are fractal.

## 1 Introduction

Since Mandelbrot[1, 2] developed the basic concept of fractal, many researchers have applied fractal theory to describe phenomena in physics, chemistry, biology, medicine and so on. Topographic images obtained from scanning tunneling microscopy [3-6] or scanning probe microscopy[7] (SPM) in general are usually used to compute fractal dimension of physical surfaces. Atomic Force Microscopy (AFM) methods have been used for molecular force measurements on a laterally smaller scale (the lateral resolution of adhesion AFM is 20nm). The ratio of the adhesion forces for different functional groups can be theoretical computed from their surface energies and compared to the experimental value obtained by analyzing [8] the image. Experimental details about AFM are discussed in [8]. Berger et al. [8] used a dimyristoylphosphatidylethanolamine (DMPE) film, made by Langmuir-Blodgett (LB) technique, at a surface pressure for which liquid-expanded (LE) and liquid-condensed (LC) phases coexist. The domain structure of the DMPE film consists of a crystalline domain (LC), for which the AFM probe tip will only interact with the hydrophobic CH3 endgroup of the DMPE molecule and an amorphous LE phase where the molecules are less densely packed and the tip will have a stronger interaction with the 12 slightly less hydrophobic CH2 groups along the hydrocarbon chain of a DMPE molecule. Berger et al.[8] measured the ratio of the adhesion forces for the two different functional groups and compared it to a theoretical evaluation. The nice fit between the two numbers demonstrated that adhesion contrast can be used to resolve the LE and LC phases. In the following, adhesion AFM images obtained from Berger et al.[8] will be analyzed in terms of fractal theory. We show that phase domain boundaries have fractal characteristics. By computing the free energy for LC domains with different shapes we will try to provide an answer to the question why are the LC domains fractal?

## 2 Theory: Free energy computation of LC domains

Liquid-condensed domain shapes can be predicted using free energy analysis. Following McConnell[9], the shape of an isolated domain results from the competition between line tension and electrostatic repulsion. If we imagine that lipid molecules (like DMPE) are interacting vertical dipoles, the excess dipole density of the liquid-condensed phase relative to the liquid-expanded phase acts to form non-compact, needle-like domains. On the other hand, there is a surface tension which acts to minimize the perimeter of the domain and form circular shapes. This surface tension arises from the excess free energy associated with the formation of the interface between the LC domain and the surrounding LE phase. Mayer and Venderlick[10] derived two simplified, yet rigorous formulations of the electrostatic energy: one being more computationally efficient for circular domains, the other for non-circular domains. Free energy is given by the sum of interfacial and electrostatic contributions:

$$F = F^L + F^E \tag{1}$$

The interfacial free energy may be expressed as $F^L = \lambda P$ where $\lambda$ is the line tension and P is the perimeter of the domain. The electrostatic energy is given by:

$$F^E = \frac{\mu^2}{2} \iint g(|r' - r|) |r' - r|^{-3} d^2r' \, d^2r \tag{2}$$

where $\mu$ is the average excess dipole density and $g(r)$ the pair distribution function. The simplest pair distribution function that obeys the conditions of being zero at small separations and approaching unity at large intermolecular separations, is the Heaviside step function:

$$g(|r' - r|) = H(|r' - r| - \delta) = 1 \; if \; |r' - r| \geq \delta \; or \tag{3}$$
$$= 0 \; if \; |r' - r| < \delta$$

where $\delta$ is a parameter representing the closest distance between molecules. McConnell[9] proposed an alternative pair distribution function given by:

$$g(|r - r'|) = \frac{|r - r'|^3}{\sqrt{|r - r'|^2 + 4\delta^2}} \tag{4}$$

To simplify the evaluation of the electrostatic energy in eq.2, one would use an approximation - so called "capacitor approximation". Mayer and Vanderlick[10] developed two simplified formulations of eq.2. One uses Green's Theorem to reduce both area integrals to contour integrals. The resulting "rigorous contour expression" (RCE) works best when applied to circular domains. The second approach uses a prudent coordinate transformation to reduce the integral from fourth to third order. This second "coordinate transformation expression" (CTE) is more computationally efficient than the first, but not in the case of circular domains. The CTE approach yields:

$$\frac{F^E}{\mu^2/2} = \frac{2\pi A}{\delta} - \int_\Omega \int_0^{2\pi} \frac{1}{\rho^*(\Phi)} d\Phi d^2 r \quad (5)$$

where $\rho(\Phi)$ is the distance from a point r to the domain boundary at polar angle $\Phi$, A is the domain area, and $\rho^*(\Phi)$ is the larger of $\rho(\Phi)$ and $\delta$. In the following, we will use eq.5 to compute free energy for fractal LC-domains and we will compare the computed energy to free energies obtained in Ref.[10] for circular domains.

## 3 Results and Discussions

We will analyze three AFM images obtained by Berger et.al.[8] for the DMPE monolayer on mica. Figure 1(a-c) shows AFM images of the DMPE monolayer for three different magnifications. Dark areas correspond to the low adhesion (LC) phase.

### 3.1 Fractal dimension versus filter intensity.

The AFM pictures were digitized and transformed into 256 gray-level images. For every gray level the fractal dimension was determined. The algorithm of this process is as follows. For every filter intensity between 0-255 we thresholded the image to obtain only 2 gray-levels (black and white). If the original pixel's gray-level is greater than the filter value, then it's converted into a black pixel. If the original gray-level is lower than the filter value, the pixel becomes white. In this simple way, we obtained a "lattice" with occupied sites (the black pixels) and unoccupied sites (white pixels). For these images, the fractal dimension of mass fractals can easily be computed using the box-counting method[12],[13],[14] (figure 2).

Fractal dimension versus filter intensity curves are presented in figure 3. All curves have the same qualitative behavior: for filter level 0 the fractal dimension is 2 (a full-black image). For the maximum filter level 255, the image is white and the fractal dimension is 0. Between these two extremes, fractal dimension varies, having average values around 1.8 for image a), 1.86 for b) and 1.9 for c).

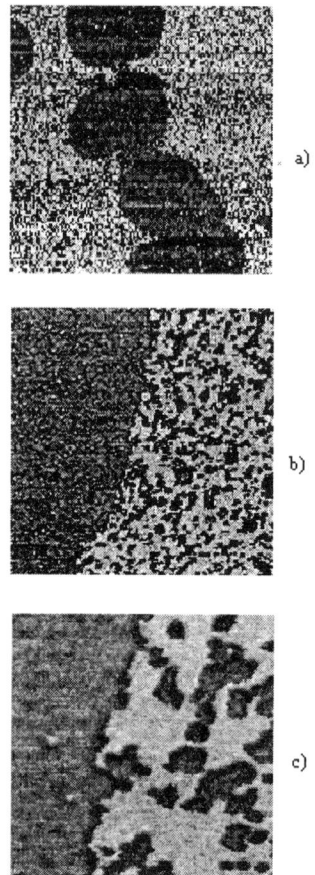

Fig. 1. AFM images of DMPE monolayer on mica at three magnifications (a-c).

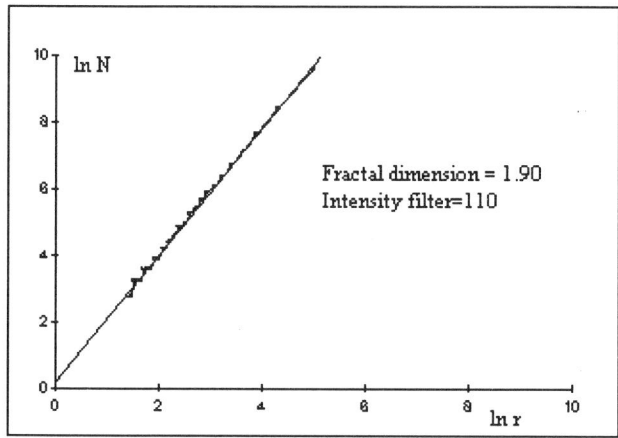

**Fig. 2.** Plot of the box-counting data

All curves share a striking feature: each has a well-defined maximum. For image a), it's at filter intensity 245, D=1.74. For image b), it's at filter intensity=226, D=2.13, and for c), at intensity=211, D=1.97. In the following discussion we show that these maxima are artifacts introduced by AFM measurement method.

Let's define:

$$f_i^{LC} = \frac{N_i^{LC}}{N} \qquad (6)$$

where $N_i^{LC}$ is the number of $CH_3$ radicals per pixels and N is the total number of active centers per pixel. Then, the gray-level of a pixel is proportional to the fraction $f_i^{LC}$ defined by eq.6. Choosing a filter value $GL_f$, we alter the image in the following way:
if $f_i^{LC} \geq GL_f$ then $f_i$=1;
if $f_i^{LC} < GL_f$ then $f_i$=0.
In this simple way, we obtain an image with an apparent fractal dimension:

$$N_f^t(R) = BR^{D_a} \qquad (7)$$

where $N_f^t(R)$ is the total number of black pixels in sphere of radius R, for the given filter intensity. Hence, by choosing the filter value and converting the image, for $i_1$ pixels we apparently increase the number of $CH_3$ radicals by

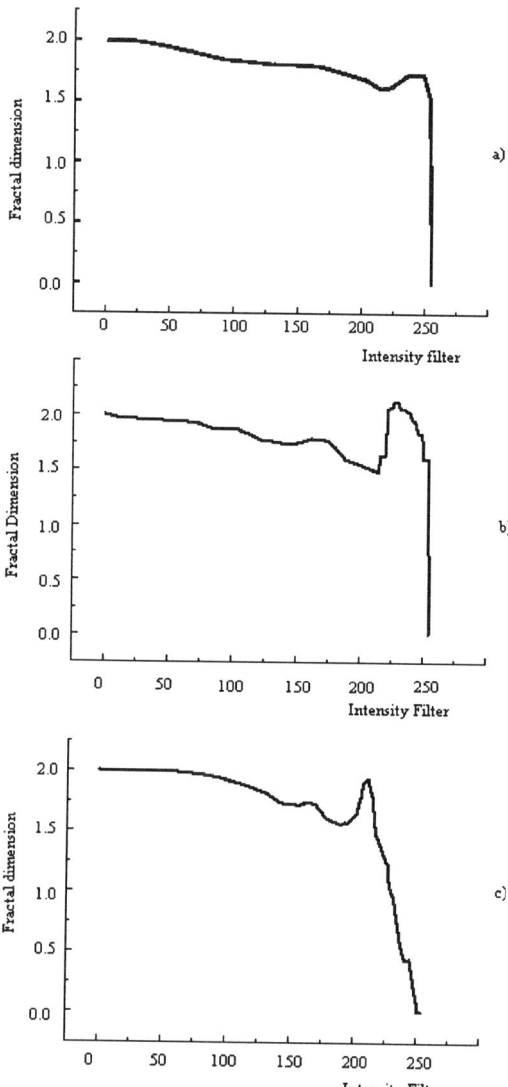

**Fig. 3.** Fractal dimension vs. filter intensity curves; a)-c) correspond to figure 1a)-c)

$$\sum_{i=1}^{i_1} N\left(1 - f_i^{LC}\right) \qquad (8)$$

and for $(i_{total} - i_1)$ pixels we apparently decrease the number of CH$_3$ radicals by:

$$\sum_{i_1}^{i_{max}} N f_i^{LC} \qquad (9)$$

This means that in sphere of radius R, the apparent number of radicals is:

$$N\ N_f^t(R) = \sum_{i=1}^{i_{max}} N f_i^{LC} + \sum_{i=1}^{i_1} N(1 - f_i^{LC}) - \sum_{i_1}^{i_{max}} N f_i^{LC} \qquad (10)$$

where the first term on the right-hand side is the real number of radicals. So:

$$N_f^t(R) = \sum_{i=1}^{i_{max}} f_i^{LC} + \sum_{i=1}^{i_1}(1 - f_i^{LC}) - \sum_{i_1}^{i_{max}} f_i^{LC} \qquad (11)$$

and therefore:

$$B\ R^{D_a} = A_0 R^D + \sum_{i=1}^{i_1}(1 - f_i^{LC}) - \sum_{i_1}^{i_{max}} f_i^{LC} \qquad (12)$$

Let's define the last two terms on the right as:

$$F(GL_f) = \sum_{i=1}^{i_1}(1 - f_i^{LC}) - \sum_{i_1}^{i_{max}} f_i^{LC} \sim \sum_{GL_i \geq GL_f}(1 - GL_i) - \sum_{GL_i < GL_f} GL_i \qquad (13)$$

The behavior of $F(GL_f)$ will determine how close to the real fractal dimension the computed, apparent fractal dimension is. In other words, if $F(GL_f)$ is close to 0, then the apparent fractal dimension will be equal to the real fractal dimension. When $F(GL_f) << 0, D_a$ is close to 0. For $F(GL_f) >> 0, D_a$ approaches 2, i.e. exactly what the graphs show.

We plotted $F(GL_f)$ for our three AFM images (figure 4 a-c). Now we can find the filter threshold value for which the $F(GL_f)$ functions are zero and compute the real fractal dimension. The results are much more consistent: D=1.82 for image a), D=1.8 for image b) and D=1.83 for image c).

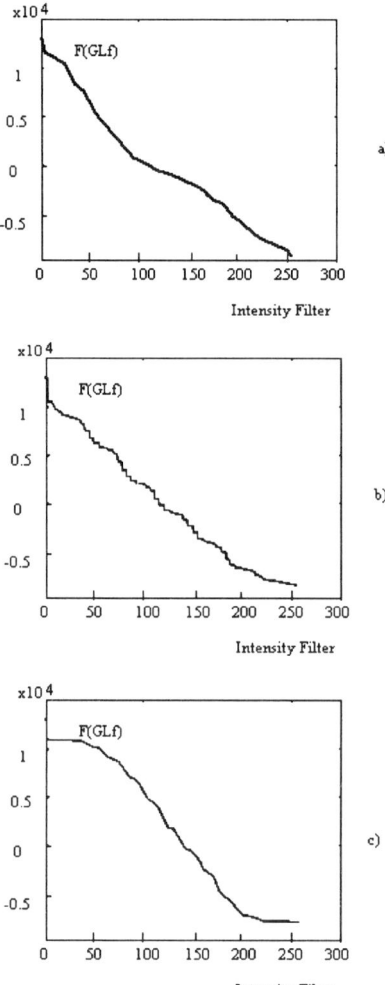

**Fig. 4.** The behavior of F(GLf); a)-c) correspond to figure 1a)-c).

## 3.2 Free energy computation.

We showed that LC domains formed by DMPE molecules on mica are fractal and how one can evaluate their fractal dimension. But why are these domains fractals? In other words, why does the DMPE monolayer form fractal islands and not circular ones, as a normal liquid would? Is there an advantage of the fractal shape over the regular one? In the following discussion we suggest an intuitive answer, based on a simple free-energy calculation. Of course, computing free energy of a such a complicated system is not an easy job, but we will simplify the problem in order to obtain at least a qualitative answer to the question above. First of all, we will use eq.5 to compute the free energy of a hypothetical LC domain with a given fractal dimension. We generated an artificial island-like fractal domain using the real part of the Weierstrass-Mandelbrot function:

$$f(x) = \sum_{n=-\infty}^{\infty} \frac{1 - \cos(b^n x)}{b^{(2-D)n}} \qquad (14)$$

where $b > 1$ for the top surface and the symmetrical function for the bottom surface. We chose this means of generating fractal shapes because it allows us to tune the fractal dimension D to any desired value.

We performed free energy calculations for eight simulated LC domains with fractal dimensions between 1.1 and 1.8 and compared these results with those obtained for an LC circular domain[10]. We chose domains with rigorously equal area and same physical properties ($\lambda, \delta, \mu$). The results are shown in Table 1. We see that the electrostatic component of the free energy decreases when fractal dimension increases, but the interfacial component increases. The quantity: $F^E/(\mu^2/2) - 2\pi A/\delta$ has the same behavior as the electrostatic energy (figure 5). The dipolar interactions between molecules decrease when domain boundaries become rough. On the other hand, the interfacial energy acts to minimize the perimeter and it can only be minimum when the domain is circular. So, the two terms that make up the free energy (electrostatic and interfacial) are competing. The minimum for the free energy is reached when the two effects are balanced. This happens for a fractal domain with D=1.3. We can say that for this simple, artificial case the minimum energy criterion requires that LC domain be fractal. This observation is a strong argument for explaining why the real, observed LC domains of the DMPE monolayer are fractals.

To take our approach one step further, we now computed free energy for 9 actual domains (figure 6), selected from the AFM image presented in figure1c, and compared them with the free energy of 9 Weierstrass-Mandelbrot domains with the same fractal dimension, same area, same line tension and same average excess dipole density. The results are presented in Table 2.

We see that the free energy for all actual domains is lower than for the Weierstrass-Mandelbrot domains. What does this mean?

Apparently, although LC domains do indeed prefer fractal arrangements in general, there is only one particular fractal shape that minimizes free energy. In other words, fractal dimension is not the only geometric parameter which controls the free energy minimum.

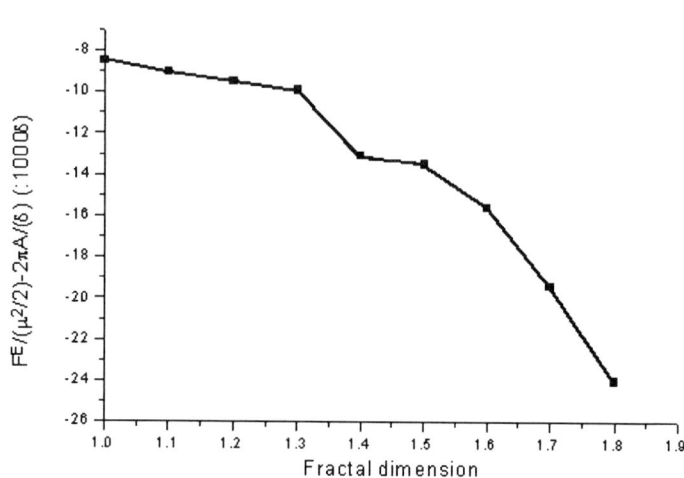

**Fig. 5.** Electrostatic component of the free energy vs. fractal dimension.

| D | $F^E (x10^2)$ | $F^L (x10^2)$ | $F(x10^2)$ | $F^E/(\mu^2/2) - 2\pi A/\delta\ (x10^2)$ |
|---|---|---|---|---|
| 1.8 | 195 | 178 | 374 | -240 |
| 1.7 | 222 | 131 | 353 | -194 |
| 1.6 | 236 | 96 | 333 | -156 |
| 1.5 | 257 | 77 | 335 | -135 |
| 1.4 | 265 | 63 | 328 | -131 |
| 1.3 | 266 | 55 | 321 | -99 |
| 1.2 | 282 | 52 | 334 | -95 |
| 1.1 | 273 | 50 | 322 | -90 |
| 1.0 | 271 | 63 | 334 | -85 |

**Table 1.** Free energy computation for LC fractal domains (Weierstrass-Mandelbrot frontier) and comparison with an LC circular domain with the same area A=10000; D=1.0 corresponds to the circular domain. Arbitrary units are assumed

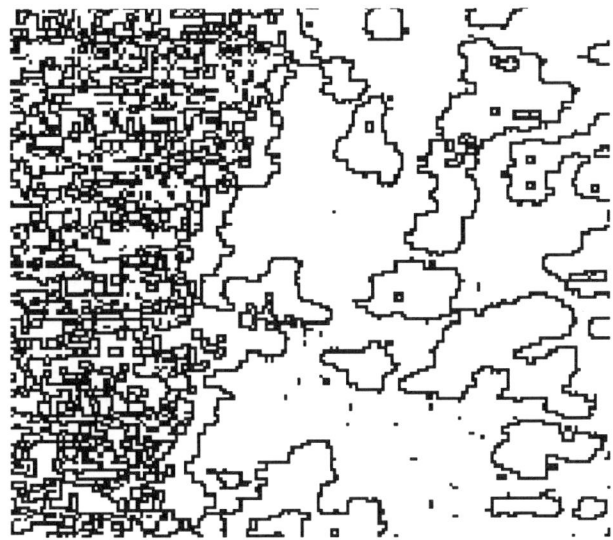

**Fig. 6.** Frontiers of LC domains from AFM image in figure 1c).

| Domain | Area | D | $F^E$ | $F^L$ | F | $F^E/(\mu^2/2) - 2\pi A/\delta$ |
|---|---|---|---|---|---|---|
| 1 | 132.0 | 1.39 | 63 | 460 | 523 | -703 |
| W-M | 132.0 | 1.39 | 133 | 579 | 712 | -526 |
| 2 | 161.0 | 1.34 | 126 | 640 | 766 | -760 |
| W-M | 161.0 | 1.34 | 270 | 611 | 881 | -637 |
| 3 | 64.0 | 1.66 | 42 | 420 | 462 | -318 |
| W-M | 64.0 | 1.66 | 40 | 359 | 398 | -276 |
| 4 | 429.5 | 1.61 | 208 | 1094 | 1302 | -2282 |
| W-M | 429.5 | 1,61 | 559 | 1380 | 1940 | -1622 |
| 5 | 808.5 | 1.12 | 602 | 1534 | 2136 | -3876 |
| W-M | 808.5 | 1.12 | 1770 | 1360 | 3130 | -1810 |
| 6 | 212.0 | 1.22 | 310 | 760 | 1070 | -713 |
| W-M | 212.0 | 1.22 | 322 | 683 | 1005 | -725 |
| 7 | 448.0 | 1.35 | 293 | 1280 | 1573 | -2229 |
| W-M | 448.0 | 1.35 | 796 | 1098 | 1890 | -1320 |
| 8 | 1819.0 | 1.41 | 1940 | 2860 | 4799 | -7756 |
| W-M | 1819.0 | 1.41 | 3852 | 2384 | 6236 | -3520 |
| 9 | 542.0 | 1.41 | 600 | 1160 | 1760 | -2220 |
| W-M | 542.0 | 1.41 | 1193 | 1309 | 2502 | -1643 |

**Table 2.** Free energy comparison between real LC fractal domains (1-9) and LC Mandelbrot-Weierstrass domains. Arbitrary units.

In order to discuss the influence of interfacial energy on the free energy minimum, we computed free energy for fractal domains with D=1.1 - 1.8 and for a circular domain, at three different values of line tension $\lambda$=5,10,20 (figure 7). We notice that free energy increases when $\lambda$ increases. When $\lambda$=20 (the interfacial term is dominant), the free energy increases when fractal dimension increases, but there is a minimum for D=1.1. For $\lambda$=10, the interfacial term becomes of the same order of magnitude as the electrostatic term, but the same dependence on fractal dimension is maintained. The minimum of free energy is obtained for D=1.3. For $\lambda$=5, the interfacial term is low, so the free energy now decreases with increasing fractal dimension. The minimum free energy is obtained for D=1.8 (or greater).

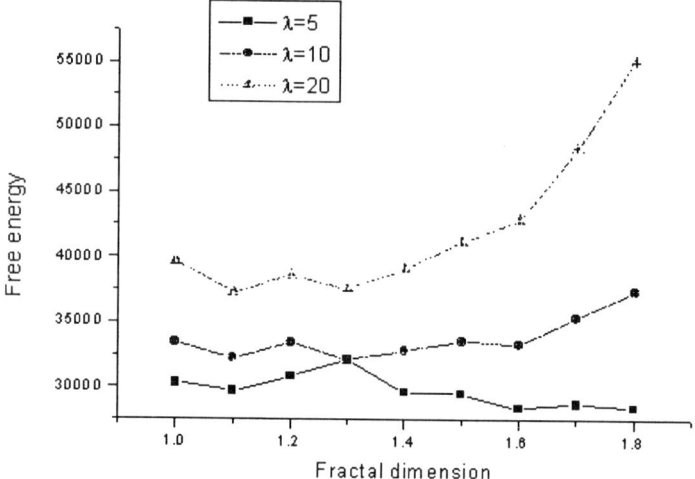

**Fig. 7.** Free energy vs. fractal dimension for three values of line tension.

We conclude that when the influence of the interfacial term decreases, the free energy minimum moves to higher fractal dimensions. For all the studied cases the minimum of the free energy is obtained for fractal domains. These observations give a qualitative understanding of LC fractal domain formation. As we saw, dipolar interactions between molecules lead to the formation of rough domain frontiers. The interaction between the LC phase and LE phase - which is represented in terms of interfacial energy - acts to minimize the domain perimeter, that thus minimize the interaction between the two phases. The competition between the two effects gives rise to the particular geometry of LC domains. A strong interaction between the LC and LE phases will lead to circular domains. On the contrary, a strong dipolar interaction in LC phase will lead to fractal domains.

# 4 Conclusions

LC domains of DMPE monolayer on mica, analyzed using adhesion-AFM, are fractal. The fractal dimension of these domains can be computed after careful image processing and removal of artifacts. A qualitative explanation of the formation of fractal domains is given in terms of free energy. In the case of strong dipolar interaction fractal LC domains have an energetic advantage over circular ones. For a better understanding of these phenomena, the free energy computation must be refined in order to include the interaction between several LC domains. Other improvements consist in adjusting the line tension, and considering not only vertical dipoles, but also freely oriented dipoles.

**Acknowledgment:** The authors wish to thank C.E.H.Berger, K.O.van der Werf, R.P.H.Kooyman, B.G.de Grooth and J.Greve for giving the permission to use AFM images from Ref.[8].

## References

1. B.B. Mandelbrot (1977): "Fractals: Form, Chance and Dimension". Freeman, San Francisco.
2. B.B. Mandelbrot (1982): "The Fractal Geometry of Nature". Freeman, San Francisco.
3. M.W. Mitchell and D.A. Bonnell (1990). J. Mater. Res., 5, 2244.
4. J.P. Carrejo, T. Thundat, L.A. Nagahara, S.M. Lindsay and A.Majumdar (1991). J. Vac. Sci. Technol. B, 9, 955.
5. D.R. Deuley (1990). J. Vac. Sci. Technol. A, 8, 603.
6. J.M. Gomez-Rodriguez, A.M. Bar,L.Vsquez, R.C. Salvarezza, J.M. Vara and A.J. Arvia (1992). J. Phys. Chem., 96, 347.
7. J.M. Williams and T.B. Beebe Jr. (1993). J. Phys. Chem, 97, 6249.
   J.M.Williams and T.P.Beebe Jr. (1993). J. Phys. Chem., 97, 6255.
8. C.E.H. Berger, K.O. van der Werf, R.P.H. Kooyman, B.G. de Grooth and J. Greve (1995). Langmuir, 11, 4188.
9. D.J. Keller, J.P. Korb and H.M. McConnell (1987). J.Phys.Chem., 91, 6417.
10. M.A. Mayer and T.K. Vanderlick (1992). Langmuir, 8, 3131.
11. H.M. McConnell and V.T. Moy (1988). J. Phys. Chem., 92, 4520.
12. P. Grassberger (1983). Phys. Lett. A, 97, 224.
13. P. Grassberger and I.Procaccia (1983). Physica 9D, 189.
14. J. Samios, P. Pfeifer, U. Mittag, M. Obert and T. Dorfmuler (1986). Chem. Phys. Lett., 128, 545.

# IMAGE COMPRESSION

# Faster Fractal Image Coding Using Similarity Search in a KL-transformed Feature Space

Jean Cardinal*

Brussels Free University
Computer Science Department
Campus de la Plaine
Bld. du Triomphe CP212
B-1050 Brussels - BELGIUM
Phone : 00-32-2-6505601
Fax : 00-32-2-6505867
jcardin@ulb.ac.be

**Abstract.** Fractal coding is an efficient method of image compression but has a major drawback: the very slow compression phase, due to a time-consuming similarity search between image blocks. A general acceleration method based on feature vectors is described, of which we can find many instances in the literature. This general method is then optimized using the well-known Karhunen-Loeve expansion, allowing optimal – in a sense to be defined – dimensionality reduction of the search space. Finally, a simple and fast search algorithm is designed, based on orthogonal range searching and avoiding the "curse of dimensionality" problem of classical best match search methods.

## 1 Introduction

### 1.1 Fractal Image Compression

The fractal compression method has been implemented for the first time in 1989 ([12]) and extensively described since then in many different publications (e.g. [9]). In fractal image compression, we try to find a mapping $W$ in the image space so that the fixed point of this mapping exists, is unique, and is as close as possible to the image $I$ we want to encode. $W$ is itself composed of $m$ different block-wise mappings $w_i, i = 1, \ldots, m$. Each mapping $w_i$ maps a so-called *domain block* onto a smaller *range block*. The set $R$ of range blocks forms a partition of the image $I$, and each range block is uniquely coded by a transformation $w_i$. There is no restriction concerning the set of domains, but these are usually taken among a set of potential domains called the *domain pool*, and are usually bigger than the ranges. The mappings $w_i$ consist in two parts: a spatial part, translating from the domain position in the image to the range position and scaling the domain size down to the range size, and a massic

---
* This work is supported by a PhD. scholarship from the National Scientific Research Fund (FNRS)

part, modifying the pixels in the block. The massic part is itself composed of two transformations: a scaling operation, multiplying all the pixel values in the block by a coefficient $s_i$, and an offset operation, adding a constant coefficient $o_i$ to each pixel value. The encoding process consists in finding for each range the best domain in the domain pool: the one that gives the least mean square error (MSE) when modified by the scaling and offset operations. Of course, the coefficients $s_i$ and $o_i$ are computed so that they minimize the error, using a simple regression formula. The decoding simply consists in iterating the mapping $W$ from any initial image. Having $|s_i| < 1, \forall i = 1, \ldots, m$ ensures the convergence during the decoding (we can however show that a weaker condition such as $|s_i| < 1.2$ is sufficient, see e.g. [9]).

The original algorithm compares each range with every domain. This method is very slow, and the compression of a single image may take hours. The simplest improvement is the use of classification schemes: dividing the domain pool in a number of classes, and comparing only the domains falling in the same class as the range. Classification schemes are numerous (e.g. [12], [9]), but none of them decreases the complexity order of the search. There are anyway some methods that achieve this goal, often with a slight quality loss (see e.g. [1])

The structure of the remainder of this paper is as follows. In section 1.2, we review the idea of feature vectors, and describe some previous work in which this idea is exploited. In sections 1.3 and 1.4, we briefly summarize the properties of the Karhunen-Loeve transform, and show how they can be used in nearest-neighbor searching problems. Section 2 describes the main contribution of this paper: a fast general acceleration technique. In section 2.1, we show how an useful representation of image blocks may be designed using the Karhunen-Loeve transform, and an appropriate search algorithm is described in 2.2. Finally experimental results are presented and commented in section 2.4.

## 1.2 A Fast General Method Using Feature Vectors

A good method to solve the problem of finding the best domain block for each range has been described in several papers ([17],[22],[13]), with different formulations and features. The general scheme is as follows:

1. find an operator mapping image blocks to *feature vectors*, so that minimizing the distance (e.g. the euclidean distance) between a feature vector corresponding to a range and a feature vector corresponding to a domain is equivalent to minimizing the collage error obtained when choosing the domain to code the range. Let us call $S_R$ and $S_D$ the feature vectors sets corresponding to the ranges and the domains, respectively;
2. for each element of $S_R$, perform a search in $S_D$ and find the closest domain feature vector. Use this domain to code the range.

Note that, since the ranges and the domains have not the same dimensionality, each block should be subsampled to the same size. Furthermore, the contractivity condition must be checked, so more than one domain should be tested in order to avoid scaling coefficients that are greater than one.

Using this method, the traditional linear search is replaced by a *nearest neighbor*, or *best match* search. This problem may be solved in $\mathcal{O}(\log |S_D|)$ average time for each query using k-d trees ([5], [3]), with a $\mathcal{O}(|S_D|.\log|S_D|)$ time preprocessing step. The overall complexity is then $\mathcal{O}((|S_D|+|S_R|).\log|S_D|)$, which is a quite good improvement compared to the original $\mathcal{O}(|S_D|.|S_R|)$.

Different approaches have been explored, using different operators to compute the feature vectors, and different structures to perform the search. In [22] and [2] the feature vectors used are the normalized DCT coefficients of the blocks, in [13], the structure used for the search is a R-tree, and the feature vectors used are the same as in [17]. We review the different methods in the following paragraphs.

**Normalization** The simplest way of finding good feature vectors might be the one described in [17], and also used in [13]. Suppose that $x$ is a vector obtained by an ordering of the pixels values of a block (e.g. in scan-line order), and $k$ the number of pixels in the block. The so-called *normalized projection operator* is defined as follows

$$\phi(x) = \frac{M.x}{\| M.x \|} \quad (1)$$

where

$$M_{ij} = \begin{cases} \frac{k-1}{k} & \text{if } i = j \\ -\frac{1}{k} & \text{if } i \neq j \end{cases} \quad (2)$$

Actually, $M$ is a mean-removing matrix: the mean value of the vector's components is substracted to each component, making it invariant to any translation whose direction is parallel to the vector $(1, 1, \ldots, 1)$ (the offset addition in the block coding transformation). In [18], this operator has been generalized to the case where a set of $p$ orthogonal fixed basis blocks was used: the matrix $M$ is then replaced by the operator projecting the vectors on the orthogonal complement of the subspace spanned by the basis blocks. Finally, the normalization operation makes the feature vector invariant to any scaling operation on the block. These vectors have two main properties:

1. the sum of their components is zero (due to the form of the matrix in 2),
2. the sum of their squared components is 1 (due to the normalization in 1).

We can then prove that minimizing the collage error is equivalent to minimizing the value $\min(d(\phi(x), \phi(y)), d(-\phi(x), \phi(y)))$ (where $x$ is the domain and $y$ the range). The geometrical interpretation of this computation has been extensively discussed in [18] and [21]. In the latter, the same kind of operator is used to eliminate useless domain blocks using an angular distance criterion.

**Frequency Domain Features.** In [4], [22] and [2], feature vectors are computed from the DCT representation of the block, as follows:

1. compute the DCT transform of the block,
2. remove the DC component,
3. normalize the other coefficients.

The DC coefficient of the transformed block is the mean intensity of the block, so that removing it is equivalent to removing the mean of the block, as performed by $M$ in the image space. It has the interesting side effect of decrementing the number of dimensions. The main advantage in this representation is that the energy packing property of the DCT allows the search algorithm to ignore the last components. This implies, however, a time-consuming frequency transform. Also note that reducing the number of dimensions by ignoring the high frequency components is a lossy speedup: the quality is worse than for a search taking all the components into account.

Extensive experiments on the use of this feature vector are described in [18], showing the obtained improvements.

### 1.3 The Discrete Karhunen-Loeve Transform

The *discrete Karhunen-Loeve transform* (KLT) is also known as *Principal Component Analysis*, *Hotelling transform* or *eigenvector transform*[1]. It has many properties that are used in the image processing field as well as in multivariate analysis.

Let $S$ be a set of vectors in a $k$-dimensional euclidean space. We define the *covariance matrix* of this set in the following way:

$$C_S = \frac{1}{|S|} \sum_{x \in S} (x - m_S).(x - m_S)^T, \qquad (3)$$

where $m_S = \frac{1}{|S|} \sum_{x \in S} x$ is called the *mean vector*. $C_S$ is a $k \times k$ matrix where each element $c_{ij}$ is the sample covariance of the $i^{th}$ and $j^{th}$ components of the vectors in $S$, and the elements $c_{ii}$ on the diagonal are the variances of the $i^{th}$ component of each vector.

$C_S$ is a real, symmetric matrix, so that it is always possible to find an orthonormal basis of eigenvectors. Let us call $A$ the $k \times k$ matrix of eigenvectors, where the $i^{th}$ column of $A$ is the eigenvector corresponding to the $i^{th}$ greatest eigenvalue of $C_S$. The K-L transform of a vector $x \in S$ is

$$y = A.(x - m_S) . \qquad (4)$$

The mean vector of the set $S' = \{y \mid y = A.(x - m_S), x \in S\}$ is zero, and the vector set $S'$ is oriented in such a way that the variance is maximized on the first coordinate axes, and minimized on the last ones. The variances on each coordinate axis are in fact the positive eigenvalues $\lambda_1, \lambda_2, \ldots, \lambda_k$, where $\lambda_1 \geq \lambda_2 \geq \ldots \geq \lambda_k$. The first axis is called the *principal component*. The KLT

---

[1] Actually, the Hotelling transform is the discrete version of the Karhunen-Loeve expansion.

is optimal in the sense that the mean square error induced by the suppression of the last vector components, those having the least variance, is minimal (see below).

## 1.4 The Use of the KLT in Best Match Searching

The KLT has been used since a long time in best match searching algorithms (see e.g. [14]). Some vector quantization algorithms project the set $S$ on its principal component and use this projection to quickly eliminate best match candidates ([15],[7]). It has also been shown that applying the KLT on a vector set before constructing a multidimensional indexing structure improves the speed of the search (see e.g. [20]). Some multidimensional tree structures based on a hierarchical space subdivision also use the principal component to partition the vector set ([19]). Actually, the KLT is very efficient in de-skewing a distribution, i.e. removing the correlation between vector components, so it is useful in strongly correlated data sets.

Before continuing, we will note some facts about distance measurement in a KL-transformed vector space. As $A$ denotes an orthonormal basis of eigenvectors, the KLT is a distance-preserving transformation. If we denote $d(x,y)$ the euclidean distance between $x \in S$ and $y \in S$, we have:

$$d(x,y) = d(x',y') \qquad (5)$$

where $x' = A.(x - m_S) \in S'$ and $y' = A.(y - m_S) \in S'$. So, if we define the set $S''$ as the projections of the vectors in $S'$ on the first $b$ coordinate axes, we can say:

$$d(x,y) \geq d(x'',y'') \qquad (6)$$

where $x''$ and $y''$ are the vectors corresponding to $x$ and $y$ in the set $S''$. This property is useful in the search methods using a projection on the principal component: if the distance between a projection of a vector $x$ on this axis and the projection of a query vector is already greater than the distance to the current best match, then $x$ does not deserve any more consideration. The principal component should be used as projection axis, since it maximizes the number of eliminated candidates. To control the amount of variance projected on the first $b$ axes, we will use the same notation than in [7], and introduce the following *preservation ratio*:

$$\frac{\sum_{i=1}^{b} \lambda_i}{\sum_{i=1}^{k} \lambda_i} \qquad (7)$$

This definition is generalized to any type of transform by replacing the eigenvalues by the variances on each axis. Another definition of the optimality of the KLT is that it maximizes the preservation ratio, i.e. there is no other orthonormal basis for which the preservation ratio is greater. When the data set is strongly correlated, we can choose $b$ so that the preservation ratio is not below a predefined threshold. Using this threshold, we can bound the difference $d(x, y) - d(x'', y'')$ on the average, and perform a search in the $b$-dimensional space.

## 2 Using the KLT in the Fractal Compression Best Match Search

### 2.1 Applying the KLT to the Feature Vectors

The computation of KL-transformed feature vectors may be carried out in four steps:

1. compute the mean-removed normalized feature vectors $\phi(x)$,
2. center the distribution, so that the mean vector is zero,
3. compute $A$,
4. multiply each feature vector by $A$.

The distribution of interest here is $S_D$, but one could imagine to use $S_R$ as well. We will introduce the notations $S'_D = \{y' \mid y' = A.(y - m_{S_D}), y \in S_D\}$, and $S'_R = \{y' \mid y' = A.(y - m_{S_R}), y \in S_R\}$.

The problem is that, even if the original image blocks form a strongly correlated vector set, the normalized feature vectors may lose this property, and the preservation ratio for a fixed $b$ is quite different in the two sets. The principal component of the non-normalized 4×4 blocks in the Lena image preserves more than 85% of the variance, but less than 55% for the feature vectors. Another important remark is that all the normalized feature vectors belong to the unit sphere, so the variances may not be as large as for the non-normalized vectors.

Another approach would be to compute the feature vectors from a KLT-based representation. We would like these KLT-based feature vectors to be invariant to the transformations used in the iterated system. This approach is more difficult than the one used for the DCT, since the KLT is a data-dependent transform. The DCT approach is also simpler when dealing with the projection part of the operator: we simply ignore the DC coefficient of the transformed vector. The equivalent operation in a KL-transformed space is much more complicated. This is why we think that the KLT should be applied in the last stage of the process, on the normalized projected feature vectors $\phi(x)$. More efficient energy packing and dimensionality reduction are expected compared to the DCT, since the KLT performs an optimal dimensionality reduction on every distribution.

Experimental results showing the preservation ratios confirm these hypothesis. Figure 1 shows the results obtained on 16K $4 \times 4$ image blocks taken from the Lena image.

A few remarks might be useful on the way the transformation may be computed efficiently. First, note that the centering step of the KLT is facultative in our application: the variances will not be affected by the position of the mean vector, so this step may be skipped, and 4 may be replaced by

$$y = A.x \qquad (8)$$

Another remark is about the compared computation time of the DCT and the KLT. The computation of $A$ has a $\mathcal{O}(k^2.|S_D|+k^3)$ time complexity, if we suppose that the matrix is computed from the $S_D$ distribution. This is a severe drawback compared to the DCT, where no such computation is needed. A good approximation of $A$ may however be obtained by subsampling the data set on which it is computed. The KL-transformation of a vector also takes $\mathcal{O}(k^2)$ operations, compared to $\mathcal{O}(k.\log k)$ for the DCT. Experimental tests carried on on 512×512 images show however that this additional computation time is not as important as the one needed for the computation of $A$.

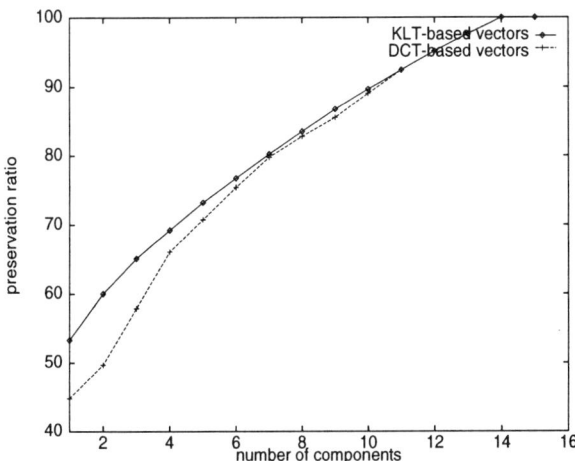

Fig. 1. Preservation ratios (in percent) for energy-packing feature vectors.

### 2.2 An Appropriate Search Algorithm

Range[2] searching consists in reporting all vectors that are contained in a query volume. For the search of the domain in the Karhunen-Loeve transformed feature space, we use a modified orthogonal range search algorithm. Orthogonal range searching in a $k$-dimensional space consists in searching only the subspace defined by the products of $k$ 1-dimensional intervals, on each of the $k$ coordinate axes. So the search is restricted in a hyper-volume bounded by two $(k-1)$-dimensional hyper-planes on each axis. It is the equivalent of a spherical range search when

---

[2] The word "range" is to be understand in its original meaning, it does not refer to range blocks.

the $L_\infty$ metric is used. Range searching and nearest neighbor searching are two closely related problems, which often use the same kind of structures: quadtrees ([8]), k-d trees ([5]), R-trees ([11]).

We outline here a general algorithm which does not use any tree structure. It is extensively described and analyzed in [16], although the same paradigm has already been exploited in some earlier algorithms ([10], [14]). The steps are shown in algorithm 1. The notation $x = (x_1, \ldots, x_k)$ is used for the vector components. This algorithm returns the set of vectors contained in the hypercube of side $2.\epsilon$ centered on the query point $Q$. This set may contain many vectors, and in that case an additional exhaustive search should be made. If the result is the empty set, then another search should be performed with a greater $\epsilon$. In [16], an efficient structure is designed to handle these operations. First, the

---

**Algorithm 1** The general range search algorithm

$S$ is the searched vector set
$Q$ is the query vector
the side of the search box is $2.\epsilon$
vector set $V$, $T$
$V \leftarrow \{x | x \in S, x_1 \in [Q_1 - \epsilon, Q_1 + \epsilon]\}$
**for** $i = 2$ to $k$ **do**
$\quad T \leftarrow \{x | x \in S, x_i \in [Q_i - \epsilon, Q_i + \epsilon]\}$
$\quad V \leftarrow V \cap T$
**end for**
**return** $V$

---

vectors are sorted along each coordinate axis: $k$ sorted lists are obtained. A mapping is maintained from the elements of each sorted lists to each vector of the original set, so that one can retrieve the index of a vector from its position in the sorted lists in constant time. This indirection is called *backward mapping*. The opposite mapping is also recorded, and is called *forward mapping*. To find the set $\{x | x \in S, x_i \in [Q_i - \epsilon, Q_i + \epsilon]\}$, two binary searches are performed, one for each limit, in the $i^{th}$ ordered list. Then we compute the intersection between the set of vectors lying in the interval and $V$, the set of vectors left in the previous step. The forward mapping allows the algorithm to use only integer comparisons for computing this intersection.

In order to suit the needs of the application, and to exploit as much as possible the properties of the transformed feature vectors, this algorithm is slightly modified. The modified version is shown in algorithm 2. In this algorithm, the trimming of the list stops as soon as the current set contains less than a predefined number $r$ of vectors. The returned vectors are not always contained in the $2.\epsilon$-sided hypercube, but the algorithm guarantees that at least $r$ vectors are returned (excepted in the case where the first set $\{x | x \in S, x_1 \in [Q_1 - \epsilon, Q_1 + \epsilon]\}$ contains less than $r$ elements, which can be made exceptional by correctly tuning the value of $\epsilon$).

**Algorithm 2** The modified range search algorithm

$S$ is the searched vector set
$Q$ is the query vector
the side of the search box is $2.\epsilon$
vector set $V, V', T$
$V \leftarrow \{x | x \in S, x_1 \in [Q_1 - \epsilon, Q_1 + \epsilon]\}$
$i \leftarrow 2$
**while** $|V| > r$ **and** $i \leq k$ **do**
  $T \leftarrow \{x | x \in S, x_i \in [Q_i - \epsilon, Q_i + \epsilon]\}$
  $V' \leftarrow V$
  $V \leftarrow V \cap T$
  $i \leftarrow i + 1$
**end while**
**if** $|V| \geq r$ **then**
  return $V$
**else**
  return $V'$
**end if**

The preprocessing step has time complexity $\mathcal{O}(k.|S|.\log|S|)$ if we use the heapsort algorithm. The cost of establishing the $Q_i \pm \epsilon$ bounds on each axis is $\mathcal{O}(k.\log|S|)$. The cost of the trimming of the lists for a single query depends on the distribution of the vectors. It has been shown in [16] that the time complexity of the original algorithm for the case where the vectors are uniformly distributed in the unit hypercube is $\mathcal{O}(|S|.(\epsilon + \frac{1}{1-\epsilon}))$ for small $\epsilon$.

It can seem strange to use an algorithm for which the complexity grows linearly with the number of domains instead of a tree-based one, logarithmic in the same variable. Two reasons motivate this choice:

1. the cost of a search in a multidimensional tree is logarithmic in the number of elements in the tree, but roughly exponential in the number of dimensions of the searched space: this is the well-known "curse of dimensionality" problem, which often makes some tree search methods useless in high dimensions. One can easily understand that by remarking that in a $k$−dimensional rectangular cell grid, a single cell has always $3^k - 1$ neighbors (two cells are neighbors if they have at least one common point), which is exponential in $k$.
2. The cost of the range search algorithm also depends on $\epsilon$, which is quite small.

As a consequence, the constant factors involved in the two complexities are of very different orders, and make the range search method attractive.

In these algorithms, the elimination of candidates begins on the first axis, and ends on the $k^{th}$ (or before in the modified version), in that order. This order is justified by the energy packing property of the vectors. We can easily show that the average number of candidates left after the $b^{th}$ iteration (including the search on the first axis) is:

$$|S| \cdot \prod_{i=1}^{b} P_i \qquad (9)$$

where

$$P_i = P\{x_i \in [Q_i - \epsilon, Q_i + \epsilon] \,|\, Q\} \qquad (10)$$

is the probability that $x_i$ lies in the interesting interval on the $i^{th}$ axis, for a given $Q$ and any $x \in S$. As the coordinate axes are sorted according to the variances of the corresponding components of the vectors, the algorithm ends up more rapidly with a small subset of the elements than for any other order, and the most important components are taken into account first.

To give an estimation of the average number of iterations of algorithm 2, we will place ourselves in a simplistic situation where the two following assertions hold:

1. the vector components $\{x_i | i = 1, \ldots, k, \; x \in S\}$ are independent, uniform random variables lying in the range $[-l_i/2, +l_i/2]$,
2. $\forall\, 1 < i \leq k$ we have $l_i = \alpha.l_{i-1}$, where $0 < \alpha < 1$.

The second assertion is an attempt to model the energy packing due to the KLT. It is fairly easy to show that the average number of iterations is[3]

$$\hat{b} = \frac{\log \frac{P_1}{\sqrt{\alpha}} + \sqrt{\log(P_1.\sqrt{\alpha})^2 - 2.\log \alpha . \log c}}{\log \alpha}, \qquad (11)$$

where $c = r/|S|$ is the desired ratio of points left at the end of the search. In the case where $\alpha = 1$, i.e. if we search in a hypercube, the solution is

$$\hat{b} = \frac{\log c}{\log P_1} - 1 . \qquad (12)$$

For a precise description of the developments, one can refer to [6].

### 2.3 Implementation

In order to show the improvements obtained, four different methods have been implemented:

1. a simple k-d tree search using the mean-removed normalized feature vectors,
2. a k-d tree search using the KL-transformed feature vectors,
3. a modified range search (algorithm 2) using the KL-transformed feature vectors,
4. a modified range search using the DCT-based feature vectors.

---
[3] the basis in which the logarithms are taken has no importance.

Only the first method finds the exact nearest neighbor in the feature vector space. The two others perform an approximated search: for the k-d tree using the KL-transformed vectors, the least significant components are ignored, and for the modified range search, the trimming of the lists should stop before the $k^{th}$ axis, and ignore as well the remaining components.

Three different range block sizes have been used: 4×4, 8×8 and 16×16. These sizes are the ones usually used in quadtree partitioning schemes, and they have been tested separately. The domain blocks are aligned on a 4×4 square lattice, and they are two times bigger than the range blocks. No square isometry is used in the transformations. All the blocks, regardless of the original size, are first down-sized to size 4×4, thus fixing $k$ =16, and the different normalizations and transform operations are performed in $\mathbb{R}^{16}$. The KL basis – as described in section 2.1 – is computed from the domain blocks distribution. Also note that in the three methods, for the sake of simplicity, we only use the vectors corresponding to $\phi(x)$, and forget the opposite key $-\phi(x)$.

The transformation coefficients are quantized, and the collage error is computed using the quantized coefficients. 5 and 7 bits are used for $s_i$ and $o_i$, respectively. We allow $s_i$ to vary between -1.2 and +1.2.

For the k-d tree methods, an optimized algorithm has been used, based on the original algorithm from Bentley ([5]), and implementing the *distance refinement* of Arya ([3]), which avoids a lot of floating point operations during the search. The search is not approximate, and the $r$ nearest neighbors are sought. In our tests, we have found $r = 10$ to be a good value for the application. This value is also the maximum number of elements contained in each bucket of the tree. For the case where the KL-transformed vectors are used, only the first $b$ components are used: as we have seen, the 90% preservation ratio is obtained with $b$ =12 on the Lena image, which is the reason why we have used this value for our tests.

We have not tested the scheme using a k-d tree with DC-transformed vectors, because we can see on figure 1 that the preservation ratio for $b$ =12 is the same as with the KLT, so the two methods should perform comparably. One can refer to [18] for this experiment.

For the modified range search method, the number of elements needed has been fixed to $r$ =20, and we use $\epsilon$ =0.3. These values have been found empirically. The number of iterations in the trimming loop of algorithm 2 using these parameters is roughly varying between 6 and 13.

The memory requirements for the KLT-based methods are the same as for the simpler feature vectors, since only the KL-transformed representations are saved.

## 2.4 Results

The three methods have been tested on 7 grey-scale images of size 512×512: Lena, Zelda, Peppers, Baboon, Goldhill, Barbara and Boat. For each block size, the overall collage error (MSE) and the average compression times in seconds are given. The time needed for the preprocessing steps (computation of the KLT basis and construction of the search structures) is also reported. The collage error is used instead of the decoded error for the sake of both simplicity and generality. The compression ratio is constant for a given block size and a coding of the transformations. In our simple quantization scheme, a transformation is coded in 26 bits, and the compression ratios for the three block sizes are respectively 1:5, 1:20 and 1:100.

Average time and error results are presented in tables 1, 2 and 3. The timings have been made on a Pentium-based machine. We can make the following conclusions:

1. the dimensionality reduction using the KLT does not introduce noticeable error,
2. this dimensionality reduction speeds up the search up to a factor of 2,
3. the modified range search method is faster than the other two, excepted for the 16×16 blocks,
4. significant MSE gains are observed for the 16×16 case,
5. the preprocessing steps in the modified range search are not prohibitive, and largely compensated by the speed gain during the search (at least for the 4×4 and 8×8 blocks),
6. the performances of the KLT-based feature vectors and the DCT-based ones are very close to each other, with a slight advantage for the DCT, mainly due to the fast preprocessing step.

Note that the block dimension used does not change the value of $k$: the search is always performed in $\mathbb{R}^{16}$. The main difference is that for a larger block size, the total number of ranges is decreasing, and the encoding is faster.

We can explain the results obtained for the bigger blocks by the fact that the distribution of the feature vectors is not the same for all the sizes: bigger blocks seem to give feature vectors with less variance. The choice of $\epsilon$ in the modified range search method should have been different for each block size, according to the distribution of the feature vectors.

We can imagine to use the new search method in a quadtree partitioning scheme. In that case, one transform is necessary for each quadtree level. The memory requirements are however the same as for a static partitioning corresponding to the last quadtree level.

**Table 1.** Results for 4×4 blocks

| Encoder | Preprocessing time (s.) | Total time (s.) | MSE |
|---|---|---|---|
| k-d tree – no transform | 5.32 | 935.92 | 37.50 |
| k-d tree – with KLT and $b=12$ | 13.18 | 417.93 | 37.83 |
| modified range search – with KLT | 15.32 | 153.79 | 38.49 |
| modified range search – with DCT | 8.61 | 153.36 | 38.49 |

Table 2. Results for 8×8 blocks

| Encoder | Preprocessing time (s.) | Total time (s.) | MSE |
|---|---|---|---|
| k-d tree – no transform | 6.32 | 186.79 | 137.09 |
| k-d tree – with KLT and $b=12$ | 13.89 | 92.21 | 140.63 |
| modified range search – with KLT | 15.75 | 73.79 | 137.37 |
| modified range search – with DCT | 9.32 | 68.21 | 137.54 |

Table 3. Results for 16×16 blocks

| Encoder | Preprocessing time (s.) | Total time (s.) | MSE |
|---|---|---|---|
| k-d tree – no transform | 9.89 | 45.93 | 265.07 |
| k-d tree – with KLT and $b=12$ | 17.32 | 33.93 | 267.85 |
| modified range search – with KLT | 18.89 | 51.36 | 257.12 |
| modified range search – with DCT | 12.32 | 44.92 | 257.04 |

## 3 Conclusion

In this paper, we have reviewed the general method of fractal compression using feature vectors, shown the improvements that could be obtained using a decorrelating rigid transform, and designed a simple algorithm using these features, faster than other methods of the same flavor. An optimized implementation of the technique, integrating an efficient block partition, could lead to competitive coder performances. Other experiments using more specific feature vectors, or wavelet domain transformations, could be fruitful too.

## 4 Acknowledgements

The author thanks Dietmar Saupe, Brendt Wohlberg, and Giuseppe Lauria for useful discussions and pieces of advice on this subject, and Efstathios Hadjidemetriou and Sameer Nene for providing the code of the algorithm described in [16]. The remarks of the referees also greatly helped to improve the quality of this text.

## References

1. E. Amram and J. Blanc-Talon (Oct. 1997): "Quick search algorithm for fractal image compression". *Proc. ICIP-97 IEEE International Conference on Image Processing*, Santa Barbara, California.
2. O. C. Au, M. L. Liou and L. K. Ma (Oct. 1997): "Fast fractal encoding in frequency domain". *Proc. ICIP-97 IEEE International Conference on Image Processing*, Santa Barbara, California.
3. S. Arya and D. Mount (1993): "Algorithms for fast vector quantization". *Proc. Data Compression Conference*, J. A. Storer and M. Cohn eds., Snowbird, Utah, IEEE Computer Society Press, 381-390.

4. K. U. Barthel (1995): *Festbilcodierung bei niedrigen bitraten unter verwendung fraktaler methoden im orts und frequenzbereich*. PhD. Thesis, Technische Universität Berlin.
5. J. L. Bentley, R. A. Finkel and J. H. Friedman (1977): "An algorithm for finding best matches in logarithmic expected time". *ACM Trans. Math. Software*, 3(3): 209-226.
6. J. Cardinal (1998): "Fractal compression using the discrete Karhunen-Loeve transform". Brussels Free University, Internal Report, http://homepages.ulb.ac.be/~jcardin.
7. C.-C. Chang, D.-C. Lin and T.-S. Chen (1997): "An improved VQ codebook search algorithm using principal component analysis". *Journal of Visual Communications and Image Representation*, Vol.8, No.1, March, pp. 27-37.
8. R. Finkel and J. L. Bentley (1974): "Quadtrees: a data structure for retrieval of composite keys". *Acta Inf.*, 4, 1, 1-9.
9. Y. Fisher (1994): *Fractal image compression - Theory and application*. Springer-Verlag, New York.
10. J. H. Friedman, F. Baskett and L. J. Shustek (Oct. 1975): "An algorithm for finding nearest neighbors". *IEEE Trans. Comput.*, vol. C-24, pp. 1000-1006.
11. A. Guttman (1984): "R-trees: a dynamic index structure for spatial searching". *Proceedings of the ACM SIGMOD International Conference on the Management of Data*, pp. 47-57.
12. A. Jacquin (August 1989): *A Fractal Theory of Iterated Markov Operators with Applications to Digital Image Coding*. PhD. thesis, Georgia Institute of Technology.
13. J. Kominek (1995): "Algorithm for fast fractal image compression". *Proceedings from IS&T/SPIE 1995 Symposium on Electronic Imaging: Science & Technology*, vol. 2419 Digital Video Compression: Algorithms and Technologies 1995.
14. R. Lee (1976): "Application of principal component analysis to multikey searching". *IEEE Transactions on Software Engineering*, SE 2(3): 185-193.
15. C. H. Lee and L. H. Chen (1995): "High-speed closest codeword search algorithms for vector quantization", *Signal Processing* 43, pp. 323-331.
16. S. A. Nene and S. K. Nayar (Sept. 1997): "A simple algorithm for nearest neighbor search in high dimensions". *IEEE Trans. on Pattern Analysis and Machine Intelligence*, vol. 19, No 9, pp. 989-1003.
17. D. Saupe (1998): "Accelerating fractal image compression by multi-dimensional nearest-neighbour search". *Proceedings DCC'95 Data Compression Conference*, J.A. Storer and M.Cohn (eds.), IEEE Comp. Soc. Press.
18. D. Saupe (1995): "Fractal image compression via nearest neighbor search". *Conf. Proc. NATO ASI Fractal Image Encoding and Analysis*, Trondheim, July 1995, Y. Fisher (ed.), Springer-Verlag, New York.
19. R. F. Sproull (1991): "Refinements to nearest-neighbor searching in k-dimensional trees". *Algorithmica*, 6, p. 579-589.
20. D. A. White and R. Jain (1997): "Algorithms and strategies for similarity retrieval". Visual Computing Laboratory, University of California, Internal Report. http://vision.ucsd.edu/papers.
21. B. E. Wohlberg and G. de Jager (1994): "On the reduction of fractal image compression encoding time". *1994 IEEE South African Symposium on Communications and Signal Processing (COMSIG'94)*, pp. 158-161.
22. B. E. Wohlberg and G. de Jager (1995): "Fast image domain fractal compression by DCT domain block matching". *Electronic Letters*, 31, 869-870.

# Can One Break the "Collage Barrier" in Fractal Image Coding?

Edward R. Vrscay[1] and Dietmar Saupe[2]

[1] Department of Applied Mathematics, University of Waterloo, Waterloo, Ontario - CANADA N2L 3G1
ervrscay@links.uwaterloo.ca
[2] Institut für Informatik, Universität Leipzig, Augustusplatz 10/11, 04109 Leipzig - GERMANY
saupe@acm.org

**Abstract.** Most fractal image coding methods rely upon *collage coding*, that is, finding a fractal transform operator $T_c$ that sends a target image $u$ as close as possible to itself. The fixed point attractor $\bar{u}_c$ of $T_c$ is generally a good approximation to $u$. However, it is well known that collage coding does not necessarily yield an optimal attractor, i.e., one for which the approximation error $||u - \bar{u}_c||$ is minimized with respect to variations in the fractal transform parameters. A number of studies have employed the "collage attractor" $\bar{u}_c$ as a starting point from which to obtain better approximations to $u$. In this paper, we show that attractors $\bar{u}$ are differentiable functions of the (affine) fractal parameters. This allows us to use gradient descent methods that search for optimal attractors in fractal parameter space, i.e., local minima of the approximation error $||u - \bar{u}||$. We report on results of corresponding computer experiments and compare them with those obtained by related (nondifferentiable) methods based on the simplex hill climbing and annealing approaches.

## 1 Introduction

Fractal image coding seeks to approximate a target image $u(x)$ with the fixed point attractor function $\bar{u}(x)$ of a contractive *fractal transform* operator $T$ that acts on a suitable metric space $(\mathcal{F}, d_\mathcal{F})$ of image functions. The *fractal code*, i.e., the parameters that define $T$, is then used to represent the target image $u$. The approximation $\bar{u}$ is generated by iteration of $T$ (*decoding*).

Given a suitable parameter space $\mathcal{P}$ of acceptable fractal codes it is generally a tedious procedure to determine the best fractal code, i.e., to determine the operator $T_{opt}$ whose fixed point $\bar{u}_{opt}$ yields the smallest possible error $d_\mathcal{F}(\bar{u}, u)$. In fact, Ruhl and Hartenstein [15] have shown that such "attractor coding" problems are NP-hard. Understandably, most, if not all, fractal coding algorithms rely upon the *Collage Theorem* [2,1]. In *collage coding*, one finds an operator $T_c$ that minimizes the *collage error* $d_\mathcal{F}(u, Tu)$. As is well known, the corresponding attractor error $d_\mathcal{F}(\bar{u}_c, u)$, where $\bar{u}_c = T\bar{u}_c$, is bounded above by a multiple of the collage error. Unfortunately, there is little theoretical knowledge about the relationship between the collage error and the minimum error.

It is also well known that collage coding is not necessarily optimal, i.e., that $\bar{u}_c \neq \bar{u}$. In fact, Ruhl and Hartenstein [15] have rigorously shown that collage coding is *not* a $\rho$-approximating algorithm: Using the notation introduced above, there exists no (finite) constant $\rho > 0$ such that

$$\frac{d_{\mathcal{F}}(u, \bar{u}_c)}{d_{\mathcal{F}}(u, \bar{u}_{opt})} < \rho \quad \forall u \in \mathcal{F}.$$

In other words, it is possible that the ratio of the collage attractor error to the optimal attractor error can be arbitrarily large. Intuitively, this is a consequence of the fact that collage coding is a "greedy algorithm" that seeks to find self-similarity of an image in one scan. Nevertheless, because of its relative simplicity and the fact that it appears to work well in most cases, collage coding continues to serve as the basis of most fractal coders.

Several studies have attempted to find attractor functions $\bar{u}$ that are better approximations to a target $u$ than the "collage attractors" $\bar{u}_c$. Indeed, these studies have typically employed the collage attractor $\bar{u}_c$ as a starting point. For example, Barthel [3] and then Lu [10] have devised "annealing schemes" that produce sequences of attractors $\bar{u}_n$ in which the fractal code for $\bar{u}_{n+1}$ is obtained from constructing a collage of pieces of $\bar{u}_n$ to approximate the target $u$. The sequences $\bar{u}_n$ are observed to provide better approximations to the target. However, there is still no rigorous theoretical basis for this method. A closely related annealing algorithm is presented by Domaszevicz and Vaishampayan [5]. On the other hand, Dudbridge and Fisher [6], using the Nelder-Mead simplex algorithm, searched the fractal code space $\mathcal{P}$ in the vicinity of the collage attractor to locate (local) minima of the approximation error $d_{\mathcal{F}}(u, \bar{u})$. Their method was applied to a rather restricted class of (separable) fractal transforms, in which four $4 \times 4$ pixel range blocks shared a common domain block.

In this paper we examine a systematic method to perform attractor optimization using the partial derivatives of attractor functions with respect to the fractal code parameters, $\partial \bar{u} / \partial \pi_k$, $k = 1, 2, \ldots, M$. Here, the parameter space $\mathcal{P}$ of acceptable fractal codes consists of vectors $\underline{\pi} = (\pi_1, \pi_2, \ldots, \pi_M)$, $M > 0$. We first establish the existence of these derivatives and show that they are attractor functions of "vector fractal transform" operators. A knowledge of these derivatives permits the computation of the gradient vector of the error function $d_{\mathcal{F}}(\bar{u}, u)$ which, in turn, allows the use of gradient descent algorithms. In principle, the method may be applied to any domain-range fractal coding scheme. However, from a complexity point of view, it is expensive, since the attractor and all derivatives must be computed at every pixel. We compare these results to those obtained by minimizing the approximation error using the Nelder-Mead simplex algorithm employed in Ref. [6] and also to those obtained using the appropriate variant of annealing schemes from Ref. [3, 11].

In a related work Withers [16] derives differentiability properties of Iterated Function Systems with probabilities whose attractors model graphs of 1D functions. Newton's method is used to compute parameters.

Finally, in our discussion of the inverse problem, we present an "Anti-Collage Theorem" which, in contrast to the infamous Collage Theorem, provides a *lower bound* to the approximation error in terms of the collage error. This result is another very simple consequence of Banach's Fixed Point Theorem.

## 2 Partial Derivatives of IFSM Attractor Functions

### 2.1 Mathematical Preliminaries

Let $(X, d)$ denote the *base* or *pixel space*, assumed to be a complete metric space. The discussion will be carried out for the continuous case, e.g., $X = [0,1]^2$, but with the understanding that it readily carries over to the discrete case, e.g., $X = \{(i,j) | 1 \leq i, j \leq n_p\}$, where $n_p$ denote the width and height of a square image in pixels. Let $\mathcal{F}(X) = \{f : X \to \mathbf{R}\}$ denote a suitable complete space of image functions with metric $d_\mathcal{F}$. Later, we shall use $\mathcal{F}(X) = \mathcal{L}^2(X)$, the space of square integrable functions on $X$ with the usual metric.

Now let $R_k \subset X$, $k = 1, 2, \ldots, N$ denote a set of *range* or *child* blocks that partition $X$, i.e., (1) $\cup_{k=1}^N R_k = X$ and (2) $R_i \cap R_j = \emptyset$ for $i \neq j$. Assume that for each range block $R_k$ are associated the following:

1. a *domain* or *parent* block, $D_k \subset X$, such that $R_k = w_k(D_k)$, where $w_k$ is a one-to-one contraction map with contraction factor $c_k \in [0,1)$,
2. an affine greyscale map $\phi_k(t) = \alpha_k t + \beta_k$, where $\alpha_k, \beta_k \in \mathbf{R}$.

In the language of [9], the above ingredients comprise an (affine) $N$-map Iterated Function System with Grey Level Maps (IFSM). Associated with such a (nonoverlapping) IFSM is a fractal transform operator $T : \mathcal{F}(X) \to \mathcal{F}(X)$ whose action is defined as follows. Given an image $u \in \mathcal{F}(X)$ then for all $x \in R_k$, $k = 1, 2, \ldots, N$,

$$v(x) = (Tu)(x) = \phi_k(u(w_k^{-1}(x)))$$
$$= \alpha_k u(w_k^{-1}(x)) + \beta_k. \quad (1)$$

The transform $T$ is defined by the fractal code parameters

$$\underline{\pi} = (\alpha_1, \ldots, \alpha_N, \beta_1, \ldots, \beta_N).$$

In addition some side information must be transmitted to the decoder in order to specify the image partition and the range-domain assignments. However, in this study, we assume a *fixed* domain-range block configuration, implying that the IFS maps $w_i$ are *fixed*. Thus, for all further discussion we can disregard this side information.

It is well known that if $|\alpha_k| < 1$, $1 \leq k \leq N$, then the operator $T$ is contractive in the space of functions $\mathcal{L}^\infty(X)$. In the space of functions $\mathcal{L}^2(X)$, a straightforward calculation shows that

$$\|Tu - Tv\|_2 \leq C\|u - v\|_2, \quad \forall u, v \in \mathcal{L}^2(X),$$

where

$$C = \sum_{k=1}^{N} c_k |\alpha_k|. \qquad (2)$$

Therefore, the condition $C < 1$ is sufficient (but not necessary) for contractivity of $T$ in $\mathcal{L}^2(X)$.

Another special type of IFSM/fractal transform, relevant below, involves "condensation" [11]. For a $u \in \mathcal{F}(X)$, define $v = Tu$ as follows: For all $x \in R_k$, $k = 1, 2, \ldots, N$,

$$v(x) = (Tu)(x) = \alpha_k u(w_k^{-1}(x)) + \theta_k(x). \qquad (3)$$

The functions $\theta_k(x)$ are known as *condensation* functions. Note that condensation functions do *not* affect the contractivity of $T$.

## 2.2 Partial Derivatives of IFSM Attractor Functions with Respect to Greyscale Map Parameters

Let us now consider an affine $N$-map IFSM and assume that its associated fractal transform $T$ in Eq. (1) is contractive in the space $\mathcal{F}(X) = \mathcal{L}^2(X)$. Therefore, there exists a unique fixed point attractor function $\bar{u} = T\bar{u}$. We now consider $\bar{u}$ as a function not only of position but also the fractal parameters, i.e., $\bar{u} = \bar{u}(x, \underline{\pi})$. Then, from Eq. (1),

$$\bar{u}(x, \underline{\pi}) = \alpha_k \bar{u}(w_k^{-1}(x), \underline{\pi}) + \beta_k, \quad x \in R_k. \qquad (4)$$

**Proposition 1.** *The attractor $\bar{u}$ is continuous with respect to the fractal parameters $\alpha_\ell$, $\beta_\ell$, $l = 1, 2, \ldots, N$.*

The continuity of IFSM attractors with respect to grey level maps $\phi_\ell$ was proved in [8], using the methods described in [4]. It is straightforward to establish the continuity in terms of the grey level parameters $\alpha_\ell$ and $\beta_\ell$. The following result, which establishes the continuity of attractor functions with respect to condensation functions, is also a simple consequence of Proposition 1.

**Proposition 2.** *Let $T_1$ and $T_2$ be contractive $N$-map IFSM operators as in Eq. (3), with condensation functions $\theta^{(1)}(x)$ and $\theta^{(2)}(x)$, respectively, and identical fractal coefficients $\alpha_k$. Let $\bar{u}_1$ and $\bar{u}_2$, respectively, denote the fixed points of these operators. Then given an $\epsilon > 0$, there exists a $\delta > 0$ such that $\|\theta^{(1)} - \theta^{(2)}\|_2 < \delta$ implies that $\|\bar{u}_1 - \bar{u}_2\|_2 < \epsilon$.*

Recall that the IFS maps $w_i$ employed in the fractal transforms are assumed to be fixed. Now define the *feasible fractal parameter space* $\mathcal{P} \subset \mathbf{R}^{2N}$ to be the set of all fractal codes $\underline{\pi} \in \mathbf{R}^{2N}$ for which the corresponding affine IFSM operators $T$ defined in Eq. (1) are contractive in $\mathcal{F}(X) = \mathcal{L}^2(X)$.

**Proposition 3.** *The set $\mathcal{P}$ is open.*

**Proof:** We prove that $\overline{\mathcal{P}} = \mathbf{R}^{2N} - \mathcal{P}$ is closed. Let $\underline{\pi}_n \in \overline{\mathcal{P}}$, $n = 1, 2, \ldots$, be a convergent sequence (in the topology of $\mathbf{R}^{2N}$) with limit $\underline{\pi}$. Each (unfeasible) fractal code vector $\underline{\pi}_n \in \overline{\mathcal{P}}$ defines a noncontractive fractal transform operator $T_n : \mathcal{L}^2(X) \to \mathcal{L}^2(X)$ with associated factor (cf. Eq. (2)) $C_n = \sum_{k=1}^{N} c_k^{1/2} |\alpha_{nk}|$. Now, for each operator $T_n$, define its "optimal" Lipschitz factor as follows,

$$L_n = \sup_{y_1 \neq y_2} \frac{\|T_n y_1 - T_n y_2\|_2}{\|y_1 - y_2\|_2}.$$

From this definition and the noncontractivity of the $T_n$, it follows that $1 \leq L_n \leq C_n$ for all $n$. From the convergence of the code vectors $\underline{\pi}_n$, it also follows that $\lim_{n \to \infty} C_n = C \geq 1$. Therefore, from Proposition 1, the fractal transform $T$ defined by the limit code vector $\underline{\pi}$ has associated factor $C$ and Lipschitz factor $L \geq 1$. Therefore $T$ is not contractive, implying that $\underline{\pi} \notin \mathcal{P}$. Thus $\overline{\mathcal{P}}$ is closed, proving the proposition. □

**Proposition 4.** *The partial derivatives of the attractor $\bar{u}$ with respect to the fractal parameters $\alpha_\ell, \beta_\ell, l = 1, 2, \ldots, N$, exist at any point $\underline{\pi} \in \mathcal{P}$.*

**Proof:** For any $\underline{\pi} \in \mathcal{P}$, the associated fractal transform $T$ is contractive. This implies that for any $u_0 \in \mathcal{L}^2$, the sequence of functions defined by $u_{n+1} = Tu_n$ converges to $\bar{u}$, that is, $\|u_n - \bar{u}\|_2 \to 0$ as $n \to \infty$. Let $u_0 = \theta$, where

$$\theta(x) = \sum_{k=1}^{N} \beta_k I_{R_k}(x)$$

and $I_S(x)$ is the characteristic function of a subset $S \subset X$. Then, for $M \geq 0$, $u_M = T^{\circ M} u_0$ is given by

$$u_M(x, \underline{\pi}) = \theta(x) + \sum_{n=1}^{M} \sum_{i_1, \ldots, i_n}^{N} \alpha_{i_1} \cdots \alpha_{i_n} \theta(w_{i_n}^{-1} \circ \cdots \circ w_{i_1}^{-1}(x)). \quad (5)$$

The $u_M$ are partial sums of an infinite series that converge, in the $\mathcal{L}^2$ metric, to $\bar{u}$. Thus, we can write

$$\bar{u}(x, \underline{\pi}) = \theta(x) + \sum_{n=1}^{\infty} \sum_{i_1, \ldots, i_n}^{N} \alpha_{i_1} \cdots \alpha_{i_n} \theta(w_{i_n}^{-1} \circ \cdots \circ w_{i_1}^{-1}(x)),$$

where the equation is understood in the $\mathcal{L}^2$ sense.

Now consider an $x \in R_k$ for some $k \in \{1, 2, \ldots, N\}$. Then the index $i_1$ in Eq. (5) must equal $k$ (in order for $w_{i_1}^{-1}(x)$ to be defined). Therefore, Eq. (5) becomes

$$u_M(x, \underline{\pi}) = \theta(x) + \alpha_k u_{M-1}(w_k^{-1}(x), \underline{\pi}), \quad x \in R_k.$$

For a given $l \in \{1, 2, \ldots, N\}$, we partially differentiate the terms in this equation with respect to $\alpha_\ell$:

$$\frac{\partial u_M}{\partial \alpha_\ell}(x, \underline{\pi}) = \alpha_k \left[ \frac{\partial u_{M-1}}{\partial \alpha_\ell}(w_k^{-1}(x), \underline{\pi}) \right] + [u_{M-1}(w_k^{-1}(x), \underline{\pi})]\delta_{kl}, \quad (6)$$

where $\delta_{kl} = 1$ if $k = l$ and zero otherwise. Define the following $N$-map IFSM operator $T_\ell$ with condensation:

$$(T_\ell u)(x) = \alpha_k u(w_k^{-1}(x)) + \xi_k(x), \quad x \in R_k, \quad 1 \leq k \leq N,$$

where $\xi_k(x) = [\bar{u}(w_k^{-1}(x))]\delta_{kl}$. Since $T$ is contractive, it follows that $T_\ell$ is contractive in $\mathcal{L}^2$. ($T$ and $T_\ell$ have identical IFS maps and fractal parameters $\alpha_k$.) Let $\bar{v}_\ell$ denote the fixed point of $T_\ell$. From Propositions 1 and 2, $\bar{v}_\ell$ is continuous with respect to the parameters $\alpha_k$, in particular, $\alpha_\ell$. We now show that $\bar{v}_\ell = \partial \bar{u}/\partial \alpha_\ell$. (In what follows, for simplicity of notation, only $x$ and $\alpha_\ell$ will be written explicitly in the list independent variables.)

Note that Eq. (6) does not correspond to a single IFSM operator with condensation. However, since the functions $u_M$ converge to $\bar{u}$, it follows, from Proposition 2, that the sequence of functions $\partial u_M/\partial \alpha_\ell$ converges to $\bar{v}_\ell$. That is, for a given $\underline{\pi} \in \mathcal{P}$ and $\epsilon_1 > 0$, there exists an $M_1 > 0$ such that

$$\left\| \frac{\partial u_M}{\partial \alpha_\ell}(x, \alpha_\ell) - \bar{v}(x, \alpha_\ell) \right\|_2 < \epsilon_1, \quad \forall M > M_1. \tag{7}$$

We denote our reference point as $\underline{\pi}^0 = (\alpha_1^0, \ldots, \alpha_N^0, \beta_1^0, \ldots, \beta_n^0) \in \mathcal{P}$. Let $N_\ell(\delta)$, $\delta > 0$, be a restricted neighbourhood of the point $\underline{\pi}^0$ in which only the component $\alpha_\ell$ is allowed to vary, i.e., $\alpha_\ell \in I_\delta = [\alpha_\ell^0 - \delta, \alpha_\ell^0 + \delta]$, such that the corresponding vectors $\underline{\pi}$ lie in $\mathcal{P}$. (The existence of such a neighbourhood is guaranteed since $\mathcal{P}$ is open.) Let $h \in \mathbf{R}$, with $|h| < \delta$. Then for each $x \in X$ there exists, by the Mean Value Theorem, a $c_M \in I_h = [\alpha_\ell^0 - h, \alpha_\ell^0 + h]$, such that

$$u_M(x, \alpha_\ell^0 + h) - u_M(x, \alpha_\ell^0) = \frac{\partial u_M}{\partial \alpha_\ell}(x, c_M)h.$$

Therefore,

$$\|u_M(x, \alpha_\ell^0 + h) - u_M(x, \alpha_\ell^0) - h\bar{v}(x, \alpha_\ell^0)\|_2$$
$$= h \left\| \frac{\partial u_M}{\partial \alpha_\ell}(x, c_M) - \bar{v}(x, \alpha_\ell^0) \right\|_2$$
$$\leq h \left\| \frac{\partial u_M}{\partial \alpha_\ell}(x, c_M) - \bar{v}(x, c_M) \right\|_2 + h\|\bar{v}(x, c_M) - \bar{v}(x, \alpha_\ell^0)\|_2$$
$$\leq h \left\| \frac{\partial u_M}{\partial \alpha_\ell}(x, c_M) - \bar{v}(x, c_M) \right\|_2$$
$$+ \max_{\alpha_\ell \in I_h} h\|\bar{v}(x, \alpha_\ell) - \bar{v}(x, \alpha_\ell^0)\|_2. \tag{8}$$

Since $I_\delta$ is closed, there exists an $\overline{M} > 0$ such that the inequality in (7) is satisfied for all $M > \overline{M}$ at all $\underline{\pi} \in N_\ell(\delta)$. Therefore, for a fixed $h \in (-\delta, \delta)$, we may take the limit $M \to \infty$ of both sides of (8) to give

$$\left\| \frac{\bar{u}(x, \alpha_\ell^0 + h) - \bar{u}(x, \alpha_\ell^0)}{h} - \bar{v}(x, \alpha_\ell^0) \right\|_2 \leq \max_{\alpha_\ell \in I_h} \|\bar{v}(x, \alpha_\ell) - \bar{v}(x, \alpha_\ell^0)\|_2.$$

Since $\bar{v}$ is continuous with respect to $\alpha_\ell$, the right side term may be made arbitrarily small by making $h$ sufficiently small, thus establishing the differentiability of $\bar{u}$ with respect to $\alpha_\ell$ at $\underline{\pi}^0$.

The differentiability of $\bar{u}$ with respect to the $\beta_\ell$ may be derived in a similar fashion. □

**Remark:** From Eq. (6) (and its analogue for differentiation with respect to $\beta_\ell$), the partial derivatives of $\bar{u}$ with respect to the fractal parameters $\alpha_\ell$ and $\beta_\ell$ may be obtained by formally differentiating both sides of Eq. (4). For a fixed $x \in R_k$:

$$\frac{\partial \bar{u}}{\partial \alpha_\ell}(x,\underline{\pi}) = \alpha_k \left[ \frac{\partial \bar{u}}{\partial \alpha_\ell}(w_k^{-1}(x),\underline{\pi}) \right] + [\bar{u}(w_k^{-1}(x),\underline{\pi})]\delta_{kl}, \tag{9}$$

$$\frac{\partial \bar{u}}{\partial \beta_\ell}(x,\underline{\pi}) = \alpha_k \left[ \frac{\partial \bar{u}}{\partial \beta_\ell}(w_k^{-1}(x),\underline{\pi}) \right] + \delta_{kl}. \tag{10}$$

Eqs. (4), (9) and (10) may be considered to define a $(2N+1)$-component "vector IFSM with condensation" that may be written in the following compact form:

$$\bar{\mathbf{u}} = \mathbf{T}\bar{\mathbf{u}},$$

where

$$\bar{\mathbf{u}}(\mathbf{x},\underline{\pi}) = \left[ \bar{u}(x,\underline{\pi}), \frac{\partial \bar{u}}{\pi_1}(x,\underline{\pi}), \ldots, \frac{\partial \bar{u}}{\pi_{2N}}(x,\underline{\pi}) \right]^T. \tag{11}$$

Now define the space $\mathcal{F}^{2N+1}(X) = \{\mathbf{f} = (f_0, f_1, \ldots, f_{2N}) \mid f_j \in \mathcal{F}(X)\}$ with associated metric $d_{\mathcal{F}^{2N+1}}(\mathbf{f},\mathbf{g}) = \max_{0 \leq j \leq 2N} d_\mathcal{F}(f_j, g_j)$. Then the vector IFSM is denoted by $\mathbf{T}: \mathcal{F}^{2N+1}(X) \to \mathcal{F}^{2N+1}(X)$. For an $f \in \mathcal{F}^{2N+1}(X)$,

$$(\mathbf{Tf})(x) = A_k \cdot \mathbf{f}(w_k^{-1}(x)) + \Theta_k(x), \quad x \in R_k.$$

The coefficients of the matrix $A_k = (a_{i,j}^{(k)})_{i,j=0,\ldots,2N}$ are

$$a_{i,j}^{(k)} = \begin{cases} \alpha_k & \text{if } i = j \\ 1 & \text{if } i = k, j = 0 \\ 0 & \text{otherwise} \end{cases}.$$

The entry "1" in the $k$-th row represents the only "mixing" of components of $\mathbf{u}$ under the action of $\mathbf{T}$. The vector $\Theta_k(x)$ represents a condensation vector composed of constant functions: $[\Theta_k(x)]^T = (\beta_k, 0, 0, \ldots, 1, \ldots, 0)$, where the "1" occurs at index $N+k$.

**Proposition 5.** *Suppose that $T$ is contractive in $(\mathcal{F}(X), d_\mathcal{F})$. Then $\mathbf{T}$ is contractive in $\mathcal{F}^{2N+1}(X)$. Its fixed point $\mathbf{u}$ is given by Eq. (11), where $\bar{u}$ is the fixed point of $T$, see Eq. (4).*

From Banach's Fixed Point Theorem, contractivity of $T$ allows the computation of its fixed point function $\bar{u}$ by means of iteration. The above proposition implies that all all partial derivatives $\partial \bar{u}/\partial \pi_\ell$ may also be computed by iteration: Begin with a "seed" $\mathbf{u}_0 \in \mathcal{F}^{2N+1}(X)$ and construct the sequence of vector functions $\mathbf{u}_{n+1} = \mathbf{T}\mathbf{u}_n$, $n \geq 0$. The reader will immediately note the complexity of such calculations: Except in special cases, $\bar{u}$ and its partial derivatives will have to be computed for all $x \in X$. This will be discussed in more detail below.

## 3  The Inverse Problem of Fractal Approximation

### 3.1  Optimal Attractor Coding vs. Collage Coding

We now consider inverse problems in $\mathcal{F}(X) = \mathcal{L}^2(X)$. As well, $v \in \mathcal{L}^2(X)$ will denote a "target" image function that we seek to approximate. Let $T$ be a fractal transform, as defined in Eq. (1), with fractal code $\underline{\pi} \in \mathcal{P}$, i.e., $T$ is contractive in $\mathcal{L}^2(X)$. We consider the squared $\mathcal{L}^2$ error in the approximation $v \approx \bar{u}$ as a function of the fractal code $\underline{\pi}$:

$$E(\underline{\pi}) = \int_X [v(x) - \bar{u}(x,\underline{\pi})]^2 dx$$
$$= \langle v - \bar{u}, v - \bar{u} \rangle.$$

The "true" inverse problem, or *attractor coding*, of $v$ is then to find $\min_{\underline{\pi}\in\mathcal{P}} E(\underline{\pi})$. Let $\underline{\pi}_{opt}$ denote a global minimum point of this error function. Assuming that the fractal transform $T_{opt}$ defined by $\underline{\pi}_{opt}$ is contractive, we denote its fixed point by $\bar{u}_{opt}$ and refer to it as the *optimal attractor*.

However, as mentioned earlier, the solution of this problem, involving the determination of optimal domain-range pairs and associated fractal parameters, is generally intractable. As such, most fractal-based methods perform *collage coding*, that is, they seek to minimize the *collage error* associated with the transform $T = T(\underline{\pi})$. We denote the squared $\mathcal{L}^2$ collage error as

$$\Delta(\underline{\pi}) = \int_X [v(x) - T(\underline{\pi})v(x)]^2 dx$$
$$= \sum_{k=1}^N \int_{R_k} [v(x) - \alpha_k v(w_k^{-1}(x)) - \beta_k]^2 dx.$$

A standard procedure is to impose stationarity conditions in order to obtain a system of two linear equations for each set of parameters $(\alpha_k, \beta_k)$, $k = 1, 2, \ldots, N$. (However, it is *not* guaranteed that the solution of these systems lies in $\mathcal{P}$.) If we let $\underline{\pi}_{col}$ denote the global minimum of $\Delta(\underline{\pi})$ then, of course,

$$E(\underline{\pi}_{opt}) \leq E(\underline{\pi}_{col}).$$

We shall refer to the fixed point of the fractal transform defined by $\underline{\pi}_{col}$ (assuming it to be contractive) as the *collage attractor*, $\bar{u}(x, \underline{\pi}_{col})$.

## 3.2 Optimizing Collage Coding

One possible compromise between solving the optimal attractor coding problem and suboptimal collage coding is to employ the collage attractor (and corresponding domain-range assignments), in particular the fractal code $\underline{\pi}_{col}$, as a starting point, varying the fractal parameters $\underline{\pi}$ in an attempt to decrease the error function $E(\underline{\pi})$ as much as possible. This was the strategy of Dudbridge and Fisher [6], who employed the Nelder-Mead simplex algorithm for some rather simple and restrictive domain-range assignments. In this scheme, the error function $E(\underline{\pi})$ is computed at strategic points.

A knowledge of the partial derivatives of $\bar{u}$ with respect to the fractal parameters permits the computation of elements of the gradient vector of $E$:

$$\frac{\partial E}{\partial \pi_\ell}(\underline{\pi}) = -2 \left\langle v - \bar{u}, \frac{\partial \bar{u}}{\partial \pi_\ell}(\underline{\pi}) \right\rangle, \quad l = 1, 2, \ldots, 2N.$$

This allows us to employ gradient-descent and related methods to search for local minima.

Practically speaking, however, the partial derivatives $\partial \bar{u}/\partial \pi_\ell(x, \underline{\pi})$ must be computed at all points (pixels) $x \in X$. In addition to an $n_p \times n_p$ matrix required to store an image, an additional $2N$ $n_p \times n_p$ matrices are needed, in general, to store the derivatives at all pixels. Borrowing from the terminology of quantum chemists, this "full configuration interaction" will compute the total rate of change of the attractor — hence the approximation error — with respect to changes in all fractal parameters $\pi_\ell$ *for a fixed set of domain-range pair assignments.* When applying a gradient descent method to minimize the error function $E$ less storage is required. It suffices to provide one additional $n_p \times n_p$ matrix to sequentially compute each component of the gradient $(\partial E/\partial \pi_1, \ldots, \partial E/\partial \pi_{2N})$.

In Section 4 we apply the conjugate gradient method to compute minima of the error function $E$. Technically we need to ensure that the gradient of $E$ is continuous. As a sketch of a proof of this property we remark that the gradient is essentially given by the attractor of a vector IFS with condensation which is continuous with respect to IFS parameters. This should be sufficient to establish continuous partial derivatives of $E$.

## 3.3 The "Anti-Collage Theorem"

We close this section by examining an interesting consequence of collage coding. For reference, we state the "Collage Theorem," a simple corollary of Banach's Fixed Point Theorem:

**Proposition 6 (Collage Theorem).** *Given $(Y, d_Y)$ a complete metric space. Let $T : Y \to Y$ be contractive with contraction factor $c_T \in [0, 1)$ and let $\bar{y}$ denote the unique fixed point of $T$. Then for any $y \in Y$,*

$$d_Y(y, \bar{y}) \leq \frac{1}{1 - c_T} d_Y(y, Ty).$$

This result, central to fractal image compression, provides an upper bound to the distance $d_Y(y,\bar{y})$, which may be viewed as the error in approximating $y$ by $\bar{y}$, in terms of the collage error $d_Y(y,Ty)$. As may or may not be well known, it may be derived by means of a simple application of the triangle inequality.

However, it may not be so well known that a slight reshuffling of the triangle inequality leads to another result that "works in the other direction," providing a lower bound. We first state a more generalized result.

**Proposition 7.** *Given $(Y, d_Y)$ a complete metric space. Let $T : Y \to Y$ be Lipschitz, i.e., there exists an $L_T \geq 0$ such that $d_Y(Ty_1, Ty_2) \leq L_T d_Y(y_1, y_2)$ for all $y_1, y_2 \in Y$. As well, assume that $\bar{y}$ is a fixed point of $T$. Then for any $y \in Y$,*

$$d_Y(y,\bar{y}) \geq \frac{1}{1+L_T} d_Y(y, Ty).$$

**Proof:** From the triangle inequality:

$$d_Y(y, Ty) \leq d_Y(y, \bar{y}) + d_Y(\bar{y}, Ty)$$
$$\leq d_Y(y, \bar{y}) + L_T d_Y(\bar{y}, y),$$

from which the desired result follows.

**Remark:** Note that in the above result, $T$ is not assumed to be contractive. Hence the fixed point $\bar{y}$ need not be unique.

We may now combine the two results for application to fractal approximation: Given a target function $v \in \mathcal{F}(X)$ consider the approximation of $v$ by the unique fixed point $\bar{u}$ of a contraction map $T : \mathcal{F}(X) \to \mathcal{F}(X)$. Then

$$\frac{1}{1+c_T} d_{\mathcal{F}}(v, Tv) \leq d_{\mathcal{F}}(v, \bar{u}) \leq \frac{1}{1-c_T} d_{\mathcal{F}}(v, Tv).$$

In other words, the collage error $d_{\mathcal{F}}(v, Tv)$ bounds the approximation error both from above and below. (In retrospect, this most interesting result is a simple consequence of the "triangle" formed by $v$, $\bar{v}$ and $Tv$.) For a fixed collage error (assuming this can be done), the spread between upper and lower bounds decreases with the contractivity factor $c_T$. On a more negative note, for nonzero collage error, there is no chance that the error can be small "by accident."

## 4 Practical Examples

We have tested the proposed method in two versions of fractal image coding. In the first one we use the approach suggested by Monro and Dudbridge [12] and explained below. This choice is motivated by the chance to compare our results with those obtained by Dudbridge and Fisher [6], who used a simplex method for attractor optimization. Also in this approach the dimensionality of the optimization problem is small, so it is a good starting point. The second series of tests were carried out using quadtree partitionings derived with the code from [7]. We provide a comparison of the results with those obtained by appropriately adjusted procedures of Barthel [3] and Lu [10].

|          | Collage attractor dB | Attractor optimization | | $\Delta$PSNR dB | $\Delta$PSNR [6] dB |
|----------|-----|------|------|------|------|
|          |     | Simplex dB (sec) | Gradient dB (sec) |  |  |
| Lena     | 29.25 | 29.87 (301) | 29.87 (229) | 0.62 | 0.35 |
| Boat     | 26.66 | 27.42 (300) | 27.42 (299) | 0.56 | 0.41 |
| Mandrill | 21.52 | 22.11 (532) | 22.08 (1500) | 0.59 | 0.33 |
| Peppers  | 29.34 | 30.02 (277) | 29.94 (591) | 0.68 | 0.33 |

**Table 1.** Results of (a) collage coding and attractor optimization using (b) simplex and (c) gradient methods, the latter two using collage coding as a starting point. All results are expressed in PSNR (dB). The final two columns list the improvement in PSNR achieved by the simplex method obtained in this study and Ref. [5], respectively.

## 4.1 Results Part 1

We first apply our method to a simple fractal transform scheme examined by Dudbridge and Fisher [6], designed to minimize the interdependency of range blocks. The following four 512 × 512 pixel images (8 bpp), used in [6] were also used in this study: *Lena, Boat, Mandrill* and *Peppers*.[1] Each image was partitioned into 4×4 pixel range blocks, with four range blocks sharing a common 8×8 pixel domain block. Therefore, for each image, the inverse problem separates into $64^2$ independent problems, each involving an 8 × 8 pixel image with four range blocks $R_k$, hence 8 fractal parameters.

As in [6], for each test image we first used collage coding to determine the fractal code $\underline{\pi}_{col}$ that minimizes the collage error. We then used this code as a starting point for a gradient-descent method. The NAG [13] subroutine E04DKF, which performs a quasi-Newton conjugate gradient minimization, was used. It was also desirable to compare these results with the non-gradient calculations of [6]. However, since some of our collage error results differed from those of [6], we have independently carried out attractor optimization using the Nelder-Mead simplex algorithm. The NAG subroutine E04CCF was used.

In all cases, the simplex and gradient methods yielded almost identical improvements. A comparison with [6] will reveal some nonnegligible differences, not only in the collage errors but also in the improvements afforded by the simplex method. In all cases, the improvements in Table 1 are greater. In both the simplex as well as the gradient algorithms, the results are quite sensitive to the settings of the tolerance/accuracy parameters as well as the maximum number of iterations (*maxiter*) allowed. Generally the best performance was obtained when the tolerance parameters for the simplex and gradient subroutines were set to $10^{-5}$ and $10^{-6}$, respectively. The parameter *maxiter* was set to 2000, which is virtually infinity.

---

[1] These images may be retrieved by anonymous ftp from the Waterloo Fractal Compression Project site links.uwaterloo.ca in the appropriate subdirectories located in ftp://links.uwaterloo.ca/pub/BragZone/GreySet2/.

|         | PCA   | Simplex      | Gradient      | $\Delta$PSNR |
|---------|-------|--------------|---------------|--------------|
|         | dB    | dB (sec)     | dB (sec)      | dB           |
| Lena    | 26.93 | 29.73 (421)  | 29.74 (288)   | 2.81         |
| Boat    | 25.08 | 27.30 (452)  | 27.32 (618)   | 2.24         |
| Mandrill| 20.85 | 22.00 (663)  | 21.97 (3333)  | 1.15         |
| Peppers | 25.97 | 29.76 (420)  | 29.56 (2888)  | 1.79         |

**Table 2.** Results of (a) piecewise constant approximation (PCA) and attractor optimization using (b) simplex and (c) gradient methods, the latter two using the PCA as a starting point. All results are expressed in PSNR (dB). The final column lists the improvement in PSNR achieved by the better of methods (b) and (c).

In Table 1 are presented the PSNR values associated with collage coding and subsequent simplex and gradient optimized attractor coding, along with the improvements in PSNR. The numbers in brackets represent the CPU time required for each calculation. (We emphasize that these numbers are presented for the purpose of comparision, since the computer codes themselves are quite unoptimized.)

In an attempt to understand how good an initial estimate is provided by collage coding, we have performed simplex and gradient optimization calculations for another set of initial conditions, namely, piecewise constant approximations to the images. In this case, all $\alpha_\ell$ are initially set to zero and the $\beta_\ell$ are simply the mean values of the range block. (Of course, in more general problems than the one studied here, there would remain the problem of assigning a domain block to each range block.) In Table 2, we present the results of these calculations. The first column gives the error associated with the initial piecewise constant approximation. The next two columns list the PSNR values of the optimized attractors obtained from, respectively, the simplex and gradient methods along with the CPU times. The final column gives the PSNR improvement yielded by the better of the two methods.

We observe that the simplex and gradient methods, using such suboptimal initial conditions, i.e., piecewise constant approximations, yield approximations that are almost as good as those found from collage attractors. The worst case is *Peppers*, for which a 0.26 dB difference is found. For the others, the discrepancy is on the order of 0.1 dB.

### 4.2 Results Part 2

In this subsection we report on the results of the gradient descent algorithm when applied to fractal encodings based on quadtree partitionings. We used the coder of Fisher [7] to produce quadtree partitions and corresponding fractal codes. The resulting scaling and offset parameters were then subject to improvement by the gradient descent method. In this case we have used the conjugate gradient algorithm from [14]. The major computational burden is the computation of the gradients required in each step, which allowed us to do experiments only with

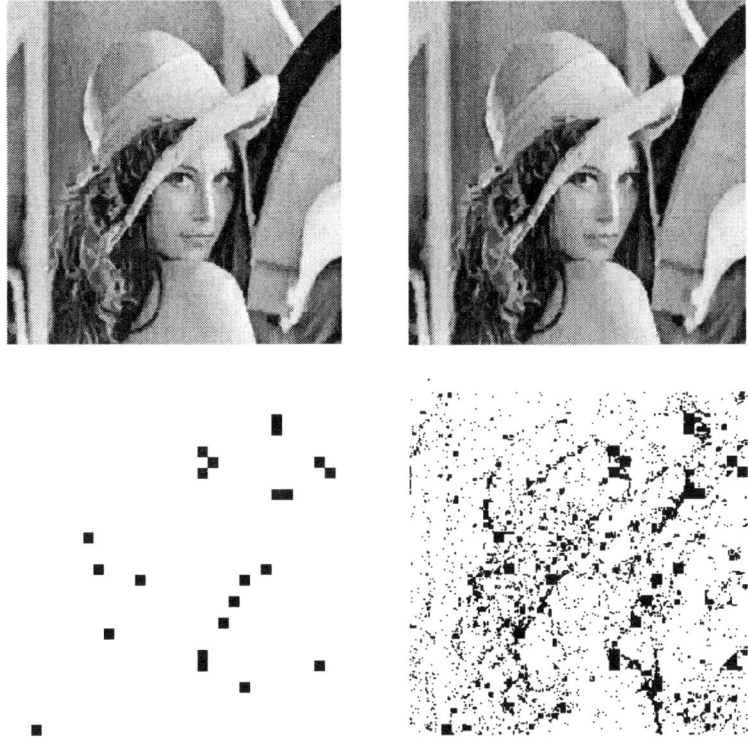

**Fig. 1.** Illustration of the results for the attractor optimization gradient descent method. The image is the 256x256 test image Lena, compressed using the fractal quadtree coder with minimal range block size of 8x8 pixels.
Top left: attractor image obtained by collage coding (26.96 dB PSNR).
Top right: attractor image after optimization (26.99 dB PSNR).
Bottom left: range blocks with new scale and offset code are shown in black.
Bottom right: pixels that are different in the optimized attractor are shown in black.

images of size 256 × 256. We considered the test image Lena and three different encodings, obtained by prescribing minimal range block sizes of 16 × 16, 8 × 8, and 4 × 4 pixels respectively.[2] The scaling and offset parameters were then modified until convergence of the conjugate gradient method, which required up to 16 steps for our setting of algorithm parameters. After convergence the obtained scaling and offset parameters were quantized for storage, where we used the same scheme as the one contained in the quadtree coder. Table 3 lists intermediate and end results. The gain obtained by the gradient descent method varied between 0.16 and 0.25 dB in PSNR. However, the necessary quantization

---

[2] The command line parameters for the quadtree coder were -t 8.000000 -m 4 -d 2 -D 2 -s 5 -o 7 -N 1.000000 -F with -M 4, -M 5, and -M 6 for minimal range block sizes of 16 × 16, 8 × 8, and 4 × 4 pixels respectively.

| min. block size | 16x16 | 8x8 | 4x4 |
|---|---|---|---|
| compression ratio | 70.2 | 23.8 | 9.8 |
| attractor errors (in dB) | 23.46 | 26.96 | 30.87 |
| step 1 | 23.57 | 27.04 | 30.93 |
| step 2 | 23.60 | 27.07 | 30.96 |
| step 3 | 23.61 | 27.10 | 31.00 |
| step 4 | 23.61 | 27.11 | 31.03 |
| step 5 | 23.62 | 27.12 | 31.06 |
| step 6 | 23.62 | 27.12 | 31.08 |
| step 7 | 23.62 | 27.12 | 31.09 |
| step 8 | 23.62 | 27.13 | 31.10 |
| step 9 | 23.62 | 27.13 | 31.10 |
| step 10 | 23.62 | 27.13 | 31.11 |
| step 11 | 23.62 | 27.13 | 31.11 |
| step 12 | — | 27.13 | 31.11 |
| step 16 | — | — | 31.12 |
| best result (in dB) | 23.62 | 27.13 | 31.12 |
| gain (in dB) | 0.16 | 0.17 | 0.25 |
| after quantization (dB) | 23.50 | 26.99 | 30.89 |
| remaining gain (in dB) | 0.04 | 0.03 | 0.02 |
| total time | 3h:40m | 6h:20m | 9h:50m |

**Table 3.** Results for the gradient descent method applied to quadtree encodings of the 256 × 256 test image Lena. The three columns correspond to encodings with different bit rates.

destroyed a large part of these gains so that improvements of only 0.02 to 0.04 dB PSNR remained. This indicates that our settings for the tolerances in the conjugate gradient methods were rather conservative. The computation times were very large.[3] In Figure 1 we display the attractor images and the ranges and pixels that have changed during the attractor optimization.

In order to compare the results with those obtained by previously published methods we implemented an annealing approach along the lines in [3, 11]. Briefly, the method proceeds by re-encoding the test image several times. In the first step the original image is used to derive the domain pool as usual, while in all subsequent iterations the attractor from the previous step is used to generate the domain pool. In this way the encoder has a better grasp of the codebook that the decoder actually uses in the attractor image. Unfortunately, there is no theory behind these heuristics and experiments show that the attractor error does not decrease monotonically as iterations proceed. This was also confirmed in our simulations, see Table 4. Since the scope of this paper includes the parameter updates, but no change in the domain assignments for the ranges, we used the annealing approach also in the corresponding version. In other words, in the first step of the annealing method both the partitioning and the domains used for each range were computed and then held fixed throughout all following iterations. The

---

[3] CPU times reported in this subsection were measured on an SGI O2 R5000 processor.

| min. block size | 16x16 | 8x8 | 4x4 |
|---|---|---|---|
| compression ratio | 70.2 | 23.8 | 9.8 |
| attractor error (in dB) | 23.46 | 26.96 | 30.87 |
| step 1 | 23.50 | 27.00 | 30.92 |
| step 2 | 23.53 | 27.01 | 30.91 |
| step 3 | 23.50 | 27.00 | 30.92 |
| step 4 | 23.52 | 27.00 | 30.93 |
| step 5 | 23.50 | 27.01 | 30.93 |
| step 6 | 23.53 | 27.01 | 30.92 |
| step 7 | 23.50 | 27.00 | 30.93 |
| step 8 | 23.52 | 27.01 | 30.92 |
| step 9 | 23.50 | 27.00 | 30.92 |
| step 10 | 23.53 | 27.00 | 30.93 |
| step 21 | 23.51 | ⋮ | ⋮ |
| step 24 |  | 27.00 | ⋮ |
| step 32 |  |  | 30.92 |
| best result (dB) | 23.53 | 27.01 | 30.93 |
| gain (in dB) | 0.07 | 0.05 | 0.06 |
| total time | 0m:35s | 0m:58s | 1m:31s |

Table 4. Results for the annealing method.

iterations were terminated after a total of ten iterations without improvement, i.e., iterations for which the attractor quality in PSNR had dropped. In this method quantization is already employed in each step, of course, and thus, no extra loss occurs as for the gradient method, which must rely on continuous parameters. Table 4 lists the results for the same test image as above. We see that the improvements are in the range between 0.05 and 0.07 dB PSNR.

# 5 Conclusions and future work

In this paper we have derived the theoretical fundamentals necessary for any application of differentiable methods for attractor error reduction in fractal image compression, namely

- the establishment of the differentiability of the attractor image as a function of its (real valued) scale and offset parameters, and
- the feasibility of gradient computation by iteration of a properly defined vector Iterated Function System with gray level Maps.

Moreover, we have implemented gradient descent algorithms for the problem and reported computational results for a few test cases. While the computer programs have demonstrated that the methods work in practice, the outcomes, however, are not promising. Although gains for the simple encoding based on the method of Dudbridge and Monro are around one half of a dB in PSNR, the conceptually less complex method using a simplex hill climbing algorithm performs just as well at the same cost in terms of computation time. Furthermore, the achievable gains for fractal coding with the quadtree method are negligible, while the competing annealing methods gave slightly better improvements at a fraction of the cost in computation time. Thus, we must conclude that gradient descent methods for parameter optimization do not provide a useful addition to conventional fractal encoding.

Even if parameter updates cannot break the "collage barrier", not all is lost. The remaining chance to achieve this goal is to also consider changing the domain assignments for the range blocks, as already proposed in [3, 11]. This, of course, is a discrete process and we cannot use differentiable methods as presented in this paper. Our current work focusses on algorithms to accomplish this without sacrificing the property of monotonically decreasing errors that we have, e.g, for the gradient descent algorithms.

# 6 Acknowledgments

We thank Raouf Hamzaoui and Hannes Hartenstein for valuable discussions as well as for some independent numerical calculations in order to verify our results. Andreas Zerbst carried out the computer experiments implementing the gradient descent and annealing methods for quadtree fractal encodings based on Yuval Fisher's public domain program [7].

This research was supported in part by a grant from the DFG Schwerpunktprogramm *Ergodentheorie, Analysis und Effiziente Simulation Dynamischer Systems tem* (DANSE).

ERV gratefully acknowledges the partial support of this research by the Natural Sciences and Engineering Council of Canada, in the form of an Operating Grant as well as a Collaborative Projects Grant.

# References

1. M.F. Barnsley (1998): *Fractals Everywhere.* Academic Press, New York.
2. M.F. Barnsley, V. Ervin, D. Hardin and J. Lancaster (1985): Solution of an inverse problem for fractals and other sets. Proc. Nat. Acad. Sci. USA **83**, 1975–1977.
3. K. U. Barthel (1996): Festbildcodierung bei niedrigen Bitraten unter Verwendung fraktaler Methoden im Orts- und Frequenzbereich. (Dissertation, Technische Universität Berlin), Wissenschaft & Technik Verlag, Berlin.
4. P. Centore and E.R. Vrscay (1994): Continuity of attractors and invariant measures for Iterated Function Systems. Canad. Math. Bull. **37**(3), 315–329.
5. J. Domaszevic and V.A. Vaishampayan (1994): Iterative collage coding for fractal compression. *Proceedings of the IEEE Intern. Conf. Image Proc.* (ICIP'94).
6. F. Dudbridge and Y. Fisher (May 1997): Attractor optimixation in fractal image encoding. Proceedings of Third "Fractals in Engineering" Conference, Arcachon, France.
7. Y. Fisher (1995): *Fractal Image Compression, Theory and Application.* Springer-Verlag, New York.

8. B. Forte and E.R. Vrscay (1995): Solving the inverse problem for function and image approximation using Iterated Function Systems. *Dyn. Cont. Disc. Imp. Syst.* **1**, 177–231.
9. B. Forte and E.R. Vrscay (1998): Inverse problem methods for generalized fractal transforms. *Fractal Image Encoding and Analysis*, ed. Y. Fisher (Springer Verlag, Heidelberg).
10. N. Lu (1997): *Fractal Imaging.* Academic Press, New York.
11. F. Mendivil and E.R. Vrscay (1997): Correspondence between fractal-wavelet transforms and Iterated Function Systems with Grey-level Maps. *Fractals in Engineering*, ed. J. Lévy Véhel, E. Lutton and C. Tricot (Springer, New York).
12. D. M. Monro and F. Dudbridge (1992): Fractal block coding of images. *Electronics Letters* **28**, 11, 1053–1055.
13. NAG Fortran Library, The Numerical Algorithms Group Ltd, Oxford, UK.
14. W. H. Press, S. A. Teukolsky, W. T. Vetterling and B. P. Flannery (1992): *Numerical Recipes in C.* Second Edition, Cambridge University Press.
15. M. Ruhl and H. Hartenstein (1997): Optimal fractal coding is NP-hard. *Proceedings of the IEEE Data Compression Conference*, ed. J. Storer and M. Cohn, Snowbird, Utah.
16. D. Withers (1989): Newton's method for fractal approximation. *Constructive Approximation* **5**, 151–170.

# Two Algorithms for Non-Separable Wavelet Transforms and Applications to Image Compression

Franklin Mendivil* and Daniel Piché

Department of Applied Mathematics
University of Waterloo
Waterloo, Ontario - CANADA N2L 3G1
mendivil@math.gatech.edu
dgpiche@mercator.math.uwaterloo.ca

**Abstract.** Previously, the use of non-separable wavelets in image processing has been hindered by the lack of a fast algorithm to perform a non-separable wavelet transform. We present two such algorithms in this paper. The first algorithm implements a periodic wavelet transform for any valid wavelet filter sequence and dilation matrices satisfying a trace condition. We discuss some of the complicating issues unique to the non-separable case and how to overcome them. The second algorithm links Haar wavelets and complex bases and uses this link to drive the algorithm. For the complex bases case, the asymmetry of the wavelet trees produced leads to a discussion of the complexities in implementing zero-tree and other wavelet compression methods. We describe some preliminary attempts at using this algorithm, with non-separable Haar wavelets, for reducing the blocking artifacts in fractal image compression.

**Keywords:** Haar, wavelet, image, compression, fractal, tiling, periodic wavelet transform, non-separable.

## 1 Introduction

Wavelets have found applications to many diverse areas of science, engineering and mathematics. A large part of their usefulness in applications is their ability to capture both scale and location information of a data signal. However, the practical application of wavelets is also driven by the existence of a fast algorithm to do a wavelet decomposition on a discrete data set. This *Fast Wavelet Transform* in many cases allows the real-time analysis of data.

For a finite length data signal, we have several options on how to analyze the data. One popular method is to periodize the signal. In this case, we think of our data as being a function on the circle $S^1$ and use wavelets on $S^1$ that are induced from wavelets on $\mathbb{R}$. There is a corresponding discrete transform called

---
* Current address: School of Math, Georgia Tech, Atlanta, GA 30332-0160

the *Discrete Periodic Wavelet Transform* that is very fast. In fact, for a data stream of length $N$, the transform has time complexity $O(N)$ (see [30]). Another method is *zero padding* where you extend your finite length data signal by zeros to make an (potentially) infinite length signal. In processing zero padded data you need only compute with the finite length signal itself, filling in the appropriate zeros where necessary.

One of the many areas in which wavelets have found applications is the area of image processing. Here their ability to extract spatially localized information is very useful in reconstructing, modifying or analyzing an image. However, most of the wavelets used in image processing (an inherently two-dimensional application) have been tensor products of wavelets from $L^2(\mathbb{R})$. While this is sufficient for many purposes, in some cases it is not. For example, when using tensor product wavelets to compress images, you often introduce artifacts from the fact that you lose information. With tensor product wavelets, these artifacts include vertical and horizontal lines in the image. This is undesirable since our eyes are particularly sensitive to errors along lines. Non-separable wavelets based on fractal or dust-like tilings introduce a natural dithering effect which helps to eliminate these linear errors.

Non-separable wavelets are wavelets that are, in some sense, intrinsic to two (or more) dimensions; they are not tensor products of wavelets on some lower dimensional space. There has been much recent activity on constructing and analyzing multidimensional non-separable wavelets (see [2, 5, 19–21]). However, multidimensional non-separable wavelets are far from being well-understood. The question of existence and properties of these wavelets is a much more delicate and intricate one than that of one dimensional wavelets. For instance, there are very few explicit constructions of these wavelets. On the other hand, certain types of wavelets, namely Haar wavelets, are fairly easy to construct. These wavelets are derived from characteristic functions.

The characterization of multidimensional Haar wavelets was given in the paper of Gröchenig and Madych [15]. These wavelets are usually non-separable and have support on fractal tilings. This leads to the idea of complex bases, which also produce fractal tilings of the complex plane.

Complex bases are a way of representing complex numbers, in a similar fashion as the decimal system is used to represent real numbers. The study of such bases began with the work of Kátai and Szabó [18]. Many results in this area, including algorithms for determining the representations, are due to Gilbert [9–11]. Gilbert also provided the connection between the fractal tiles of complex bases and iterated function systems. This allowed the development of the long division algorithm for complex bases [14].

In this paper we present two recent algorithms to perform non-separable wavelet transforms, one a periodic transform and one a zero padding transform. The periodic transform requires that the dilation matrix satisfy a trace condition but will work with any valid wavelet filter sequence for the given dilation matrix. Thus it will work for wavelets more general than Haar wavelets. The zero padding algorithm is based on a new link between complex bases and wavelets, which enables an understanding of the translation of the Mallat algorithm to the language of complex bases. This algorithm only works for Haar wavelets. However, the dilation matrices associated with complex bases do not in general satisfy the trace condition. Thus, while the two algorithms have a substantial area of overlap, they complement each other in that neither one generalizes the other.

In some sense, our periodic transform is simply the transform on the $n$ dimensional torus induced by the discrete wavelet transform on $\mathbb{R}155^n$ and periodicity. Unlike the one dimensional case, however, this turns out to have surprising complications so that it doesn't work out as simply. We indicate in this paper how these difficulties may be overcome.

After presenting each algorithm, we give some preliminary experiments in using non-separable Haar wavelets combined with a fractal-wavelet transform in an attempt to reduce the blocking artifacts which are common in conventional fractal image compression. In the case of the periodic transform, we use the Haar wavelets primarily because of a lack of suitable wavelet filters for other smoother wavelets. However, the benefits of non-separable wavelets are evident even in the simple Haar wavelet case.

## 2 Wavelets

In this section, we summarize the necessary background of wavelet analysis. For a more complete account of the wavelet theory the reader is referred to [6, 15] and of the fractal compression theory see [8, 22, 29].

### 2.1 Background

To start, let $L^2(\mathbb{R}^n)$ be the space of all square Lebesgue integrable functions from $\mathbb{R}^n$ to $\mathbb{R}$. Recall the following definitions:

**Definition 1.** *A matrix $A$ on $\mathbb{R}^n$ is an* acceptable dilation *for $\mathbb{Z}^n$ if $A\mathbb{Z}^n \subset \mathbb{Z}^n$ and if $|\lambda| > 1$ for each eigenvalue $\lambda$ of $A$.*

Throughout this paper we will let $q = |\det A|$. The properties of an acceptable dilation imply that $q$ is an integer $\geq 2$.

Let $A$ be an acceptable dilation on $L^2(\mathbb{R}^n)$, $f \in L^2(\mathbb{R}^n)$ and $x, y \in \mathbb{R}^n$. Define the unitary dilation operator $U_A$ by

$$U_A f(x) = |\det A|^{-1/2} f(A^{-1}x)$$

and the translation operator $\tau_y$ by

$$\tau_y f(x) = f(x - y).$$

Then for each $i \in \mathbb{Z}$ and $j \in \mathbb{Z}^n$ let $f_{i,j} \equiv U_A^{-i} \tau_j f$. Hence,

$$f_{i,j}(x) = q^{i/2} f(A^i x - j).$$

Using this notation, $f_{0,0} = f$.

We are interested in wavelet bases given by translation by integers. By a basis we will mean a (orthonormal) Hilbert space basis.

**Definition 2.** *A wavelet basis $B$, associated with an acceptable dilation $A$, is a basis of $L^2(\mathbb{R}^n)$ whose members are $A$ dilates and $\mathbb{Z}^n$ translates of a finite orthonormal set $S = \{\psi^1, \ldots, \psi^m\} \subset L^2(\mathbb{R}^n)$, where $m \in \mathbb{N}^+$. More precisely,*

$$B = \{\psi_{i,j}^l : l = 1, \ldots, m;\ i \in \mathbb{Z};\ j \in \mathbb{Z}^n\},$$

*where $\psi_{i,j}^l = q^{i/2}\psi^l(A^i x - j)$. The elements of $S$ are called the* basic (mother) wavelets.

Consider the definition of a multiresolution analysis as given in [15] where the lattice $\Gamma = \mathbb{Z}^n$.

**Definition 3.** *Let $A$ be an acceptable dilation for $\mathbb{Z}^n$. A* multiresolution analysis (MRA) *associated with $A$ is a sequence of closed subspaces $(V_i)_{i \in \mathbb{Z}}$ of $L^2(\mathbb{R}^n)$, satisfying*

*i)* $V_i \subset V_{i+1},\ \forall i \in \mathbb{Z}$
*ii)* $\cup_{i \in \mathbb{Z}} V_i = L^2(\mathbb{R}^n)$
*iii)* $V_i = U_A^{-i} V_0,\ \forall i \in \mathbb{Z}$
*iv)* $\tau_j V_0 = V_0,\ \forall j \in \mathbb{Z}^n$
*v) there is a function $\phi \in V_0$, called the* scaling function,
  *such that $\{\tau_j \phi : j \in \mathbb{Z}^n\}$ is a basis for $V_0$.*

These properties imply that $\{\phi_{i,j} : j \in \mathbb{Z}^n\}$ is a basis for $V_i$ for each $i \in \mathbb{Z}$. Since $\phi \in V_0 \subset V_1$, we obtain the *dilation equation*:

$$\phi(x) = \sum_{j \in \mathbb{Z}^n} h_j \phi_{1,j}(x)$$
$$= \sum_{j \in \mathbb{Z}^n} h_j |\det A|^{1/2} \phi(Ax - j),\ \forall x \in \mathbb{R}^n$$

where $h_j = \langle \phi, \phi_{1,j} \rangle, \forall j \in \mathbb{Z}^n$.

Given a multiresolution analysis we define, for each $i \in \mathbb{Z}$, the space $W_i$ as the orthogonal complement of $V_i$ in $V_{i+1}$: $W_i = V_{i+1} \ominus V_i$. Thus, it follows that $W_i = U_A^{-i} W_0$ and that $L^2(\mathbb{R}^n) = \bigoplus_{i \in \mathbb{Z}} W_i$.

Recalling the result of Meyer [27] that there exist $q-1$ functions $\psi^1, \ldots, \psi^{q-1}$ such that $\{\tau_j \psi^l : j \in \mathbb{Z}^n;\ l = 1, \ldots, q-1\}$ is a basis for $W_0$, then the set

$$\{\psi_{i,j}^l : l = 1, \ldots, q-1\ ; i \in \mathbb{Z}\ ; j \in \mathbb{Z}^n\}$$

is a wavelet basis for $L^2(\mathbb{R}^n)$. Since $W_0 \subset V_1$, we get a dilation equation for each of the $\psi^l$, $l = 1, \ldots, q-1$:

$$\psi^l = \sum_{j \in \mathbb{Z}^n} g_j^l \phi_{1,j},$$

where $g_j^l = \langle \psi^l, \phi_{1,j} \rangle, \forall j \in \mathbb{Z}^n$. The coefficients $h_j$ and $g_j^l$ are called the *filter coefficients* of the scaling function and wavelet functions respectively.

Finding the scaling function $\phi$ and the functions $\psi^l$ may be extremely difficult in general. However, in the case where the scaling functions are *characteristic functions on self-similar lattice tilings*, such basic wavelets can always be found.

## 2.2 Self-Similar Lattice Tilings

Let $Q, R \subset \mathbb{R}^n$ be Lebesgue measurable. Denote the characteristic function of $Q$ by $\chi_Q$ and write $|Q|$ to denote its Lebesgue measure. Write $Q \simeq R$ if $|Q \backslash R| = |R \backslash Q| = 0$. We also state the following definitions.

**Definition 4.** *A set $Q$ is said to* tile $\mathbb{R}^n$ *by integer translates if*

i) $\bigcup_{k \in \mathbb{Z}^n} (Q + k) \simeq \mathbb{R}^n$ and
ii) $Q \cap (Q + k) \simeq \emptyset$, $\forall k \in \mathbb{Z}^n \backslash \{0\}$.

It can be shown that such a set $Q$ must have Lebesgue measure 1.

*Example 1.* The set $Q = [0, 1]$ tiles $\mathbb{R}$ by integer translates.

**Definition 5.** *A set containing a unique element from each coset is called a complete residue system.*

One can show that a complete residue system for $\mathbb{Z}^n / A\mathbb{Z}^n$ contains $q = |\det A|$ elements.

*Example 2.* Letting $A=10$ in $\mathbb{Z}$ one complete residue system for $\mathbb{Z}/A\mathbb{Z}$ is $\{0, 1, \ldots, 9\}$. This follows since each integer is equivalent to a unique element of this set, modulo 10.

For our applications, we will be particularly interested in the case of Haar wavelets. Haar wavelets are wavelets where the scaling function is the characteristic function of some set.

**Definition 6.** *We say that a scaling function $\phi$ for a MRA of $L^2(\mathbb{R}^n)$ is a* Haar scaling function *if $\phi = \chi_Q$ for some measurable subset $Q$ of $\mathbb{R}^n$.*

The fact that the scaling function has to satisfy the dilation equation means that its support has to be a *self-similar tile*. In [15], Gröchenig and Madych completely characterized these scaling functions and their supports. The following theorems summarize their results.

**Theorem 1.** *Suppose $A$ is an acceptable dilation for $\mathbb{Z}^n$ and let $Q$ be a measurable subset of $\mathbb{R}^n$. The function $\phi = |Q|^{-1/2} \chi_Q$ is the scaling function for a multiresolution analysis associated with $A$ if and only if the following conditions are satisfied:*

i) $Q$ tiles $\mathbb{R}^n$ by integer translates.

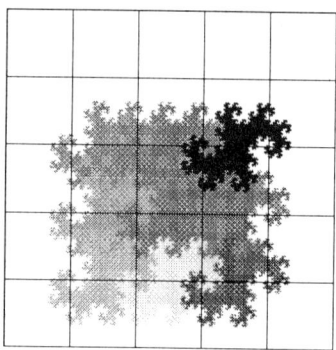 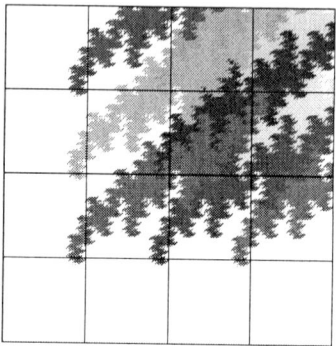

Fig. 1. Examples of self-similar tiles.

ii) $AQ \simeq \bigcup_{k \in K}(Q+k)$ for some complete residue system $K$ of $\mathbb{Z}^n/A\mathbb{Z}^n$.
iii) $Q \simeq C$ for some compact subset $C$ of $\mathbb{R}^n$.

Since $Q$ satisfies ii) it is called *self-similar in the affine sense*. These properties imply that the filter coefficients $h_j$ of $\phi$ are identically $q^{-1/2}$ for $j \in K$ and zero otherwise. Figure 1 illustrates examples of two different self-similar tilings. The first tile (the *twin dragon* tile) is generated by the matrix and complete residue system (also called the *digit set*)

$$\begin{pmatrix} 1 & -1 \\ 1 & 1 \end{pmatrix} \quad \{(0,0),(0,1)\}$$

while the second tile is generated by the matrix and digit set

$$\begin{pmatrix} 1 & 1 \\ -1 & 2 \end{pmatrix} \quad \{(0,0),(1,0),(2,0)\}.$$

In both figures, we show nine translations of the basic tile. Since the scaling function is a characteristic function, all the non-zero coefficients in the dilation equation have the value 1. The tile is the attractor of the IFS $\{A^{-1}(x) + A^{-1}d : d \in K\}$.

**Theorem 2.** *Let $K$ be a complete residue system of $\mathbb{Z}^n/A\mathbb{Z}^n$. Then there exists a unique solution of $\phi = \sum_{k \in K} q^{-1/2}\phi_{1,k}$ in $L^1(\mathbb{R}^n)$, up to multiplication by a constant. Furthermore, this solution has support in the compact set*

$$Q = \left\{ \sum_{i=1}^{\infty} A^{-i}k_i \ : \ k_i \in K \right\}.$$

**Theorem 3.** *Let $A$ be an acceptable dilation for $\mathbb{Z}^n$ and let $Q \subset \mathbb{R}^n$. Then the function $\phi = \chi_Q$ is the scaling function of a MRA associated with $A$ if and only if $|Q| = 1$ and $Q$ is of the form given in Theorem 2 for some complete residue system $K$ of $\mathbb{Z}^n/A\mathbb{Z}^n$.*

Gröchenig and Madych also characterized the wavelets for such MRA.

**Theorem 4.** *Let $\phi = \chi_Q$ be the scaling function for a MRA of $L^2(\mathbb{R}^n)$ associated with $A$ and let $K = \{k_1, \ldots, k_q\}$ be the complete residue system generating $Q$. Let $U = (u_{ij})$ be a unitary $q \times q$ matrix, with $u_{1j} = q^{-1/2}$, $j = 1, \ldots, q$. For $i = 1, \ldots, q-1$ define*

$$\psi^i = \sum_{j=1}^{q} u_{i+1\,j} \phi_{1,k_j}.$$

*Then $\{\tau_j \psi^i : i = 1, \ldots, q-1 \,; j \in \mathbb{Z}^n\}$ is a basis for $W_0$. Conversely, any set of basic wavelets for a MRA associated with $\phi = \chi_Q$ must arise in such a way.*

An immediate corollary is that the set

$$\{\psi^l_{i,j} : l = 1, \ldots, q-1 \,; i \in \mathbb{Z}; j \in \mathbb{Z}^n\}$$

is a basis for $L^2(\mathbb{R}^n)$.

*Example 3.* For $q = 2$, there are only two possible matrices $U$:

$$U = \begin{pmatrix} 1/\sqrt{2} & 1/\sqrt{2} \\ \pm 1/\sqrt{2} & \mp 1/\sqrt{2} \end{pmatrix}.$$

Another example, slightly different than the one in [15] is the following.

*Example 4.* For $q \geq 3$, define the unitary matrix $U = (u_{ij})$, by letting $u_{1j} = q^{-1/2}$ and

$$u_{ij} = \sqrt{\frac{2}{q}} \cos \frac{(i-1)(2j-1)\pi}{2q},$$

for $i = 2, \ldots, q$ and $j = 1, \ldots, q$.

The reader is referred to [15] for further details and examples.

### 2.3 Reconstruction and Decomposition Algorithm

The strength of the multiresolution analysis method lies in the reconstruction and decomposition algorithms, discovered initially by Mallat [23]. These algorithms are a fundamental component of wavelet analysis applied to signal and image processing. They will be reviewed here only briefly, mostly for notational purposes.

Suppose $(V_i)_{i \in \mathbb{Z}} \subset L^2(\mathbb{R}^n)$ is a MRA and that $f \in V_{i+1} = V_i \oplus W_i$. We have two bases: one for $V_{i+1}$ and one for $V_i \oplus W_i$. Therefore,

$$f = \sum_{z \in \mathbb{Z}^n} s_{i+1,z} \phi_{i+1,z}$$

$$= \sum_{j \in \mathbb{Z}^n} s_{i,j} \phi_{i,j} + \sum_{l=1}^{q-1} \sum_{j \in \mathbb{Z}^n} w^l_{i,j} \psi^l_{i,j},$$

where the *scaling coefficients* $s_{m,n} = \langle f, \phi_{m,n} \rangle$ and the *wavelet coefficients* $w_{m,n}^l = \langle f, \psi_{m,n}^l \rangle$. By using the dilation equations we obtain the identities

$$s_{i,j} = \sum_{z \in \mathbb{Z}^n} h_{z-Aj} s_{i+1,z} \quad \text{and} \quad w_{i,j}^l = \sum_{z \in \mathbb{Z}^n} g_{z-Aj}^l s_{i+1,z}, \qquad (1)$$

which give the decomposition of higher resolution scaling coefficients into lower resolution scaling coefficients and wavelet coefficients. The reconstruction algorithm is given by

$$s_{i+1,z} = \sum_{j \in \mathbb{Z}^n} h_{z-Aj} s_{i,j} + \sum_{l=1}^{q-1} \sum_{j \in \mathbb{Z}^n} g_{z-Aj}^l w_{i,j}^l.$$

As is well known, this procedure yields a tree structure for the wavelet coefficients.

## 3  Fractal Image Compression

A fractal representation of an image tries to use self-similarity within a picture to encode the picture. The usual way of doing this in a compression scheme is to break the image up into a regular grid of large blocks (called the *source blocks* or *parent blocks*) and a regular grid of small blocks (called the *target blocks* or the *child blocks*). Then for each child block we search among all the parent blocks for the block that is the closest match. We usually allow some kind of affine modification of the image values on the block. If we represent the image as a function $f$ then the usual transformation is of the form $\alpha f(\cdot) + \beta$.

The above "fractal block" algorithm often introduces undesirable blocking artifacts in the reconstructed image. These artifacts arise because the image is treated as a collection of disjoint child blocks. Mixed fractal-wavelet methods were introduced in an attempt to reduce these blocking artifacts.

The *Fractal-Wavelet Transform* (see [26, 29]) is based on the observation that the wavelet coefficients have a very natural tree structure, as illustrated in the following diagram (depicting the situation where the dilation is by a factor of 2).

| $w_{0,0}$ | | | |
|---|---|---|---|
| $w_{1,0}$ | | $w_{1,1}$ | |
| $B_{20}$ | $B_{21}$ | $B_{22}$ | $B_{23}$ |

In this diagram, each entry $B_{ij}$ represents a tree of infinite length. Using this notation, an example of a simple type of Fractal-Wavelet Transform is the transformation $W$ defined by:

$$W_1 : B_{10} \to B_{20}, \quad W_2 : B_{11} \to B_{21},$$

and

$$W_3 : B_{10} \to B_{22}, \quad W_4 : B_{11} \to B_{23},$$

with associated multipliers $\alpha_i$, $1 \leq i \leq 4$. Diagrammatically, this IFSW transforms $B_{00}$ into

| $w_{0,0}$ | | | |
|---|---|---|---|
| $w_{1,0}$ | | $w_{1,1}$ | |
| $\alpha_1 B_{10}$ | $\alpha_2 B_{11}$ | $\alpha_3 B_{10}$ | $\alpha_4 B_{11}$ |

where $|\alpha_i| < 1/\sqrt{2}$. The restrictions on the $\alpha_i$ follow from the condition that the wavelet coefficient sequence $w_{i,j}$ is square summable.

The map $W$ has a unique fixed point function $\bar{u}$ whose wavelet coefficients decay geometrically according to the $\alpha_i$'s.

All the previous work on fractal-wavelet methods have used separable wavelets. This was mainly due to the lack of an algorithm to do the wavelet transform for non-separable wavelets. In this paper, we concentrate on using non-separable wavelets.

## 4 Periodic Non-Separable Wavelet Transform

In this section, we present a brief discussion of the algorithm for the discrete periodic wavelet transform for non-separable wavelets (for a more complete description and discussion of the algorithm see the paper [24]). Our algorithm is general in the sense that you can use *any* valid scaling function filter and wavelet filters for an appropriate dilation matrix. However, in our applications we concentrate on the case of Haar wavelets. We do this mainly because very few wavelet filters for smooth non-separable wavelets are known [2].

For simplicity, we restrict ourselves to two dimensions. The extension to higher dimensions is reasonably straightforward.

We assume that we are given a dilation matrix $A$ and that our signal is of size $det(A)^N$ by $det(A)^N$ for some $N$. Furthermore, we make the assumption that

$$trace(A) = 0 \bmod det(A) \qquad (2)$$

and that the only solution to $A(a,0)^T = (0,0)^T$ is $a = 0$. It is possible to weaken these assumptions, but the algorithm is slightly more complicated.

The basic idea is that (1) tells us how to filter and downsample our given signal, so we implement these equations. We use the proper arithmetic on the subscripts to keep track of where to subsample and how to convolve. However, this is not entirely straightforward.

First, the data is downsampled by a factor of $det(A)$ on each iteration of the wavelet analysis algorithm. This may make the storage and indexing of the data complicated. For example, if the image is square and $det(A) = 2$, then after one iteration you no longer have a square. This also complicates the periodization (performing the arithmetic correctly). The indexing problem can be solved by using $A$ to help with the indexing from one "level" to the next. The periodization problem (performing the correct arithmetic) can be solved by noting that after one subsampling we are on the sublattice $A(\Lambda) \cong \Lambda/ker(A)$, where

$\Lambda = \mathbb{Z}_{det(A)^N} \times \mathbb{Z}_{det(A)^N}$ is the lattice (module) of integers modulo $det(A)^N$. Thus, we perform our arithmetic modulo $det(A)^N$ and then reduce modulo $ker(A)$.

This leads to the second problem. In general, the subsampling lattices $A^n(\Lambda)$ get more complicated with each iteration. Thus, in order to perform the correct arithmetic on the $n^{th}$ stage it would be necessary to reduce modulo $ker(A^n)$. It would be necessary to know $ker(A^n)$ for all $n$ to do this and this reduction would make the algorithm very slow.

Finally, suppose $A$ has an eigenvector $v$ over $\mathbb{Z}_{det(A)}$ associated with an eigenvalue $\lambda$ which is invertible modulo $det(A)$. Then $det(A)^{N-1}v$ is an eigenvector of $A$ over $\Lambda$ with eigenvalue $\lambda$. So after repeatedly sub-sampling the lattice $\Lambda$ using $A$, we will eventually get to a stage where our sub-sampling points are no longer changing; the wavelet analysis algorithm cannot be completed. The only way around this problem is to restrict our possible dilation matrices to those that have no such eigenvectors. It is easy to see that if $A$ has an invertible eigenvalue modulo $det(A)^N$ then $A$ must necessarily have an invertible eigenvalue modulo $det(A)$. Thus it is sufficient to check for eigenvalues modulo $det(A)$.

Our assumption given in equation (2) eliminates these problems. Since $A$ is a $2 \times 2$ matrix, the characteristic polynomial of $A$ is

$$\lambda^2 - trace(A)\lambda + det(A).$$

If $trace(A) = 0 \mod det(A)$, then $A$ has no non-invertible eigenvalues modulo $det(A)$, and thus has no associated eigenvectors. Furthermore, by the Cayley-Hamilton Theorem (which remains true for matrices over commutative rings, see [4]) we have

$$A^2 \equiv 0 \mod det(A)$$

which means that each entry of $A^2$ is a multiple of $det(A)$ so that

$$A^2(\Lambda) \subset det(A)\Lambda.$$

In fact, this equation is an equality. Thus, while $A(\Lambda)$ might be some arbitrary submodule of $\Lambda$, $A^2(\Lambda)$ sits in $\Lambda$ very simply. In order to do arithmetic in $A^2(\Lambda)$ we simply perform our arithmetic modulo $det(A)^{N-1}$ in each coordinate! Thus, we can go from our "square" arrays of data to "rectangular" and then back to "square".

Here is an example to illustrate these ideas.

*Example 5.* Let $A$ be the dilation matrix

$$A = \begin{pmatrix} 2 & 1 \\ 1 & 4 \end{pmatrix}.$$

Then $A(2,1)^T \equiv 6(2,1)^T \mod 7$ so $A$ has an eigenvector modulo 7 corresponding to the invertible eigenvalue 6.

In Figure 2 we use the lattice $\Lambda = \mathbb{Z}_{49} \times \mathbb{Z}_{49}$ and show $A(\Lambda)$ and $A^2(\Lambda)$. For this example $A^n(\Lambda) = A^2(\Lambda)$ for $n \geq 2$.

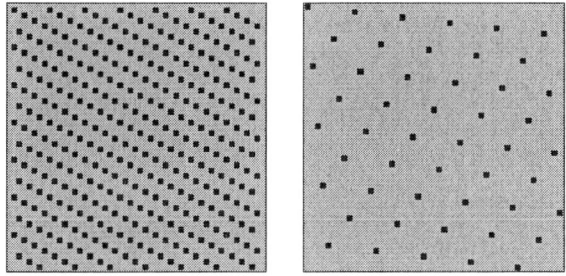

**Fig. 2.** $A(\Lambda)$ and $A^n(\Lambda)$ for $n \geq 2$

## Application to Image Compression

The motivation for using non-separable wavelets in image processing is that any "blocking" artifacts created by these wavelets would be less noticeable to the human eye. This is especially true if we use Haar wavelets. In this case, using separable wavelets creates very noticeable blockiness in the image since separable Haar wavelets are characteristic functions of squares. When we use non-separable wavelets, this blockiness is considerably reduced.

Figure 3 illustrates the basic difference between the use of separable and non-separable Haar wavelets. The first image uses the separable Haar wavelets and represents truncating the bottom 2 levels of the wavelet tree while the second uses Haar wavelets based on the twin dragon tiling (the first tiling in Figure (1) and represents truncating the bottom 4 levels. The information content is similar in the two pictures since for the separable Haar wavelets each level represents downsampling by a factor of 4 ($det(A) = 4$ since $A = 2I$) while for the non-separable one each level represents downsampling by a factor of 2.

In Figure 4 we illustrate the use of Haar wavelets based on the second tiling in Figure 1. In this case we have truncated the finest three levels of the wavelet tree (which corresponds to subsampling by a factor of nine, since $det(A) = 3$). The second image in Figure 4 is the result of using the wavelet analysis and synthesis algorithm for wavelets whose dilation matrix $A$ does not satisfy the assumption in equation (2). This matrix is the same one whose subsampling lattice is shown in Figure 2.

Figure 5 shows the result of two types of encoding: zero-tree and fractal-wavelet. First we performed a wavelet decomposition using the twin dragon Haar wavelets. Next we traversed the wavelet tree truncating the tree when the total energy of the branch was less than some specified threshold. The image on the left is the image reconstructed from this pruned wavelet tree. The image on the right shows the result of fractal-wavelet compression using the twin dragon Haar wavelets. We first computed the wavelet tree for the lenna image and store the first 12 levels of the tree. Next we used these levels to predict the higher resolution levels in the manner described in Section 3.

Fig. 3. Left: Separable Haar 4:1. Right: Twin-dragon Haar 4:1.

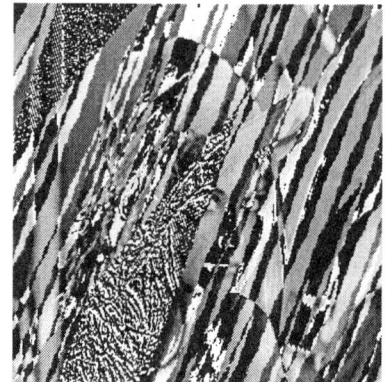

Fig. 4. Left: Non-separable Haar. Right: Improper subsampling.

Fig. 5. Left: Zero-tree encoding. Right: Fractal-Wavelet encoding.

This periodic transform has the benefit of generating a full tree so that fractal type methods are easily implemented. However, periodizing the data sometimes introduces undesirable wrap-around effects on the edges of the image. Another benefit of this transform is that we can use smooth wavelets. For the purposes of illustrating the algorithm it is sufficient to use Haar wavelets; however, an area of future research is using more general smooth non-separable wavelets.

## 5 Complex Bases

Now we turn to a discussion of the wavelet transform algorithm for Haar wavelets based on the connection between self-similar lattice tilings and complex bases. In order to do this, we first give some background on complex bases. We use a base and digits from an integer set to represent the numbers in a given system. For more on the theory of complex bases, we refer the reader to some papers by Gilbert [10–12]. Most of the details for the results that follow can be found there.

### 5.1 Background

The concept of representing numbers in a base is a very simple and familiar one. In real numbers, the most common examples are the decimal and binary systems.

*Example 6.* Representing positive real numbers with $\mathbb{N}$ as the integers.
  Base 10, digits $\{0, 1, 2, \ldots, 9\}$
$$193_{10} = 1 \cdot 10^2 + 9 \cdot 10^1 + 3 \cdot 10^0.$$
  Base 2, digits $\{0, 1\}$
$$11010_2 = 1 \cdot 2^4 + 1 \cdot 2^3 + 0 \cdot 2^2 + 1 \cdot 2^1 + 0 \cdot 2^0 = 26_{10}.$$

The same idea carries over to complex numbers. In this case, the integers are the Gaussian integers:

**Definition 7.** *The set of* Gaussian integers, *denoted by* $\mathbb{Z}[i]$, *is the set of complex numbers of the form* $a + bi$, *where* $a, b \in \mathbb{Z}$. *The* norm *of a Gaussian integer* $a + bi$ *is defined as*
$$\mathrm{norm}(a + bi) \equiv a^2 + b^2.$$

**Definition 8.** *A valid base is a pair* $(b, D)$ *where* $b$ *is a Gaussian integer and* $D \subset \mathbb{Z}[i]$, *such that* $0 \in D$ *and every integer* $z$ *can be represented uniquely as a sum of powers of* $b$, *with coefficients in* $D$. *More precisely, each* $z \in \mathbb{Z}[i]$ *can be written uniquely as*
$$z = \sum_{j=0}^{t} a_j b^j, \text{ where } a_j \in D \text{ and } t \in \mathbb{N}.$$

If $z$ has this form, write $z = (a_t a_{t-1} \cdots a_1 a_0)_b$. This is called base $b$ positional notation. The integer $b$ is often referred to as the base, and the set $D$ is called the digit set. When the base is understood, the subscript $b$ in the positional notation is often omitted.

A result below will show that $(-1+i, \{0,1\})$ is a valid base.

**Example 7.** We can expand 3 in $(-1+i, \{0,1\})$:

$$3 = (-1+i)^3 + (-1+i)^2 + (-1+i)^0 = (1101)_{-1+i}.$$

**Definition 9.** Let $(b, D)$ be a valid base. If $z \in \mathbb{C}$ has the form

$$z = \sum_{j=-\infty}^{t} a_j b^j, \text{ where } a_j \in D \text{ and } t \in \mathbb{N},$$

then the radix expansion of $z$ in base $(b, D)$ is defined as

$$z = (a_t a_{t-1} \cdots a_1 a_0 . a_{-1} a_{-2} \cdots)_b.$$

The point between the digits $a_0$ and $a_{-1}$ is called the radix point. The string of digits to the left of the radix point is called the integer part of $z$, while the string to the right is called the radix part. Another name for the radix expansion of a complex number is the address of the number.

**Example 8.** The complex number $(-1-8i)/5$ can be expressed in base $(-1+i, \{0,1\})$ as

$$(-1-8i)/5 = (111.\overline{10})_{-1+i}.$$

where $\overline{10}$ indicates that the string 10 is repeated indefinitely. This can be seen by considering a geometric series in $(-1+i)^{-1}$.

**Proposition 1.** *(Gilbert [11]) If $(b, D)$ is a valid base, then $D$ is a complete residue system for $\mathbb{Z}[i]$ modulo $b$ and hence contains norm($b$) elements.*

**Theorem 5.** *(Gilbert [13]) Each $z \in \mathbb{C}$ has an infinite radix expansion in a valid base. However, this expansion may not necessarily be unique.*

It therefore makes sense to define the *fundamental tile* of a valid base.

**Definition 10.** Given a valid base $(b, D)$, define the fundamental tile $T(b, D)$ as the set of complex numbers with zero integer part in the base.

By Theorem 5, $\mathbb{C} = \cup_{z \in \mathbb{Z}[i]} (T(b, D) + z)$. The following result shows that there are many valid bases.

**Theorem 6.** *(Davio, Deschamps and Gossart [7]) Given any $b \in \mathbb{Z}[i]$ with modulus larger than one, except 2 and $1 \pm i$, there exists a complete residue system $D$ such that $(b, D)$ is a valid base for $\mathbb{C}$.*

The following result gives an entire class of complex bases.

**Theorem 7.** *(Kátai and Szabó [18]) If $b \in \mathbb{Z}[i]$, with norm $N$ and $D = \{0, 1, 2, \ldots, N-1\}$, then $(b, D)$ is a valid base for $\mathbb{C}$ iff $b = -n \pm i$, for some positive integer $n$.*

Further generalizations can be found in [9, 17].

**Corollary 1.** *The pair $(-1 + i, \{0, 1\})$ is a valid base.*

### 5.2 Representation in a Complex Base

There are various algorithms for determining the representation of Gaussian integers in a valid base. These are due to Gilbert [9, 12–14].

---

(Gilbert [13]) Let $(b, D)$ be a valid base. Since $D$ is a complete residue system for $\mathbb{Z}[i]$ modulo $b$ then, given $z \in \mathbb{Z}[i]$, there exist unique integers $q_j \in \mathbb{Z}[i]$ and $a_j \in D$, $j = 1 \ldots t$, $t \in \mathbb{N}^+$ such that

$$z = q_1 b + a_0$$
$$q_1 = q_2 b + a_1$$
$$\vdots$$
$$q_t = 0 b + a_t.$$

Hence, $z = (a_t \ldots a_1 a_0)_b$.

---

*Example 9.* Let $z = 5 + 12i$. Find the address of $z$ in $(-2 + i, \{0, 1, 2, 3, 4\})$.
Using the Division Algorithm

$$5 + 12i = (2 - 5i)(-2 + i) + 4$$
$$2 - 5i = (-1 + 2i)(-2 + i) + 2$$
$$-1 + 2i = 2(-2 + i) + 3$$
$$2 = 0(-2 + i) + 2$$

Hence the address of $5 + 12i$ is (2324).

A fast algorithm, called the Clearing Algorithm, also exists for finding the expansion of integers in bases of the form $(b, D)$ where $D \subset \mathbb{Z}$. For simplicity, we consider only bases $(-n + i, \{0, 1, \ldots, n^2\})$, $n \in \mathbb{N}^+$. The reader is referred to [9] for the general result.

*Example 10.* Determine the expansion of $5 + 12i$ in base $(-2 + i, \{0, 1, 2, 3, 4\})$.
The minimal polynomial of $b = -2 + i$ is $x^2 + 4x + 5$. Hence, $b^2 + 4b + 5 = 0$ and, by abuse of notation, we can write this as $(1\ 4\ 5)_b = 0$. (Note that 5 is not in the digit set.)

Begin with the expansion $5 + 12i = 12b + 29 = (12\ 29)_b$. Then, we clear the polynomial in $\mathbb{Z}[b]$ as follows:

(Gilbert [9]) Consider the valid base $(-n+i, \{0, 1, \ldots, n^2\})$ and let $p(x)$ be the minimal polynomial of $b = -n + i$. Thus

$$p(x) = x^2 + 2nx + n^2 + 1.$$

Then the representation of any integer $z \in \mathbb{Z}[i]$ in the base can be obtained as follows. Begin with $z = m(b) = a_k b^k + \cdots + a_1 b + a_0$, an expansion of $z$ in powers of $b$ with integer coefficients. For instance, any Gaussian integer $c+id$ can be expanded in powers of $b = -n + i$ as $c + id = db + (c + nd)$. Consider this expansion as an element $m(b)$ of the polynomial ring $\mathbb{Z}[b]$. Let $r$ be the least integer such that $a_r \notin D$. If no such $r$ exists, then $m(b)$ is the unique expansion of $z$ in the base. We call such a polynomial *clear*.

If $r$ exists, then let $s$ be the integer such that $0 \leq a_r + s(n^2 + 1) \leq n^2$. Add $sb^r$ times $p(b)$ to $m(b)$. Remember that we perform this operation in the polynomial ring $\mathbb{Z}[b]$. Call this new polynomial $m_1(b)$.

However, $p(-n + i) = 0$, thus $m(b)$ and $m_1(b)$ are equal in $\mathbb{C}$. Hence, $m_1(b)$ is an expansion of $z$ in $\mathbb{Z}[b]$. In addition, the coefficient of the $r$-th power of $b$ in $m_1(b)$ is a digit. We say that $a_r$ has been *cleared*.

Repeat the process of clearing coefficients, by induction on $r$, until a clear polynomial is obtained. The resulting polynomial is the expansion of $z$ in $(b, D)$. This process must terminate after a finite number of steps.

|   |   | 12 | 29 |
|---|---|----|----|
|   | -5 | -20 | -25 |
|   | -5 | -8 | 4 |
| 2 | 8 | 10 |   |
| 2 | 3 | 2 | 4 |

Hence, the expansion of $5 + 12i$ in base $(-2 + i, \{0, 1, 2, 3, 4\})$ is $(2324)$. A quick calculation verifies that this is indeed correct.

# 6 Linking Complex Bases to Wavelets

## 6.1 Addressing

The theory of MRA and complex bases are related. Multiplication by the base $b = c + id$ in $\mathbb{C}$ is equivalent to multiplication by the acceptable dilation

$$A = \begin{pmatrix} c & -d \\ d & c \end{pmatrix}$$

in $\mathbb{R}^2$, via the natural association $s + it \leftrightarrow (s, t)^T$. Here, $\det A = c^2 + d^2 = \text{norm}(b)$ and $\lambda = c \pm id$. The fundamental tile $T(b, D)$ for a valid base $(b, D)$ is simply the basic tile $Q$ of $A$, where the complete residue system $K$ is identified with $D$ through the natural association. Therefore,

$$T(b, D) = \left\{ \sum_{j=-\infty}^{-1} a_j b^j : a_j \in D \right\} \equiv \left\{ \sum_{i=1}^{\infty} A^{-i} k_i : k_i \in K \right\} = Q.$$

## 6.2 Reconstruction and Decomposition Algorithm

Consider the reconstruction and decomposition algorithms for a Haar MRA associated with an acceptable dilation $A$ for $\mathbb{Z}^2$, with complete residue system $K$. Assume $A$ is associated with a complex base $(b, D)$ in the sense of Section 6.1. Rewriting the equation for the wavelet decomposition, we have

$$s_{i,j} = \sum_{z \in \mathbb{Z}^2} h_{z-Aj} s_{i+1,z}$$
$$= \sum_{k \in K} h_k s_{i+1, Aj+k},$$

since $h_k = 0$ when $k \notin K$. We can translate this identity to the language of complex bases as

$$s_{i,j} = \sum_{d \in D} h_d s_{i+1, bj+d},$$

where the scaling filters and coefficients are re-indexed, via the natural association given above.

By looking at the decomposition algorithm in the language of complex bases, we can consider $bj + d$ as the address of $j$ shifted to the left by one place, with $d$ added as the zero-th order digit. We can then think of the index $i$ as the length of the address of $j$. The scaling coefficient associated with the point $j$ is thus a linear combination of the coefficients of the points whose addresses are one digit longer than that of $j$, and start with the address of $j$. In other words, letting $j$ have address $(a_t \ldots a_1)$ in base $b$ we have

$$s_{t, (a_t \ldots a_1)} = \sum_{d \in D} h_d s_{t+1, (a_t \ldots a_1 d)}.$$

This gives a precise and **implementable** tree structure for the coefficients. A similar relation holds for the reconstruction algorithm.

## 6.3 Application to Images

We present here an example of this process applied to images.

*Example 11.* Perform the decomposition of the function $f$ in base $(-1+i, \{0,1\})$, where $f$ is the *simple function* given by the following diagram:

| 1 | 2 |
|---|---|
| 0 | 1 |

By this representation, we take the boxes to be unit squares in $\mathbb{C}$, with the left hand corner at 0. The value of the function on each square is the number written inside the square.

The addresses of the values are

| Integer | Address | Value |
|---------|---------|-------|
| 0       | 0000    | 0     |
| 1       | 0001    | 1     |
| i       | 0011    | 1     |
| 1+i     | 1110    | 2     |

The decomposition begins at level 4, the length of the longest address. Choose the standard Haar wavelet basis as given in Example 3. We then obtain the scaling and wavelet trees using the decomposition algorithm. The lowest level on the scaling tree represents the initial values of the function. Going down the left branch in a tree represents the digit 0. The right represents the digit 1. For example, the value of the function on the square $i$, which is at address 0011, is placed in the node down the scaling tree left, left, right, right. We normalize the scaling filters in this example for the purposes of clarity. The correct tree would reduce the values by $\sqrt{2}$ at each level, beginning with level 3. Empty nodes are set to zero.

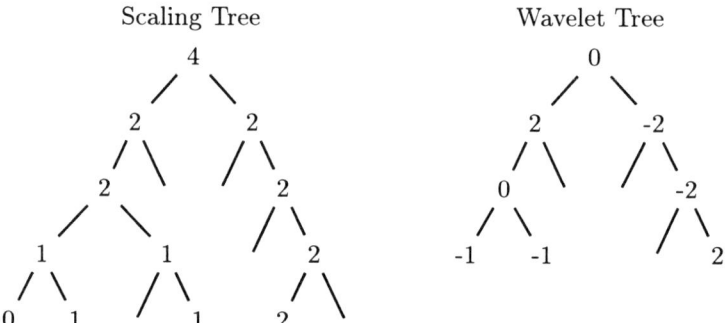

Notice the number of empty nodes in the trees. This demonstrate the asymmetrical nature of the trees generated by this process. This occurs since the support of the original function sits inside the principal tile, but is not the entire tile; it is a consequence of the zero padding of the function.

*Example 12.* An illustration of the above method applied to a 512 × 512 image of Lena is shown in Figure 6. The wavelet decomposition was performed on this image, using the complex base $(2 + i, \{0, 1, i, -i, -2 - 3i\})$. The wavelets are given by the 5×5 unitary matrix of Example 4. A threshold of 50 was used to set individual wavelet coefficients to zero. In this case, 94% of the coefficients were set to zero. The tile of this base is disconnected and dust-like. It is shown in Figure 7 along with the first basic wavelet.

The implementation of compression methods on wavelet trees of complex bases might not be trivial. Current compression methods, such as the zero-tree (see [3]), rely on extremely symmetric and balanced distributions of the coefficients in the expansion tree. The trees created by complex bases are tremendously unbalanced, as is seen in Example 11, above. This asymmetry is much greater with larger numbers of points such as in images. There may also be

**Fig. 6.** Wavelet compression on dust. Left: Original 512 × 512 of Lena. Right: Reconstruction of Lena from pruned wavelet tree. PSNR=27.22 dB.

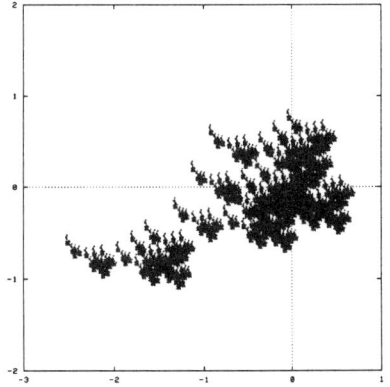 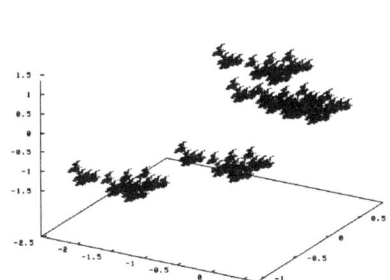

**Fig. 7.** Fundamental tile of the base $(2 + i, \{0, 1, i, -i, -2 - 3i\})$ with a basic wavelet.

fractal-wavelet transforms (see [26, 29]) which could utilize the structure of the complex bases. This could lead to a general fractal-wavelet transform for nonseparable wavelets.

However, it is possible that the asymmetry of the trees does not permit the construction of a viable compression method. If this is the case one could investigate compression on translations of the image which result in more symmetric trees. It may even be necessary to assume that the image is exactly the tile by mapping it onto the tile of an appropriate size. Computational complexity of such methods may result in algorithms which are slower than the standard ones.

*Acknowledgment.* The first author was supported in part by a Natural Sciences and Engineering Research Council of Canada (NSERC) Collaborative Grant (C. Tricot (Ecole Polytechnique, Montreal), E.R. Vrscay (Waterloo), J. Levy-Vehel (INRIA, Rocquencourt) and B. Forte (Verona)) in the form of a Postdoctoral Fellowship. The support of an NSERC Postgraduate Scholarship is gratefully acknowledged by the second author. The authors also thank Professors William Gilbert and Edward Vrscay, as well as the members of the Waterloo "fractalators" (http://links.uwaterloo.ca) for their most helpful advice and discussions.

# References

1. M. F. Barnsley (1988): *Fractals Everywhere*. Academic Press, New York.
2. E. Belogay and Yang Wang. Arbitrarily smooth orthogonal nonseparable wavelets in $\mathbb{R}^2$. To appear in *SIAM Math. Analysis*.
3. J. J. Benedetto and M. W. Frazier, editors (1994): *Wavelets: Mathematics and Applications*. Studies in Advanced Mathematics, CRC Press.
4. W. C. Brown (1993): *Matrices over Commutative Rings*. Marcel Dekker, New York.
5. C. Cabrelli, C. Heil, and U. Molter. Accuracy of lattice translates of several multidimensional refinable functions. To appear in *J. Approximation Theory*.
6. I. Daubechies (1992): *Ten Lectures on Wavelets*. SIAM Press, Philadelphia, PA.
7. M. Davio, J. P. Deschamps and C. Gossart (May 1978): *Complex Arithmetic*. Philips MBLE Research Lab. Report R369.
8. Yuval Fisher (1995): *Fractal Image Compression: Theory and Applications*. Springer-Verlag, New York.
9. W. Gilbert (1981): Radix representations of quadratic fields. *J. Math. Anal. Appl.*, 83, 264–274.
10. W. Gilbert (1982): Fractal geometry derived from complex bases. *Math. Intelligencer*, 4(2), 78–86.
11. W. Gilbert (1982): Geometry of radix representations. In C. Davis, B. Grünbaum and F. A. Sherk, editors, *The Geometric Vein, The Coxeter Festschrift*, Springer-Verlag, 129–139.
12. W. Gilbert (1984): Arithmetic in complex bases. *Mathematics Magazine*, 57(2), 77–81.
13. W. Gilbert (May 1994): Complex based number systems. Unpublished.
14. W. Gilbert (1996): The division algorithm in complex bases. *Canadian Mathematical Bulletin*, 39(1), 47–54.

15. K. Gröchenig and W. R. Madych (1992): Multiresolution analysis, haar bases, and self-similar tilings of $R^n$. *IEEE Trans. Inform. Theory*, 39, 556–568.
16. J. E. Hutchinson (1981): Fractals and Self-similarity. *Indiana Univ. Math. J.* **30**, 713–747.
17. I. Kátai and B. Kovács (1981): Canonical number systems in imaginary quadratic fields. *Acta Math. Acad. Sci. Hungaricae*, 37, 159–164.
18. I. Kátai and J. Szabó (1975): Canonical number systems for complex integers. *Acta. Sci. Math. (Szeged)*, 37, 255–260.
19. J. Kovačević and M. Vetterli (1995): Nonseparable two- and three-dimensional wavelets. *IEEE Trans. on Signal Processing* **43**, 1269–1273.
20. J. C. Lagarias and Yang Wang (1995): Haar-type orthonormal wavelet basis in $\mathbb{R}^2$. *J. Fourier Analysis and Appl.*, **2**, 1–14.
21. Kevin Leeds (1997): Dilation Equations with Matrix Dilations. PhD Thesis, Department of Math, Georgia Tech.
22. N. Lu (1997): *Fractal Imaging*. Academic Press.
23. S. Mallat (1989): Multiresolution approximation and wavelet orthonormal bases of $L^2(\mathbb{R})$. *Trans. AMS*, 315, 69–88.
24. F. Mendivil (1998): A discrete periodic wavelet transform for non-separable wavelets. Preprint.
25. F. Mendivil (1998): The application of a fast non-separable discrete periodic wavelet transform to fractal image compression. Preprint. *Submitted to Fractals in Engineering 99*.
26. F. Mendivil and E. R. Vrscay (1997): Correspondence between fractal-wavelet transforms and iterated function systems with grey level maps. In E. Lutton, C. Tricot and J. Lévy Véhel, editors, *L'Ingénieur et les fractales*, Springer Verlag, 54–64.
27. Y. Meyer (1987): *Ondelettes, fonctions splines et analyses graduées*. Rapport CEREMADE n. 8703.
28. C. A. Micchelli and H. Prautzsch (1989): Uniform refinement of curves. *Linear Algebra Appl.*, **114/115**, 841–870.
29. E. R. Vrscay (1998): A new class of fractal-wavelet transforms for image representation and compression. *Can. J. Elect. Comp. Eng.*, 23, 69–83.
30. G. Strang and T. Nguyen (1996): *Wavelets and Filter Banks*. Wellesley-Cambridge Press, Wellesley.
31. Yang Wang (1997): Self-Affine Tiles. Preprint, to appear in *Proc. Chinese Univ. of Hong Kong Workshop on Wavelet*.